Partial Differential Equations of Mathematical Physics

Partial Differential Equations of Mathematical Physics

Second Edition

Tyn Myint-U

NORTH HOLLAND
New York • Oxford

Elsevier North Holland, Inc.
52 Vanderbilt Avenue, New York, New York 10017

Distributors outside the United States and Canada:

Thomond Books
(A Division of Elsevier/North-Holland Scientific Publishers, Ltd.)
P.O. 85
Limerick, Ireland

First edition published in 1973 by Elsevier North Holland, Inc.

Library of Congress Cataloging in Publication Data

Tyn Myint-U
Partial differential equations of mathematical physics.

Bibliography: p.
Includes index.
1. Differential equations, Partial. I. Title.
QA374.T94 1980 515′.353 79-19310
ISBN 0-444-00352-5

Desk Editor Philip Schafer
Design Edmée Froment
Production Manager Joanne Jay
Compositor Science Typographers, Inc.
Printer Haddon Craftsmen

Manufactured in the United States of America

To
U and Mrs. Hla Din
U and Mrs. Thant

Contents

Preface

The theory of partial differential equations has been one of the most important fields of study in mathematics. This is essentially due to the frequent occurrence of partial differential equations in many branches of physics, engineering, and other sciences. With much interest and great demand for applications in diverse areas of sciences, several excellent books on linear partial differential equations have been published. This book is written for the purpose of presenting an approach based mainly on the mathematical problems and their related solutions, and also of formulating a course appropriate for all students of mathematical sciences. The primary concern, therefore, is not with the general theory, but to provide students with the fundamental concepts, the underlying principles, and the techniques and methods of solution of partial differential equations.

An attempt has been made to present a clear and concise exposition of the mathematics used in analyzing a variety of problems. With this in mind, the chapters are carefully arranged to enable students to view the material to be studied in an orderly perspective. For example, theorems in the chapters on Fourier series and eigenvalue problems are explicitly mentioned, whenever possible, to avoid confusion with their use in the development of principles of partial differential equations. A wide range of problems in mathematical physics with various boundary conditions has been included to improve student understanding.

This book is in part based on lectures given at Manhattan College. It is used by advanced undergraduate or beginning graduate students in applied mathematics, physics, engineering, and other sciences. The prerequisite for its study is a standard calculus sequence with elementary ordinary differential equations.

The first chapter is concerned with an introduction to partial differential equations. The second chapter deals with the mathematical models corresponding to the physical events that yield the three basic types of partial differential equations. The third chapter constitutes an account of the classification of second-order equations with two independent variables, and in addition, illustrates the determination of the general solution for a class of relatively simple equations.

Cauchy's problem, Goursat problem and the initial-boundary value problems involving hyperbolic equations of the second order are presented in the fourth chapter. The fifth chapter contains a full treatment of Fourier series essential for the further study of partial differential equations.

Separation of variables is one of the simplest and is the most widely used method of solving partial differential equations. Its basic concept and the separability conditions necessary for its application are described in the sixth chapter, followed by some well-known problems of mathematical physics with a detailed analysis of each problem. In the seventh chapter, self-adjoint eigenvalue problems are treated in depth building on their introduction in the preceding chapter. In addition, Green's function and its application to eigenvalue problems and boundary value problems for ordinary differential equations are presented.

Boundary value problems and the maximum principle are discussed in the eighth chapter, whereas more involved higher dimensional problems and the method of eigenfunctions are treated in the ninth chapter. The tenth chapter deals with the basic concepts and the construction of Green's function and its application to boundary value problems. In the final chapter, the fundamental properties and the techniques of Fourier and Laplace transforms are introduced.

The chapters on the mathematical models, Fourier series and eigenvalue problems are self-contained, and hence these chapters can be omitted for those students who have prior knowledge of the subjects. The exercises are the integral part of the text and range from simple to more difficult problems. Answers to most exercises are given at the end of the book. For students wishing to have further insight into the subject matter, detailed references are listed in the Bibliography.

In this second edition, many changes are made to improve the exposition and clarity. Major changes are made to add some materials which are considered to be important for many students. The specific modifications are:

1. The chapter on the Cauchy problem has been reviewed with an emphasis on wave propagation. Examples on initial-boundary value problems, the characteristic initial value problem, and the Goursat problem have been added.
2. Method of separation of variables has been extended by including the method of eigenfunctions for nonhomogeneous problems.

3. The chapter on eigenvalue problems has been reorganized and has added new sections relating to singular Sturm–Liouville system. Green's function and its application to boundary value problems for ordinary differential equations have also been included.
4. Many new examples and exercises have been added for further understanding.

The author wishes to express his gratitude to those who had assisted him in the preparation of the first edition and to the readers who have indicated corrections and offered many valuable suggestions and comments. The author also wishes to thank Mr. Kenneth Bowman and the staff of Elsevier-North Holland, Inc. for their kind help and cooperation.

Tyn Myint-U

New York, NY

Partial Differential Equations of Mathematical Physics

CHAPTER ONE

Introduction

1.1 Basic Concepts and Definitions

A differential equation that contains, in addition to the dependent variable and the independent variables, one or more partial derivatives of the dependent variable is called a *partial differential equation*. In general, it may be written in the form

$$f(x,y,\ldots,u,u_x,u_y,\ldots,u_{xx},u_{xy},\ldots)=0 \tag{1.1.1}$$

involving several independent variables $x,y,\ldots$, an unknown function u of these variables and the partial derivatives[1] $u_x,u_y,\ldots,u_{xx},u_{xy},\ldots$ of the function. Here Eq. (1.1.1) is considered in a suitable domain D of the n-dimensional space R^n in the independent variables $x,y,\ldots$. We seek functions $u=u(x,y,\ldots)$ which satisfy Eq. (1.1.1) identically in D. Such functions, if they exist, are called *solutions* of Eq. (1.1.1). From these many possible solutions we attempt to select a particular one by introducing suitable additional conditions.

For instance,

$$\begin{gathered} uu_{xy}+u_x=y \\ u_{xx}+2yu_{xy}+3xu_{yy}=4\sin x \\ (u_x)^2+(u_y)^2=1 \\ u_{xx}-u_{yy}=0 \end{gathered} \tag{1.1.2}$$

---

[1]Subscripts on dependent variables denote differentiations, e.g.,

$$u_x=\partial u/\partial x \qquad u_{xy}=\partial^2u/\partial y\partial x$$

are partial differential equations. The functions

$$u(x,y)=(x+y)^3$$
$$u(x,y)=\sin(x-y)$$

are solutions of the last equation of (1.1.2) as can easily be verified.

The *order* of a partial differential equation is the order of the highest-ordered partial derivative appearing in the equation. For example,

$$u_{xx}+2xu_{xy}+u_{yy}=e^{y}$$

is a second-order partial differential equation, and

$$u_{xxy}+xu_{yy}+8u=7y$$

is a third-order partial differential equation.

A partial differential equation is said to be *linear* if it is linear in the unknown function and all its derivatives with coefficients depending only on the independent variables; it is said to be *quasilinear* if it is linear in the highest-ordered derivative of the unknown function. For example, the equation

$$yu_{xx}+2xyu_{yy}+u=1$$

is a second-order linear partial differential equation whereas

$$u_x u_{xx}+xuu_y=\sin y$$

is a second-order quasilinear partial differential equation. The equation which is not linear is called a *nonlinear* equation.

We shall be primarily concerned with linear second-order partial differential equations frequently arising in problems of mathematical physics. The most general second-order linear partial differential equation in n independent variables has the form

$$\sum_{i,j=1}^{n} A_{ij}u_{x_ix_j}+\sum_{i=1}^{n} B_iu_{x_i}+Fu=G \tag{1.1.3}$$

where we assume without loss of generality that $A_{ij}=A_{ji}$. We also assume that B_i, F, and G are functions of the n independent variables x_i.

If G is identically zero, the equation is said to be *homogeneous*; otherwise it is *nonhomogeneous*.

The general solution of a linear ordinary differential equation of nth order is a family of functions depending on n independent arbitrary constants. In the case of partial differential equations the general solution depends on arbitrary functions rather than on arbitrary constants. To illustrate this, consider the equation

$$u_{xy}=0$$

If we integrate this equation with respect to y, we obtain

$$u_x(x,y)=f(x)$$

A second integration with respect to x yields

$$u(x,y)=g(x)+h(y)$$

where $g(x)$ and $h(y)$ are arbitrary functions.

Suppose u is a function of three variables, x, y, and z. Then for the equation

$$u_{yy}=2$$

one finds the general solution

$$u(x,y,z)=y^2+yf(x,z)+g(x,z)$$

where f and g are arbitrary functions of two variables x and z.

We recall that in the case of ordinary differential equations, the first task is to find the general solution, and then a particular solution is determined by finding the values of arbitrary constants from the prescribed conditions. But, for partial differential equations, selecting a particular solution satisfying the supplementary conditions from the general solution of a partial differential equation may be as difficult as, or even more difficult than, the problem of finding the general solution itself. This is so because the general solution of a partial differential equation involves arbitrary functions; the specialization of such a solution to the particular form which satisfies supplementary conditions requires the determination of these arbitrary functions, rather than merely the determination of constants.

For linear homogenous ordinary differential equations of order n, a linear combination of n linearly independent solutions is a solution. Unfortunately, this is not true, in general, in the case of partial differential equations. This is due to the fact that the solution space of every homogenous linear partial differential equation is infinite dimensional. For example, the partial differential equation

$$u_x-u_y=0 \tag{1.1.4}$$

can be transformed into the equation

$$2u_\eta=0$$

by the transformation of variables

$$\xi=x+y$$
$$\eta=x-y$$

The general solution is

$$u(x,y)=f(x+y)$$

where $f(x+y)$ is an arbitrary function. Thus, we see that each of the functions

$$(x+y)^n$$
$$\sin n(x+y)$$
$$\cos n(x+y)$$
$$\exp n(x+y), \qquad n=1,2,3,\ldots$$

is a solution of Eq. (1.1.4). The fact, that a simple equation, such as (1.1.4), yields infinitely many solutions, is an indication of an added difficulty which must be overcome in the study of partial differential equations. Thus, we generally prefer to determine directly the particular solution of a partial differential equation satisfying prescribed supplementary conditions.

1.2 Mathematical Problems

A problem consists of finding an unknown function of a partial differential equation satisfying appropriate supplementary conditions. These conditions may be initial and/or boundary conditions. For example

$$\begin{array}{llll} P.D.E. & u_t - u_{xx} = 0 & 0 < x < l, & t > 0 \\ I.C. & u(x,0) = \sin x & 0 \leqslant x \leqslant l & \\ B.C. & u(0,t) = 0 & & t \geqslant 0 \\ B.C. & u(l,t) = 0 & & t \geqslant 0 \end{array}$$

is a problem which consists of a partial differential equation and three supplementary conditions. The equation describes the heat conduction in a rod of length l. The last two conditions are called the *boundary conditions* which describe the function at two prescribed boundary points. The first condition is known as the *initial condition* which prescribes the unknown function $u(x,t)$ throughout the given region at some initial time t, in this case $t=0$. This problem is known as the *initial-boundary value problem.* Mathematically speaking, the time and the space coordinates are regarded as some independent variables. In this respect, the initial condition is merely a point prescribed on the t-axis and the boundary conditions are prescribed, in this case, as two points on the x-axis. Initial conditions are usually prescribed at a certain time $t=t_0$ or $t=0$, but it is not customary to consider the other end point of a given time interval.

In many cases, in addition to prescribing the unknown function, other conditions such as their derivatives are specified on the boundary.

In considering the problem of unbounded domain, the solution can be determined uniquely by prescribing initial conditions only. The problem is called the initial value problem.[2] The solution of such a problem may be interpreted physically as the solution unaffected by the boundary conditions at infinity. For problems affected by the boundary at infinity, boundedness conditions on the behavior of solutions at infinity must be prescribed.

---

[2]The mathematical definition is given in Chapter 4.

A mathematical problem is said to be properly posed if it satisfies the following requirements:

1. Existence: There is at least one solution.
2. Uniqueness: There is at most one solution.
3. Continuity: The solution depends continuously on the data.

The first requirement is an obvious logical condition, but we must keep in mind that we cannot simply state that the mathematical problem has a solution just because the physical problem has a solution. We may well be erroneously developing a mathematical model, say, consisting of a partial differential equation whose solution may not exist at all. The same can be said about the uniqueness requirement. In order to really reflect the physical problem that has a unique solution, the mathematical problem must have a unique solution.

The last requirement is a useful one. In practice, small errors occur in the process of measurements. Thus, for the mathematical problem to represent a physical phenomenon, a small variation of the given data should lead to at most a small change in the solution.

1.3 Linear Operators

An operator is a mathematical rule which when applied to a function produces another function. For example, in the expressions

$$L[u]=\frac{\partial^2 u}{\partial x^2}+\frac{\partial^2 u}{\partial y^2}$$

$$M[u]=\frac{\partial^2 u}{\partial x^2}-\frac{\partial u}{\partial x}+x\frac{\partial u}{\partial y}$$

$L=\partial^2/\partial x^2+\partial^2/\partial y^2$ and $M=\partial^2/\partial x^2-\partial/\partial x+x(\partial/\partial y)$ are called the differential operators.

An operator is said to be *linear* if it satisfies the following:

1. A constant c may be taken outside the operator:
$$L[cu]=cL[u] \tag{1.3.1}$$
2. The operator operating on the sum of two functions gives the sum of the operator operating on the individual functions:
$$L[u_1+u_2]=L[u_1]+L[u_2] \tag{1.3.2}$$

We may combine (1) and (2) to express as

$$L[c_1u_1+c_2u_2]=c_1L[u_1]+c_2L[u_2] \tag{1.3.3}$$

where c_1 and c_2 are constants. This can be extended to a finite number of functions. If $u_1,u_2,\ldots,u_k$ are k functions and $c_1,c_2,\ldots,c_k$ are k constants, then by repeated application of (1.3.3)

$$L\left[\sum_{j=1}^{k} c_j u_j\right] = \sum_{j=1}^{k} c_j L[u_j] \tag{1.3.4}$$

We may now define the sum of two linear differential operators formally. If L and M are two linear operators, then the sum of L and M is defined as

$$(L+M)[u] = L[u] + M[u] \tag{1.3.5}$$

where u is a sufficiently differentiable function. It can be readily shown that $L+M$ is also a linear operator.

The product of two linear differential operators L and M is the operator which produces the same result as is obtained by the successive operations of the operators L and M on u, that is,

$$LM[u] = L(M[u]) \tag{1.3.6}$$

in which we assume that $M[u]$ and $L(M[u])$ are defined. It can be readily shown that LM is also a linear operator.

Linear differential operators satisfy the following:

1. $L+M=M+L$ (commutative) (1.3.7)
2. $(L+M)+N=L+(M+N)$ (associative) (1.3.8)
3. $(LM)N=L(MN)$ (associative) (1.3.9)
4. $L(M+N)=LM+LN$ (distributive) (1.3.10)

For linear differential operators with constant coefficients,

5. $LM=ML$ (commutative) (1.3.11)

EXAMPLE 1.3.1. Let $L=\dfrac{\partial^2}{\partial x^2}+x\dfrac{\partial}{\partial y}$ and $M=\dfrac{\partial^2}{\partial y^2}-y\dfrac{\partial}{\partial y}$

$$\begin{aligned}
LM[u] &= \left(\frac{\partial^2}{\partial x^2}+x\frac{\partial}{\partial y}\right)\left(\frac{\partial^2 u}{\partial y^2}-y\frac{\partial u}{\partial y}\right)\\
&= \frac{\partial^4 u}{\partial x^2\partial y^2}-y\frac{\partial^3 u}{\partial x^2\partial y}+x\frac{\partial^3 u}{\partial y^3}-xy\frac{\partial^2 u}{\partial y^2}-x\frac{\partial u}{\partial y}\\
ML[u] &= \left(\frac{\partial^2}{\partial y^2}-y\frac{\partial}{\partial y}\right)\left(\frac{\partial^2 u}{\partial x^2}+x\frac{\partial u}{\partial y}\right)\\
&= \frac{\partial^4 u}{\partial y^2\partial x^2}+x\frac{\partial^3 u}{\partial y^3}-y\frac{\partial^3 u}{\partial y\partial x^2}-xy\frac{\partial^2 u}{\partial y^2}
\end{aligned}$$

which shows that $LM \neq ML$.

Now let us consider a linear second-order partial differential equation. In the case of two independent variables, such an equation takes the form

$$A(x,y)u_{xx}+B(x,y)u_{xy}+C(x,y)u_{yy} \\ +D(x,y)u_x+E(x,y)u_y+F(x,y)u=G(x,y) \qquad (1.3.12)$$

where A, B, C, D, E, and F are the coefficients, and G is the nonhomogeneous term.

If we denote

$$L=A\frac{\partial^2}{\partial x^2}+B\frac{\partial^2}{\partial x\partial y}+C\frac{\partial^2}{\partial y^2}+D\frac{\partial}{\partial x}+E\frac{\partial}{\partial y}+F,$$

then Eq. (1.3.12) may be written in the form

$$L[u]=G \qquad (1.3.13)$$

Very often the square bracket is omitted and one simply writes

$$Lu=G$$

Let $v_1, v_2, \ldots, v_n$ be n functions which satisfy

$$L[v_j]=G_j, \quad j=1,2,\ldots,n$$

and let $w_1, w_2, \ldots, w_n$ be n functions which satisfy

$$L[w_j]=0, \quad j=1,2,\ldots,n.$$

If we let

$$u_j=v_j+w_j,$$

then the function

$$u=\sum_{j=1}^{n} u_j$$

satisfies

$$L[u]=\sum_{j=1}^{n} G_j$$

This is called the *principle of superposition*.

In particular, if v is a particular solution of (1.3.13), that is, $L[v]=G$ and w is a solution of the associated homogeneous equation, that is, $L[w]=0$, then $u=v+w$ is a solution of $L[u]=G$.

1.4 Superposition

We may express supplementary conditions using the operator notation. For instance, the initial boundary value problem

$$u_{tt}-c^2u_{xx}=G(x,t) \qquad 0<x<l, \qquad t>0 \qquad (1.4.1)$$

$$u(x,0)=g_1(x) \qquad 0\leqslant x\leqslant l$$

$$u_t(x,0)=g_2(x) \qquad 0\leqslant x\leqslant l$$
$$u(0,t)=g_3(t) \qquad t\geqslant 0$$
$$u(l,t)=g_4(t) \qquad t\geqslant 0$$

may be written in the form

$$\begin{aligned} L[u]&=G \\ M_1[u]&=g_1 \\ M_2[u]&=g_2 \\ M_3[u]&=g_3 \\ M_4[u]&=g_4 \end{aligned} \tag{1.4.2}$$

where g_i are the prescribed functions and the subscripts on operators are assigned arbitrarily.

Now let us consider the problem

$$\begin{aligned} L[u]&=G \\ M_1[u]&=g_1 \\ M_2[u]&=g_2 \\ &\vdots \\ M_n[u]&=g_n \end{aligned} \tag{1.4.3}$$

By virtue of the linearity of the equation and the supplementary conditions, we may divide problem (1.4.3) into a series of problems as follows:

$$\begin{array}{cccc} L[u_1]=G & L[u_2]=0 & & L[u_n]=0 \\ M_1[u_1]=0 & M_1[u_2]=g_1 & & M_1[u_n]=0 \\ M_2[u_1]=0 & M_2[u_2]=0 & \cdots & M_2[u_n]=0 \\ \vdots & \vdots & & \vdots \\ M_n[u_1]=0 & M_n[u_2]=0 & \cdots & M_n[u_n]=g_n \\ (1.4.4) & (1.4.5) & & (1.4.6) \end{array}$$

Then the solution of (1.4.3) is given by

$$u=\sum_{i=1}^{n} u_i \tag{1.4.7}$$

Let us consider one of the subproblems, say, (1.4.5). Suppose we find a sequence of functions $\phi_1,\phi_2,\ldots,$ which may be finite or infinite, satisfying the homogeneous system

$$\begin{aligned} L[\phi_i] &= 0 \\ M_2[\phi_i] &= 0 \\ &\vdots \\ M_n[\phi_i] &= 0, \quad i = 1,2,3,\ldots \end{aligned} \tag{1.4.8}$$

and suppose we can express g_1 in terms of the series

$$g_1 = c_1 M_1[\phi_1] + c_2 M_1[\phi_2] + \ldots \tag{1.4.9}$$

Then the linear combination

$$u_1 = c_1\phi_1 + c_2\phi_2 + \ldots \tag{1.4.10}$$

is the solution of problem (1.4.5). In the case of an infinite number of terms in the linear combination (1.4.10), we require that the infinite series be uniformly convergent and sufficiently differentiable and all the series $N_k(u_1)$ where $N_0 = L$, $N_j = M_j$ for $j = 1,2,\ldots,n$ converge uniformly.

Exercises

1. For each of the following state whether the partial differential equation is linear, quasilinear or nonlinear. If it is linear, state whether it is homogeneous or nonhomogeneous, and give its order.

a. $u_{xx} + xu_y = y$

b. $uu_x - 2xyu_y = 0$

c. $u_x^2 + uu_y = 1$

d. $u_{xxxx} + 2u_{xxyy} + u_{yyyy} = 0$

e. $u_{xx} + 2u_{xy} + u_{yy} = \sin x$

f. $u_{xxx} + u_{xyy} + \log u = 0$

g. $u_{xx}^2 + u_x^2 + \sin u = e^y$

2. Verify that the functions

$$u(x,y) = x^2 - y^2$$
$$u(x,y) = e^x \sin y$$

are solutions of the equation

$$u_{xx} + u_{yy} = 0$$

3. Show that $u = f(xy)$ where f is an arbitrary differentiable function, satisfies

$$xu_x - yu_y = 0$$

and verify that the functions $\sin(xy)$, $\cos(xy)$, $\log(xy)$, e^{xy}, and $(xy)^3$ are solutions.

4. Show that $u=f(x)g(y)$ where f and g are arbitrary twice differentiable functions, satisfies

$$uu_{xy}-u_x u_y=0$$

5. Determine the general solution of the differential equation

$$u_{yy}+u=0$$

6. Find the general solution of

$$u_{xy}+u_x=0$$

by setting $u_x=v$.

7. Find the general solution of

$$u_{xx}-4u_{xy}+3u_{yy}=0$$

by assuming the solution to be in the form $u(x,y)=f(\lambda x+y)$ where λ is an unknown parameter.

8. Find the general solution of

$$u_{xx}-u_{yy}=0$$

CHAPTER TWO

Mathematical Models

2.1 The Classical Equations

The three basic types of second-order partial differential equations are:

a. The wave equation

$$u_{tt}-c^2(u_{xx}+u_{yy}+u_{zz})=0$$

b. The heat equation

$$u_t-k(u_{xx}+u_{yy}+u_{zz})=0$$

c. The Laplace equation

$$u_{xx}+u_{yy}+u_{zz}=0$$

Many problems in mathematical physics reduce to the solving of partial differential equations, in particular, to the partial differential equations listed above. We will begin our study of these equations by first examining in detail the mathematical models representing physical problems.

2.2 The Vibrating String

One of the most important problems in mathematical physics is the vibration of a stretched string. Simplicity and frequent occurrence in many branches of mathematical physics make it a classic example in the theory of partial differential equations.

Let us consider a stretched string of length l fixed at the end points. The problem here is to determine the equation of motion which characterizes the position $u(x,t)$ of the string at time t after an initial disturbance is given.

In order to obtain a simple equation we make the following assumptions:

1. The string is flexible and elastic, that is the string cannot resist bending moment and thus the tension in the string is always in the direction of the tangent to the existing profile of the string.
2. There is no elongation of a single segment of the string, and hence by Hooke's law, the tension is constant.
3. The weight of the string is small compared with the tension in the string.
4. The deflection is small compared with the length of the string.
5. The slope at any point of the displaced string is small compared with unity.
6. There is only pure transverse vibration.

We consider a differential element of the string. Let T be the tension at the end points as shown in Fig. 2.2.1. The forces acting on the element of the string in the vertical direction are

$$T\sin\beta - T\sin\alpha$$

By Newton's second law of motion, the resultant force is equal to the mass times the acceleration. Hence,

$$T\sin\beta - T\sin\alpha = \rho\Delta s u_{tt} \tag{2.2.1}$$

where ρ is the density and Δs is the small arc length of the string. Since the slope of the displaced string is small, we have

$$\Delta s \simeq \Delta x$$

Since the angles α and β are small

$$\sin\alpha \simeq \tan\alpha \qquad \sin\beta \simeq \tan\beta$$

Thus Eq. (2.2.1) becomes

$$\tan\beta - \tan\alpha = \frac{\rho\Delta x}{T} u_{tt} \tag{2.2.2}$$

But, from calculus we know that

$$\tan\alpha = (u_x)_x$$

and

$$\tan\beta = (u_x)_{x+\Delta x}$$

at time t. Equation (2.2.2) may thus be written as

$$\frac{1}{\Delta x}\left[(u_x)_{x+\Delta x} - (u_x)_x\right] = \frac{\rho}{T} u_{tt}$$

In the limit as Δx approaches zero, we find

$$u_{tt} = c^2 u_{xx} \tag{2.2.3}$$

where $c^2 = T/\rho$. This is called the one-dimensional wave equation.

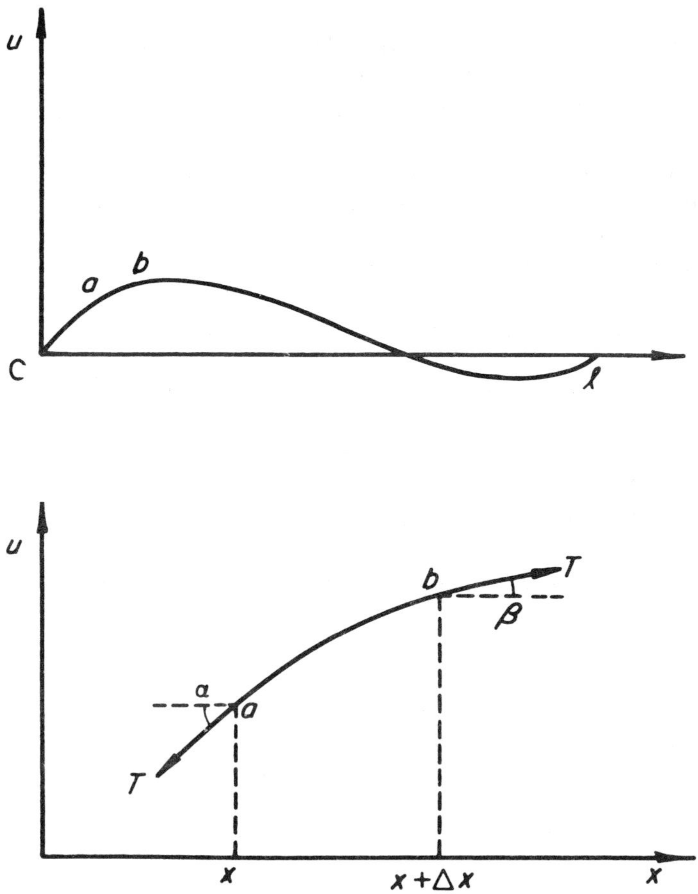

Figure 2.2.1

If there is an external force f per unit length acting on the string, Eq. (2.2.3) assumes the form

$$u_{tt} = c^2 u_{xx} + f^*, \qquad f^* = f/\rho \tag{2.2.4}$$

where f may be pressure, gravitation, resistance, and so on.

2.3 The Vibrating Membrane

The equation of the vibrating membrane occurs in a great number of problems in mathematical physics. Before we derive the equation for the vibrating membrane we make certain simplifying assumptions as in the case of the vibrating string:

1. The membrane is flexible and elastic, that is, the membrane cannot resist bending moment and the tension in the membrane is always in the direction of the tangent to the existing profile of the membrane.
2. There is no elongation of a single element of the membrane, and hence by Hooke's law, the tension is constant.
3. The weight of the membrane is small compared with the tension in the membrane.
4. The deflection is small compared with the minimal diameter of the membrane.
5. The slope at any point of the displaced membrane is small compared with unity.
6. There is only pure transverse vibration.

We consider a small element of the membrane. Since the deflection and slope are small, the area of the element is approximately equal to $\Delta x\Delta y$. If T is the tensile force per unit length, then the forces acting on the sides of the element are $T\Delta x$ and $T\Delta y$, as shown in Fig. 2.3.1.

The forces acting on the element of the membrane in the vertical direction are

$$T\Delta x\sin\beta - T\Delta x\sin\alpha + T\Delta y\sin\delta - T\Delta y\sin\gamma$$

Since the slopes are small, sines of the angles are approximately equal to their tangents. Thus the resultant force becomes

$$T\Delta x(\tan\beta-\tan\alpha)+T\Delta y(\tan\delta-\tan\gamma)$$

By Newton's second law of motion, the resultant force is equal to the mass times the acceleration. Hence

$$T\Delta x(\tan\beta-\tan\alpha)+T\Delta y(\tan\delta-\tan\gamma)=\rho\Delta A u_{tt} \tag{2.3.1}$$

where ρ is the mass per unit area, $\Delta A \simeq \Delta x\Delta y$ is the area of this element, and u_{tt} is computed at some point in the region under consideration. But from calculus, we have

$$\begin{aligned}\tan\alpha&=u_y(x_1,y)\\ \tan\beta&=u_y(x_2,y+\Delta y)\\ \tan\delta&=u_x(x,y_1)\\ \tan\gamma&=u_x(x+\Delta x,y_2)\end{aligned}$$

where x_1 and x_2 are the values of x between x and $x+\Delta x$, and y_1 and y_2 are the values of y between y and $y+\Delta y$. Substituting these values in (2.3.1), we obtain

$$T\Delta x\left[u_y(x_2,y+\Delta y)-u_y(x_1,y)\right]+T\Delta y\left[u_x(x+\Delta x,y_2)-u_x(x,y_1)\right]=\rho\Delta x\Delta y u_{tt}$$

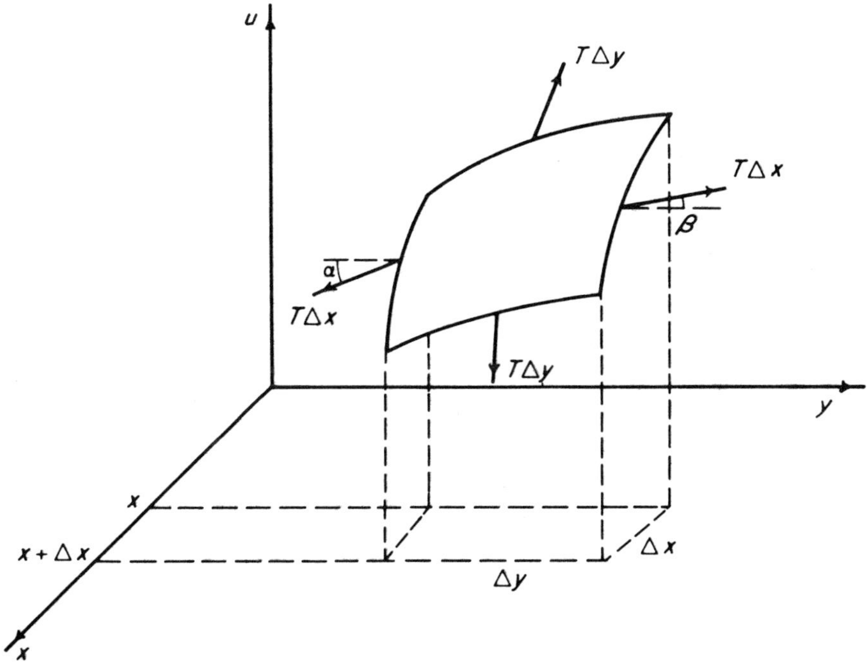

Figure 2.3.1

Division by $\rho\Delta x\Delta y$ yields

$$\frac{T}{\rho}\left[\frac{u_y(x_2,y+\Delta y)-u_y(x_1,y)}{\Delta y}+\frac{u_x(x+\Delta x,y_2)-u_x(x,y_1)}{\Delta x}\right]=u_{tt} \tag{2.3.2}$$

In the limit as Δx approaches zero and Δy approaches zero, we obtain

$$u_{tt}=c^2(u_{xx}+u_{yy}) \tag{2.3.3}$$

where $c^2=T/\rho$. This equation is called the two-dimensional wave equation.

If there is an external force f per unit area acting on the membrane, Eq. (2.3.3) takes the form

$$u_{tt}=c^2(u_{xx}+u_{yy})+f^* \tag{2.3.4}$$

where $f^*=f/\rho$

2.4 Waves in Elastic Medium

If a small disturbance is originated at a point in an elastic medium, neighboring particles are set into motion, and the medium is under a state of strain. We consider such states of motion extend in all directions. We assume that the displacements of the medium are small and that we are not concerned with translation or rotation of the medium as a whole.

Let the body under investigation be homogeneous and isotropic. Let ΔV be a differential volume of the body, and let the stresses acting on the faces of the volume be $\tau_{xx}, \tau_{yy}, \tau_{zz}, \tau_{xy}, \tau_{xz}, \tau_{yx}, \tau_{yz}, \tau_{zx}, \tau_{zy}$. The first three stresses are called the normal stresses and the rest are called the shear stresses. (See Fig. 2.4.1)

We shall assume that the stress tensor τ_{ij} is symmetric,[3] that is,

$$\tau_{ij} = \tau_{ji} \quad i \neq j \quad i,j = x,y,z \tag{2.4.1}$$

Neglecting the body forces, the sum of all the forces acting on the volume element in the x-direction is

$$\left[(\tau_{xx})_{x+\Delta x} - (\tau_{xx})_x\right]\Delta y \Delta z + \left[(\tau_{xy})_{y+\Delta y} - (\tau_{xy})_y\right]\Delta z \Delta x + \left[(\tau_{xz})_{z+\Delta z} - (\tau_{xz})_z\right]\Delta x \Delta y$$

By Newton's law of motion this resultant force is equal to the mass times the acceleration. Thus we obtain

$$\left[(\tau_{xx})_{x+\Delta x} - (\tau_{xx})_x\right]\Delta y \Delta z + \left[(\tau_{xy})_{y+\Delta y} - (\tau_{xy})_y\right]\Delta z \Delta x + \left[(\tau_{xz})_{z+\Delta z} - (\tau_{xz})_z\right]\Delta x \Delta y = \rho \Delta x \Delta y \Delta z u_{tt} \tag{2.4.2}$$

where ρ is the density of the body and u is the displacement component in the x-direction. Hence, in the limit as ΔV approaches zero, we obtain

$$\frac{\partial \tau_{xx}}{\partial x} + \frac{\partial \tau_{xy}}{\partial y} + \frac{\partial \tau_{xz}}{\partial z} = \rho \frac{\partial^2 u}{\partial t^2} \tag{2.4.3}$$

Similarly, the following two equations corresponding to y and z directions are obtained:

$$\frac{\partial \tau_{yx}}{\partial x} + \frac{\partial \tau_{yy}}{\partial y} + \frac{\partial \tau_{yz}}{\partial z} = \rho \frac{\partial^2 v}{\partial t^2} \tag{2.4.4}$$

$$\frac{\partial \tau_{zx}}{\partial x} + \frac{\partial \tau_{zy}}{\partial y} + \frac{\partial \tau_{zz}}{\partial z} = \rho \frac{\partial^2 w}{\partial t^2} \tag{2.4.5}$$

where v and w are the displacement components in the y and z directions respectively.

---

[3] The condition of the rotational equilibrium of the volume element.

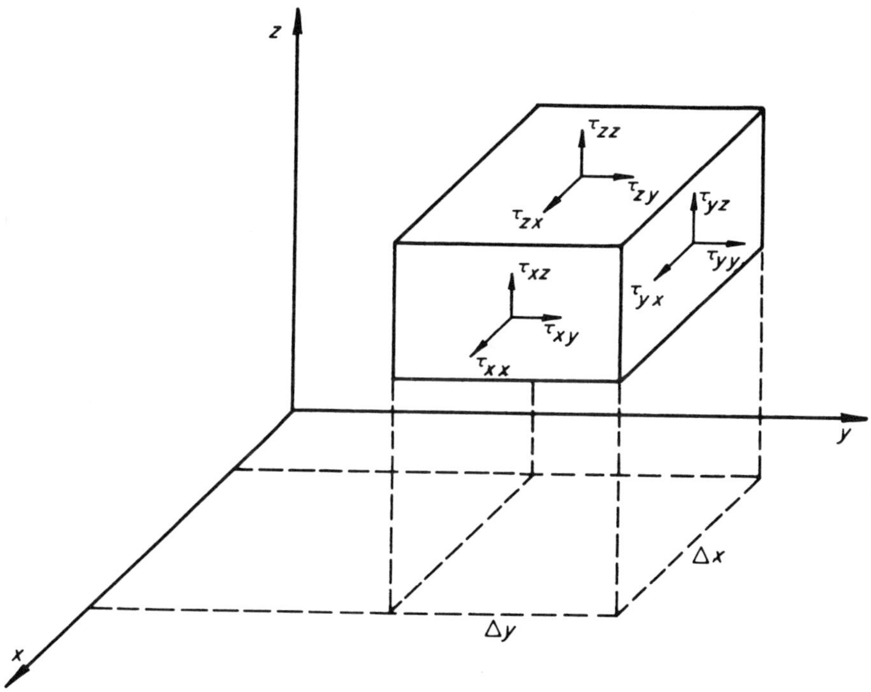

Figure 2.4.1

We may now define linear strains [40] as

$$
\begin{aligned}
\varepsilon_{xx} &= \frac{\partial u}{\partial x} & \varepsilon_{yz} &= \frac{\partial w}{\partial y} + \frac{\partial v}{\partial z} \\
\varepsilon_{yy} &= \frac{\partial v}{\partial y} & \varepsilon_{zx} &= \frac{\partial u}{\partial z} + \frac{\partial w}{\partial x} \\
\varepsilon_{zz} &= \frac{\partial w}{\partial z} & \varepsilon_{xy} &= \frac{\partial v}{\partial x} + \frac{\partial u}{\partial y}
\end{aligned}
\tag{2.4.6}
$$

in which $\varepsilon_{xx}, \varepsilon_{yy}, \varepsilon_{zz}$ represent unit elongations and $\varepsilon_{yz}, \varepsilon_{zx}, \varepsilon_{xy}$ represent unit shearing strains.

In the case of an isotropic body, generalized Hooke's law takes the form

$$
\begin{aligned}
\tau_{xx} &= \lambda\theta + 2\mu\varepsilon_{xx} & \tau_{yz} &= \mu\varepsilon_{yz} \\
\tau_{yy} &= \lambda\theta + 2\mu\varepsilon_{yy} & \tau_{zx} &= \mu\varepsilon_{zx} \\
\tau_{zz} &= \lambda\theta + 2\mu\varepsilon_{zz} & \tau_{xy} &= \mu\varepsilon_{xy}
\end{aligned}
\tag{2.4.7}
$$

where $\theta = \varepsilon_{xx} + \varepsilon_{yy} + \varepsilon_{zz}$ and λ and μ are Lame's constants.

Expressing stresses in terms of displacements we have

$$\begin{aligned}\tau_{xx} &= \lambda\theta + 2\mu\frac{\partial u}{\partial x}\\ \tau_{xy} &= \mu\left(\frac{\partial v}{\partial x} + \frac{\partial u}{\partial y}\right)\\ \tau_{xz} &= \mu\left(\frac{\partial w}{\partial x} + \frac{\partial u}{\partial z}\right)\end{aligned} \tag{2.4.8}$$

By differentiating Eqs. (2.4.8), we obtain

$$\begin{aligned}\frac{\partial \tau_{xx}}{\partial x} &= \lambda\frac{\partial \theta}{\partial x} + 2\mu\frac{\partial^2 u}{\partial x^2}\\ \frac{\partial \tau_{xy}}{\partial y} &= \mu\frac{\partial^2 v}{\partial x \partial y} + \mu\frac{\partial^2 u}{\partial y^2}\\ \frac{\partial \tau_{xz}}{\partial z} &= \mu\frac{\partial^2 w}{\partial x \partial z} + \mu\frac{\partial^2 u}{\partial z^2}\end{aligned} \tag{2.4.9}$$

Substitution of Eqs. (2.4.9) into Eq. (2.4.3) yields

$$\lambda\frac{\partial \theta}{\partial x} + \mu\left(\frac{\partial^2 u}{\partial x^2} + \frac{\partial^2 v}{\partial x \partial y} + \frac{\partial^2 w}{\partial x \partial z}\right) + \mu\left(\frac{\partial^2 u}{\partial x^2} + \frac{\partial^2 u}{\partial y^2} + \frac{\partial^2 u}{\partial z^2}\right) = \rho\frac{\partial^2 u}{\partial t^2} \tag{2.4.10}$$

We note that

$$\frac{\partial^2 u}{\partial x^2} + \frac{\partial^2 v}{\partial x \partial y} + \frac{\partial^2 w}{\partial x \partial z} = \frac{\partial}{\partial x}\left(\frac{\partial u}{\partial x} + \frac{\partial v}{\partial y} + \frac{\partial w}{\partial z}\right) = \frac{\partial \theta}{\partial x}$$

and introduce the notation

$$\boldsymbol{\Delta} = \nabla^2 = \frac{\partial^2}{\partial x^2} + \frac{\partial^2}{\partial y^2} + \frac{\partial^2}{\partial z^2}$$

The symbol Δ or ∇^2 is called the Laplace operator. Hence, Eq. (2.4.10) becomes

$$(\lambda + \mu)\frac{\partial \theta}{\partial x} + \mu\nabla^2 u = \rho\frac{\partial^2 u}{\partial t^2} \tag{2.4.11}$$

In a similar manner, we obtain the other two equations which are

$$(\lambda + \mu)\frac{\partial \theta}{\partial y} + \mu\nabla^2 v = \rho\frac{\partial^2 v}{\partial t^2} \tag{2.4.12}$$

$$(\lambda + \mu)\frac{\partial \theta}{\partial z} + \mu\nabla^2 w = \rho\frac{\partial^2 w}{\partial t^2} \tag{2.4.13}$$

In vector form, the equations of motion assume the form

$$(\lambda+\mu)\,\mathrm{grad}\,\mathrm{div}\,\mathbf{u}+\mu\nabla^2\mathbf{u}=\rho\mathbf{u}_{tt} \tag{2.4.14}$$

where $\mathbf{u}=u\mathbf{i}+v\mathbf{j}+w\mathbf{k}$ and $\theta=\mathrm{div}\,\mathbf{u}$.

(i) If $\mathrm{div}\,\mathbf{u}=0$, the general equation becomes

$$\mu\nabla^2\mathbf{u}=\rho\mathbf{u}_{tt}$$

or

$$\mathbf{u}_{tt}=c^2\nabla^2\mathbf{u} \tag{2.4.15}$$

where the velocity c of propagated wave is

$$c=\sqrt{\mu/\rho}$$

This is the case of an equivoluminal wave propagation since the volume expansion θ is zero for waves moving with this velocity. Sometimes these waves are called waves of distortion because the velocity of propagation depends on μ and ρ; the shear modulus μ characterizes the distortion and rotation of the volume element.

(ii) When $\mathrm{curl}\,\mathbf{u}=\mathbf{0}$, the identity

$$\mathrm{curl}\,\mathrm{curl}\,\mathbf{u}=\mathrm{grad}\,\mathrm{div}\,\mathbf{u}-\nabla^2\mathbf{u}$$

gives

$$\mathrm{grad}\,\mathrm{div}\,\mathbf{u}=\nabla^2\mathbf{u}$$

Then the general equation becomes

$$(\lambda+2\mu)\nabla^2\mathbf{u}=\rho\mathbf{u}_{tt}$$

or

$$\mathbf{u}_{tt}=c^2\nabla^2\mathbf{u} \tag{2.4.16}$$

where the velocity of propagation is

$$c=\sqrt{\frac{\lambda+2\mu}{\rho}}$$

This is the case of an irrotational or dilatational wave propagation, since $\mathrm{curl}\,\mathbf{u}=0$ describes irrotational motion. Equations (2.4.15) and (2.4.16) are called the three-dimensional wave equations.

In general the *wave equation* may be written as

$$u_{tt}=c^2\nabla^2 u \tag{2.4.17}$$

where the Laplace operator may be one, two or three dimensional. The importance of the wave equation stems from the fact that this type of equation arises in many physical problems; for example: sound wave in space, electrical vibration in a conductor, and torsional oscillation of a rod.

2.5 Conduction of Heat in Solids

We consider a domain D^* bounded by a closed surface B^*. Let $u(x,y,z,t)$ be the temperature at a point (x,y,z) at time t. If the temperature is not constant, heat flows from places of higher temperature to places of lower temperature. Fourier law states that the rate of flow is proportional to the gradient of the temperature. Thus the velocity of the heat flow in an isotropic body is

$$\mathbf{v} = -K \operatorname{grad} u \tag{2.5.1}$$

where K is a constant, called the thermal conductivity of the body.

Let D be an arbitrary domain bounded by a closed surface B in D^*. Then the amount of heat leaving D per unit time is

$$\iint_B v_n \, ds$$

where $v_n = \mathbf{v} \cdot \mathbf{n}$ is the component of $\mathbf{v}$ in the direction of the outer unit normal $\mathbf{n}$ of B. Thus by Gauss theorem (Divergence theorem)

$$\begin{aligned}\iint_B v_n \, ds &= \iiint_D \operatorname{div}(-K \operatorname{grad} u)\, dx\, dy\, dz \\ &= -K \iiint_D \nabla^2 u \, dx\, dy\, dz \end{aligned}\tag{2.5.2}$$

But the amount of heat in D is given by

$$\iiint_D \sigma \rho u \, dx\, dy\, dz \tag{2.5.3}$$

where ρ is the density of the material of the body, σ is its specific heat. Assuming that integration and differentiation are interchangeable, the rate of decrease of heat in D is

$$-\iiint_D \sigma \rho \frac{\partial u}{\partial t} \, dx\, dy\, dz \tag{2.5.4}$$

Since the rate of decrease of heat in D must be equal to the amount of heat leaving D per unit time, we have

$$-\iiint_D \sigma \rho u_t \, dx\, dy\, dz = -K \iiint_D \nabla^2 u \, dx\, dy\, dz$$

or

$$\iiint_D \left[\sigma \rho u_t - K \nabla^2 u\right] dx\, dy\, dz = 0 \tag{2.5.5}$$

for an arbitrary D in D^*. We assume that the integrand is continuous. If we suppose that the integrand is not zero at a point (x_0, y_0, z_0) in D, then by continuity, the integrand is not zero in the small region surrounding the point (x_0,y_0,z_0). Continuing in this fashion we extend the region encompassing D. Hence the integral must be nonzero. This contradicts (2.5.5). Thus, the integrand is zero everywhere, that is,

$$u_t = K\nabla^2 u \tag{2.5.6}$$

where $k = K/\sigma\rho$. This is known as the *heat equation*.

This type of equation appears in a great variety of problems in mathematical physics; for example: the concentration of diffusing material, the motion of tidal wave in a long channel, and the transmission in electrical cables.

2.6 The Gravitational Potential

In this section we shall derive one of the most well-known equations in the theory of partial differential equations, the Laplace equation.

We consider two particles of masses m and M, at P and Q as shown in Fig. 2.6.1. Let r be the distance between them. Then according to Newton's law of gravitation, a force proportional to the product of their masses, and inversely proportional to the square of the distance between them, is given in the form

$$F = G\frac{mM}{r^2} \tag{2.6.1}$$

where G is the gravitational constant.

It is customary in potential theory to choose the unit of force so that $G = 1$. Thus F becomes

$$F = \frac{mM}{r^2} \tag{2.6.2}$$

If $\mathbf{r}$ represents the vector PQ, the force per unit mass at Q due to the mass at P may be written as

$$\mathbf{F} = \frac{-m\mathbf{r}}{r^3} = \nabla\left(\frac{m}{r}\right) \tag{2.6.3}$$

which is called the intensity of the gravitational field of force.

We suppose a particle of unit mass moves under the attraction of the particle of mass m at P from infinity up to Q. The work done by the force $\mathbf{F}$ is

$$\int_\infty^r \mathbf{F}\cdot d\mathbf{r} = \int_\infty^r \nabla\left(\frac{m}{r}\right)\cdot d\mathbf{r} = \frac{m}{r} \tag{2.6.4}$$

This is called the potential at Q due to the particle at P. We denote this by

$$V = -\frac{m}{r} \tag{2.6.5}$$

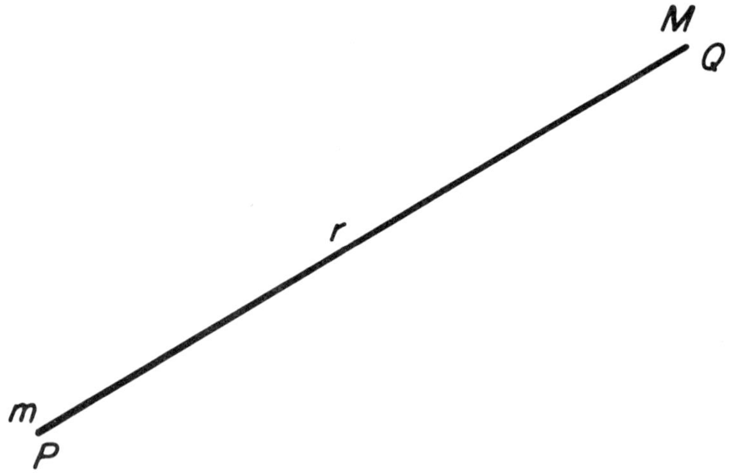

Figure 2.6.1

Hence the intensity of force at P is

$$\mathbf{F} = \nabla\left(\frac{m}{r}\right) = -\nabla V \tag{2.6.6}$$

We shall now consider a number of masses $m_1, m_2, \ldots, m_n$, whose distances from Q are $r_1, r_2, \ldots, r_n$, respectively. Then the force of attraction per unit mass at Q due to the system is

$$\mathbf{F} = \sum_{k=1}^{n} \nabla \frac{m_k}{r_k} = \nabla \sum_{k=1}^{n} \frac{m_k}{r_k} \tag{2.6.7}$$

The work done by the forces acting on a particle of unit mass is

$$\int_{\infty}^{r} \mathbf{F} \cdot d\mathbf{r} = \sum_{k=1}^{n} \frac{m_k}{r_k} = -V \tag{2.6.8}$$

The potential satisfies

$$\nabla^2 V = -\nabla^2 \sum_{k=1}^{n} \frac{m_k}{r_k} = -\sum_{k=1}^{n} \nabla^2 \frac{m_k}{r_k} = 0, \; r_k \neq 0 \tag{2.6.9}$$

In the case of a continuous distribution of mass in some volume R, we have, as in Fig. 2.6.2

$$V(x,y,z) = \int\int\int_R \frac{\rho(\xi,\eta,\zeta)}{r}\, dR \tag{2.6.10}$$

where $r = \sqrt{(x-\xi)^2 + (y-\eta)^2 + (z-\zeta)^2}$ and Q is outside the body. It immediately follows that

$$\nabla^2 V = 0 \tag{2.6.11}$$

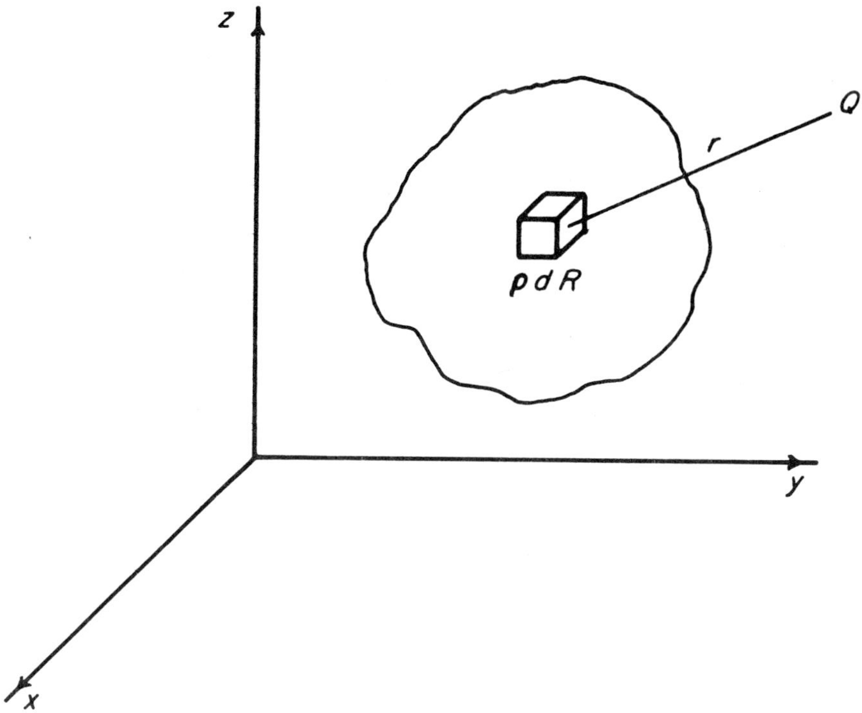

Figure 2.6.2

This equation is called the *Laplace equation*, also known as the potential equation. It appears in many physical problems, such as those of electrostatic potentials, potentials in hydrodynamics, and harmonic potentials in the theory of elasticity. We observe that the Laplace equation can be viewed as the special case of the heat and the wave equations when the dependent variables involved are independent of time.

Exercises

1. Show that the equation of motion of a *long string* is

$$u_{tt} = c^2 u_{xx} - g$$

where g is the gravitational acceleration.

2. Derive the *damped wave equation* of a string

$$u_{tt} + a u_t = c^2 u_{xx}$$

where the damping force is proportional to the velocity and a is a constant. Considering a restoring force proportional to the displacement of a string, show that the resulting equation is

$$u_{tt}+au_t+bu=c^2u_{xx}$$

where b is a constant. This equation is called the *telegraph equation.*

3. Consider the transverse vibration of a uniform beam. Adopting Euler's beam theory, the moment M at a point can be written as

$$M=-EIu_{xx}$$

where EI is called the flexural rigidity, E is the elastic modulus, and I is the moment of inertia of the cross section of the beam. Show that the transverse motion of the beam may be described by

$$u_{tt}+c^2u_{xxxx}=0$$

where $c^2=EI/\rho A$, ρ is the density and A is the cross-sectional area of the beam.

4. Derive the deflection equation of a thin elastic plate

$$\nabla^4 u=q/D$$

where q is the uniform load per unit area, D is the flexural rigidity of the plate, and

$$\nabla^4 u=u_{xxxx}+2u_{xxyy}+u_{yyyy}$$

5. Derive the one dimensional heat equation

$$u_t=ku_{xx}$$

Assuming that heat is also lost by radioactive exponential decay of the material in the bar, show that the above equation becomes

$$u_t=ku_{xx}+he^{-\alpha x}$$

where h and α are constants.

6. Starting from *Maxwell's equations* in electrodynamics, show that in a conducting medium, electric intensity **E**, magnetic intensity **H**, and current density **J**, satisfy

$$\nabla^2\mathbf{X}=\mu\varepsilon\mathbf{X}_{tt}+\mu\sigma\mathbf{X}_t$$

where **X** represents **E**, **H**, and **J**, and μ is the magnetic inductive capacity, ε is the electric inductive capacity and σ is the electric conductivity.

7. Derive the *continuity equation*

$$\rho_t+\operatorname{div}(\rho\mathbf{u})=0$$

and the *Euler's equations of motion*

$$\rho[\mathbf{u}_t+(\mathbf{u}\cdot\operatorname{grad})\mathbf{u}]+\operatorname{grad}p=0$$

in fluid dynamics.

8. In the derivation of the Laplace equation (2.6.11) the potential at Q which is outside the body is ascertained. Now determine the potential at Q when it is inside the body, and show that it satisfies the *Poisson equation*

$$\nabla^2 u = -4\pi\rho$$

where ρ is the density of the body.

9. Setting $U = e^{ikt}u$ in the wave equation $U_{tt} = \nabla^2 U$ and setting $U = e^{-k^2 t}u$ in the heat equation $U_t = \nabla^2 U$, show that $u(x,y,z)$ satisfies the *Helmholtz equation*

$$\nabla^2 u + k^2 u = 0$$

CHAPTER THREE

Classification of Second-order Equations

3.1 Second-order Equations in Two Independent Variables

The general linear second-order partial differential equation in one dependent variable u may be written as

$$\sum_{i,j}^{n} A_{ij}u_{x_ix_j} + \sum_{i}^{n} B_iu_{x_i} + Fu = G \tag{3.1.1}$$

in which, we assume $A_{ij} = A_{ji}$ and A_{ij}, B_i, F, and G are real-valued functions defined in some region of the space $(x_1, x_2, \ldots, x_n)$.

Here we shall be concerned with second order equations in the dependent variable u and the independent variables x,y. Hence Eq. (3.1.1) can be put in the form

$$Au_{xx} + Bu_{xy} + Cu_{yy} + Du_x + Eu_y + Fu = G \tag{3.1.2}$$

where the coefficients are functions of x and y and do not vanish simultaneously. We shall assume that the function u and the coefficients are twice continuously differentiable in some domain R.

The classification of second-order equations[4] is based upon the possibility of reducing Eq. (3.1.2) by a coordinate transformation to *canonical* or *standard* form at a point. An equation is said to be hyperbolic, parabolic,

[4]The classification of partial differential equations is suggested by the classification of the quadratic equation in analytic geometry. The equation

$$Ax^2 + Bxy + Cy^2 + Dx + Ey + F = 0$$

is hyperbolic, parabolic, or elliptic accordingly as $B^2 - 4AC$ is positive, zero, or negative.

or elliptic at a point (x_0,y_0) according as

$$B^2(x_0,y_0)-4A(x_0,y_0)C(x_0,y_0) \tag{3.1.3}$$

is positive, zero, or negative. If this is true at all points, then the equation is said to be hyperbolic, parabolic, or elliptic in a domain. In the case of two independent variables, a transformation can always be found to reduce the given equation to the canonical form in a given domain. However, in the case of several independent variables, it is not, in general, possible to find such a transformation.

To transform Eq. (3.1.2) to a canonical form we make a change of independent variables. Let the new variable be

$$\begin{aligned}\xi&=\xi(x,y)\\ \eta&=\eta(x,y)\end{aligned} \tag{3.1.4}$$

Assuming that ξ and η are twice continuously differentiable and that the Jacobian

$$J=\begin{vmatrix}\xi_x & \xi_y\\ \eta_x & \eta_y\end{vmatrix} \tag{3.1.5}$$

is nonzero in the region under consideration, then x and y can be determined uniquely from the system (3.1.4). Let x and y be twice continuously differentiable functions of ξ and η. Then we have

$$\begin{aligned}u_x&=u_\xi\xi_x+u_\eta\eta_x\\ u_y&=u_\xi\xi_y+u_\eta\eta_y\\ u_{xx}&=u_{\xi\xi}\xi_x^2+2u_{\xi\eta}\xi_x\eta_x+u_{\eta\eta}\eta_x^2+u_\xi\xi_{xx}+u_\eta\eta_{xx}\\ u_{xy}&=u_{\xi\xi}\xi_x\xi_y+u_{\xi\eta}(\xi_x\eta_y+\xi_y\eta_x)+u_{\eta\eta}\eta_x\eta_y+u_\xi\xi_{xy}+u_\eta\eta_{xy}\\ u_{yy}&=u_{\xi\xi}\xi_y^2+2u_{\xi\eta}\xi_y\eta_y+u_{\eta\eta}\eta_y^2+u_\xi\xi_{yy}+u_\eta\eta_{yy}\end{aligned} \tag{3.1.6}$$

Substituting these values in Eq. (3.1.2) we obtain

$$A^*u_{\xi\xi}+B^*u_{\xi\eta}+C^*u_{\eta\eta}+D^*u_\xi+E^*u_\eta+F^*u=G^* \tag{3.1.7}$$

where

$$\begin{aligned}A^*&=A\xi_x^2+B\xi_x\xi_y+C\xi_y^2\\ B^*&=2A\xi_x\eta_x+B(\xi_x\eta_y+\xi_y\eta_x)+2C\xi_y\eta_y\\ C^*&=A\eta_x^2+B\eta_x\eta_y+C\eta_y^2\\ D^*&=A\xi_{xx}+B\xi_{xy}+C\xi_{yy}+D\xi_x+E\xi_y\\ E^*&=A\eta_{xx}+B\eta_{xy}+C\eta_{yy}+D\eta_x+E\eta_y\\ F^*&=F\\ G^*&=G\end{aligned} \tag{3.1.8}$$

The resulting equation (3.1.7) is in the same form as the original equation (3.1.2) under the general transformation (3.1.4). The nature of the equation remains invariant under such a transformation if the Jacobian does not vanish. This can be seen from the fact that the sign of the discriminant does not alter under the transformation, that is,

$$B^{*2}-4A^{*}C^{*}=J^{2}(B^{2}-4AC) \tag{3.1.9}$$

which can be easily verified. It should be noted here that the equation can be of a different type at different points of the domain. But for our purpose we shall assume that the equation under consideration is of the single type in a given domain.

The classification of Eq. (3.1.2) depends on the coefficients $A(x,y)$, $B(x,y)$, and $C(x,y)$ at a given point (x,y). We shall, therefore, rewrite Eq. (3.1.2) as

$$Au_{xx}+Bu_{xy}+Cu_{yy}=H \tag{3.1.10}$$

where

$$H=H(x,y,u,u_x,u_y)$$

and Eq. (3.1.7) as

$$A^{*}u_{\xi\xi}+B^{*}u_{\xi\eta}+C^{*}u_{\eta\eta}=H^{*} \tag{3.1.11}$$

where

$$H^{*}=H^{*}(\xi,\eta,u,u_{\xi},u_{\eta})$$

3.2 Canonical Forms

In this section, we shall consider the problem of reducing Eq. (3.1.10) to canonical form.

We suppose first none of A, B, C, is zero. Let ξ and η be the new variables such that the coefficients A^{*} and C^{*} in Eq. (3.1.11) vanish. Thus, from (3.1.8), we have

$$A^{*}=A\xi_x^2+B\xi_x\xi_y+C\xi_y^2=0$$

$$C^{*}=A\eta_x^2+B\eta_x\eta_y+C\eta_y^2=0$$

These two equations are of the same type and hence we may write them in the form

$$A\zeta_x^2+B\zeta_x\zeta_y+C\zeta_y^2=0 \tag{3.2.1}$$

in which ζ stands for either of the functions ξ or η. Dividing through by ζ_y^2, Eq. (3.2.1) becomes

$$A\left(\frac{\zeta_x}{\zeta_y}\right)^2+B\left(\frac{\zeta_x}{\zeta_y}\right)+C=0 \tag{3.2.2}$$

Along the curve $\zeta = \text{constant}$, we have

$$d\zeta = \zeta_x dx + \zeta_y dy = 0$$

Thus,

$$\frac{dy}{dx} = -\frac{\zeta_x}{\zeta_y}, \tag{3.2.3}$$

and therefore, Eq. (3.2.2) may be written in the form

$$A\left(\frac{dy}{dx}\right)^2 - B\left(\frac{dy}{dx}\right) + C = 0 \tag{3.2.4}$$

the roots of which are

$$\frac{dy}{dx} = \left(B + \sqrt{B^2 - 4AC}\,\right)/2A \tag{3.2.5}$$

$$\frac{dy}{dx} = \left(B - \sqrt{B^2 - 4AC}\,\right)/2A \tag{3.2.6}$$

These equations which are known as the *characteristic equations*, are the ordinary differential equations for families of curves in the xy-plane along which $\xi = \text{constant}$ and $\eta = \text{constant}$. The integrals of Eqs. (3.2.5) and (3.2.6) are called the characteristic curves. Since the equations are first order ordinary differential equations, the solutions may be written as

$$\phi_1(x,y) = c_1, \quad c_1 = \text{constant}$$
$$\phi_2(x,y) = c_2, \quad c_2 = \text{constant}$$

Hence the transformation

$$\xi = \phi_1(x,y)$$
$$\eta = \phi_2(x,y)$$

will transform Eq. (3.1.10) to a canonical form.

(A) Hyperbolic Type

If $B^2 - 4AC > 0$, then integrations of Eqs. (3.2.5) and (3.2.6) yield two real and distinct families of characteristics. Eq. (3.1.11) reduces to

$$u_{\xi\eta} = H_1 \tag{3.2.7}$$

where $H_1 = H^*/B^*$. It can be easily shown that $B^* \neq 0$. This form is called the *first canonical form of the hyperbolic equation*.

Now if the new independent variables

$$\alpha = \xi + \eta$$
$$\beta = \xi - \eta \tag{3.2.8}$$

are introduced, then Eq. (3.2.7) is transformed into

$$u_{\alpha\alpha} - u_{\beta\beta} = H_2(\alpha, \beta, u, u_\alpha, u_\beta) \tag{3.2.9}$$

This form is called the *second canonical form of the hyperbolic equation.*

(B) Parabolic Type

In this case, we have $B^2 - 4AC = 0$, and Eqs. (3.2.5) and (3.2.6) coincide. Thus, there exists one real family of characteristics, and we obtain only a single integral $\xi = \text{constant}$ (or $\eta = \text{constant}$).

Since $B^2 = 4AC$ and $A^* = 0$, we find that

$$A^* = A\xi_x^2 + B\xi_x\xi_y + C\xi_y^2 = \left(\sqrt{A}\,\xi_x + \sqrt{C}\,\xi_y\right)^2 = 0$$

From this it follows that

$$\begin{aligned} B^* &= 2A\xi_x\eta_x + B(\xi_x\eta_y + \xi_y\eta_x) + 2C\xi_y\eta_y \\ &= 2\left(\sqrt{A}\,\xi_x + \sqrt{C}\,\xi_y\right)\left(\sqrt{A}\,\eta_x + \sqrt{C}\,\eta_y\right) \\ &= 0 \end{aligned}$$

for arbitrary values of $\eta(x,y)$ which is functionally independent of $\xi(x,y)$; for instance, if $\eta = y$, the Jacobian does not vanish in the domain of parabolicity.

Division of Eq. (3.1.11) by C^* yields

$$u_{\eta\eta} = H_3(\xi, \eta, u, u_\xi, u_\eta) \quad C^* \neq 0 \tag{3.2.10}$$

This is called the *canonical form of the parabolic equation.*

Eq. (3.1.11) may also assume the form

$$u_{\xi\xi} = H_3^*(\xi, \eta, u, u_\xi, u_\eta) \tag{3.2.11}$$

if we choose $\eta = \text{constant}$ as the integral of Eq. (3.2.5)

(C) Elliptic Type

For an equation of the elliptic type, we have $B^2 - 4AC < 0$. Consequently the quadratic equation (3.2.4) has no real solutions, but it has two complex conjugate solutions which are continuous complex-valued functions of the real variables x and y. Thus, in this case, there are no real characteristic curves. However, if the coefficients A, B and C are analytic functions[5] of x and y, then one can consider Eq. (3.2.4) for complex x and y.

---

[5]A function of two real variables x, y is said to be analytic in a certain domain if in some neighborhood of every point (x_0, y_0) of this domain, the function can be represented as a Taylor series in the variables $(x - x_0)$ and $(y - y_0)$.

Since ξ and η are complex, we introduce the new real variables

$$\alpha = \frac{1}{2}(\xi + \eta)$$

$$\beta = \frac{1}{2i}(\xi - \eta) \tag{3.2.12}$$

so that

$$\xi = \alpha + i\beta$$

$$\eta = \alpha - i\beta \tag{3.2.13}$$

First, we transform Eq. (3.1.10). We then have

$$A^{**}(\alpha,\beta)u_{\alpha\alpha} + B^{**}(\alpha,\beta)u_{\alpha\beta} + C^{**}(\alpha,\beta)u_{\beta\beta} = H_4(\alpha,\beta,u,u_\alpha,u_\beta) \tag{3.2.14}$$

in which the coefficients assume the same form as the coefficients in Eq. (3.1.11). With the use of transformation (3.2.13), the equations $A^* = C^* = 0$ become

$$\left(A\alpha_x^2 + B\alpha_x\alpha_y + C\alpha_y^2\right) - \left(A\beta_x^2 + B\beta_x\beta_y + C\beta_y^2\right)$$
$$+ i\left[2A\alpha_x\beta_x + B(\alpha_x\beta_y + \alpha_y\beta_x) + 2C\alpha_y\beta_y\right] = 0$$
$$\left(A\alpha_x^2 + B\alpha_x\alpha_y + C\alpha_y^2\right) - \left(A\beta_x^2 + B\beta_x\beta_y + C\beta_y^2\right)$$
$$- i\left[2A\alpha_x\beta_x + B(\alpha_x\beta_y + \alpha_y\beta_x) + 2C\alpha_y\beta_y\right] = 0$$

or

$$(A^{**} - C^{**}) + iB^{**} = 0$$
$$(A^{**} - C^{**}) - iB^{**} = 0$$

These equations are satisfied, if and only if,

$$A^{**} = C^{**} \quad \text{and } B^{**} = 0$$

Hence, Eq. (3.2.14) transforms into

$$A^{**}u_{\alpha\alpha} + A^{**}u_{\beta\beta} = H_4(\alpha,\beta,u,u_\alpha,u_\beta)$$

Dividing through by A^{**}, we obtain

$$u_{\alpha\alpha} + u_{\beta\beta} = H_5(\alpha,\beta,u,u_\alpha,u_\beta) \tag{3.2.15}$$

where $H_5 = H_4/A^{**}$. This is called the *canonical form of the elliptic equation.*

EXAMPLE 3.2.1. Consider the equation

$$y^2u_{xx} - x^2u_{yy} = 0$$

Here

$$A = y^2 \quad B = 0 \quad C = -x^2$$

Thus,

$$B^2-4AC=4x^2y^2>0$$

The equation is hyperbolic everywhere except on the coordinate axes $x=0$ and $y=0$. From the characteristic equations (3.2.5) and (3.2.6), we have

$$\frac{dy}{dx}=\frac{x}{y}$$

$$\frac{dy}{dx}=-\frac{x}{y}$$

After integration of these equations, we obtain

$$\frac{1}{2}y^2-\frac{1}{2}x^2=c_1$$

$$\frac{1}{2}y^2+\frac{1}{2}x^2=c_2$$

The first of these curves is a family of hyperbolas

$$\frac{1}{2}y^2-\frac{1}{2}x^2=c_1,$$

and the second is a family of circles

$$\frac{1}{2}y^2+\frac{1}{2}x^2=c_2$$

To transform the given equation to the canonical form, we consider

$$\xi=\frac{1}{2}y^2-\frac{1}{2}x^2$$

$$\eta=\frac{1}{2}y^2+\frac{1}{2}x^2$$

From the relations (3.1.6), we have

$$\begin{aligned}
u_x&=u_\xi\xi_x+u_\eta\eta_x\\
&=-xu_\xi+xu_\eta\\
u_y&=u_\xi\xi_y+u_\eta\eta_y\\
&=yu_\xi+yu_\eta\\
u_{xx}&=u_{\xi\xi}\xi_x^2+2u_{\xi\eta}\xi_x\eta_x+u_{\eta\eta}\eta_x^2+u_\xi\xi_{xx}+u_\eta\eta_{xx}\\
&=x^2u_{\xi\xi}-2x^2u_{\xi\eta}+x^2u_{\eta\eta}-u_\xi+u_\eta\\
u_{yy}&=u_{\xi\xi}\xi_y^2+2u_{\xi\eta}\xi_y\eta_y+u_{\eta\eta}\eta_y^2+u_\xi\xi_{yy}+u_\eta\eta_{yy}\\
&=y^2u_{\xi\xi}+2y^2u_{\xi\eta}+y^2u_{\eta\eta}+u_\xi+u_\eta
\end{aligned}$$

Thus, the given equation assumes the canonical form

$$u_{\xi\eta}=\frac{\eta}{2(\xi^2-\eta^2)}u_\xi-\frac{\xi}{2(\xi^2-\eta^2)}u_\eta$$

EXAMPLE 3.2.2. Consider the partial differential equation

$$x^2 u_{xx} + 2xyu_{xy} + y^2 u_{yy} = 0$$

In this case, the discriminant is

$$B^2 - 4AC = 4x^2y^2 - 4x^2y^2 = 0$$

The equation is therefore parabolic everywhere. The characteristic equation is

$$\frac{dy}{dx} = \frac{y}{x},$$

and hence the characteristics are

$$\frac{y}{x} = c$$

which is the equation of a family of straight lines.

Consider the transformation

$$\xi = \frac{y}{x}$$

$$\eta = y$$

where η is chosen arbitrarily. The given equation is then reduced to the canonical form

$$y^2 u_{\eta\eta} = 0$$

Thus,

$$u_{\eta\eta} = 0 \quad \text{for } y \neq 0$$

EXAMPLE 3.2.3. The equation

$$u_{xx} + x^2 u_{yy} = 0$$

is elliptic everywhere except on the coordinate axis $x = 0$ because

$$B^2 - 4AC = -4x^2 < 0, \ x \neq 0$$

The characteristic equations are

$$\frac{dy}{dx} = ix$$

$$\frac{dy}{dx} = -ix$$

Integration yields

$$2y - ix^2 = c_1$$

$$2y + ix^2 = c_2$$

Thus, if we write

$$\xi = 2y - ix^2$$

$$\eta = 2y + ix^2,$$

and hence

$$\alpha=\frac{1}{2}(\xi+\eta)=2y$$

$$\beta=\frac{1}{2i}(\xi-\eta)=-x^2,$$

we obtain the canonical form

$$u_{\alpha\alpha}+u_{\beta\beta}=-\frac{1}{2\beta}u_\beta$$

It should be remarked here that a given partial differential equation may be of different type in a different domain. Thus, for example *Tricomi's equation*

$$u_{xx}+xu_{yy}=0 \tag{3.2.16}$$

is elliptic for $x>0$ and hyperbolic for $x<0$, since $B^2-4AC=-4x$. For a detail treatment, see Hellwig [19].

3.3 Equations with Constant Coefficients

In the case of an equation with real constant coefficients, the equation is of the single type at all points in the domain. This is because the discriminant B^2-4AC is a constant.

From the characteristic equation

$$\frac{dy}{dx}=(B+\sqrt{B^2-4AC}\,)/2A \tag{3.3.1}$$

and

$$\frac{dy}{dx}=(B-\sqrt{B^2-4AC}\,)/2A,$$

we can see that the characteristics

$$y=\left(\frac{B+\sqrt{B^2-4AC}}{2A}\right)x+c_1$$

$$y=\left(\frac{B-\sqrt{B^2-4AC}}{2A}\right)x+c_2 \tag{3.3.2}$$

are two families of straight lines. Consequently, the characteristics coordinates take the form

$$\xi=y-\lambda_1 x$$

$$\eta=y-\lambda_2 x \tag{3.3.3}$$

where

$$\lambda_{1,2} = \frac{B \pm \sqrt{B^2 - 4AC}}{2A} \tag{3.3.4}$$

The linear second-order partial differential equation with constant coefficients may be written as

$$Au_{xx} + Bu_{xy} + Cu_{yy} + Du_x + Eu_y + Fu = G(x,y) \tag{3.3.5}$$

(A) Hyperbolic Equation

When $B^2 - 4AC > 0$, the equation is of hyperbolic type, in which case the characteristics form two distinct families.

Using (3.3.3), Eq. (3.3.5) becomes

$$u_{\xi\eta} = D_1 u_\xi + E_1 u_\eta + F_1 u + G_1(\xi, \eta) \tag{3.3.6}$$

where D_1, E_1, and F_1 are constants. Here, since the coefficients are constants, the lower order terms are expressed explicitly.

When $A = 0$, Eq. (3.3.1) does not hold. In this case, the characteristic equation may be put in the form

$$-B(dx/dy) + C(dx/dy)^2 = 0$$

which may again be rewritten as

$$dx/dy = 0$$

$$-B + C(dx/dy) = 0$$

Integration gives

$$x = c_1$$

$$x = (B/C)y + c_2$$

where c_1 and c_2 are integration constants. Thus, the characteristic coordinates are

$$\xi = x$$

$$\eta = x - (B/C)y \tag{3.3.7}$$

Under this transformation, Eq. (3.3.5) reduces to the canonical form

$$u_{\xi\eta} = D_1^* u_\xi + E_1^* u_\eta + F_1^* u + G_1^*(\xi, \eta) \tag{3.3.8}$$

where D_1^*, E_1^* and F_1^* are constants.

(B) Parabolic Equation

When $B^2 - 4AC = 0$, the equation is of parabolic type in which case, only one real family of characteristics exists. From Eq. (3.3.4), we find that

$$\lambda_1 = \lambda_2 = (B/2A)$$

so that the single family of characteristics is given by

$$y = (B/2A)x + c_1$$

where c_1 is an integration constant. Thus, we have

$$\xi = y - (B/2A)x$$
$$\eta = hy + kx \tag{3.3.9}$$

where η is chosen arbitrarily such that the Jacobian of the transformation is not zero, and h and k are constants.

With the proper choice of the constants h and k in the transformation (3.3.9), Eq. (3.3.5) reduces to

$$u_{\eta\eta} = D_2 u_\xi + E_2 u_\eta + F_2 u + G_2(\xi, \eta) \tag{3.3.10}$$

where D_2, E_2, and F_2 are constants.

If $B = 0$, we can see at once from the relation

$$B^2 - 4AC = 0$$

that C or A vanishes. The given equation is then already in the canonical form. Similarly, in the other cases when A or C vanishes B vanishes. The given equation is then in canonical form.

(C) Elliptic Equation

When $B^2 - 4AC < 0$, the equation is of elliptic type. In this case, the characteristics are complex conjugates.

The characteristic equations yield

$$y = \lambda_1 x + c_1$$
$$y = \lambda_2 x + c_2 \tag{3.3.11}$$

where λ_1 and λ_2 are complex numbers. Accordingly, c_1 and c_2 are allowed to take on complex values. Thus,

$$\xi = y - (a + ib)x$$
$$\eta = y - (a - ib)x \tag{3.3.12}$$

where $\lambda_{1,2} = a \pm ib$ in which a and b are real constants, and $a = B/2A$ and $b = \sqrt{4AC - B^2}\ /2A$.

Introduce the new variables

$$\alpha = \frac{1}{2}(\xi + \eta) = y - ax$$

$$\beta = \frac{1}{2i}(\xi - \eta) = -bx \tag{3.3.13}$$

Application of this transformation readily reduces Eq. (3.3.5) to the canonical form

$$u_{\alpha\alpha} + u_{\beta\beta} = D_3 u_\alpha + E_3 u_\beta + F_3 u + G_3(\alpha, \beta) \tag{3.3.14}$$

where D_3, E_3 and F_3 are constants.

We note that $B^2-4AC<0$, so neither A nor C is zero.

EXAMPLE 3.3.1. Consider the equation

$$4u_{xx}+5u_{xy}+u_{yy}+u_x+u_y=2$$

Since $A=4$, $B=5$, $C=1$, and $B^2-4AC=9>0$, the equation is hyperbolic. Thus the characteristic equations take the form

$$\frac{dy}{dx}=1$$

$$\frac{dy}{dx}=\frac{1}{4}$$

and hence the characteristics are

$$y=x+c_1$$

$$y=(x/4)+c_2$$

The transformation

$$\xi=y-x$$

$$\eta=y-(x/4)$$

therefore reduces the given equation to the canonical form

$$u_{\xi\eta}=\tfrac{1}{3}u_\eta-\tfrac{8}{9}$$

This is the first canonical form.

The second canonical form may be obtained by the transformation

$$\alpha=\xi+\eta$$

$$\beta=\xi-\eta$$

as

$$u_{\alpha\alpha}-u_{\beta\beta}=\tfrac{1}{3}u_\alpha-\tfrac{1}{3}u_\beta-\tfrac{8}{9}$$

EXAMPLE 3.3.2. The equation

$$u_{xx}-4u_{xy}+4u_{yy}=e^y$$

is parabolic since $A=1$, $B=-4$, $C=4$ and $B^2-4AC=0$. Thus, we have from Eq. (3.3.9)

$$\xi=y+2x$$

$$\eta=y$$

in which η is chosen arbitrarily. By means of this mapping, the equation transforms into

$$u_{\eta\eta}=\tfrac{1}{4}e^\eta$$

EXAMPLE 3.3.3. Consider the equation

$$u_{xx}+u_{xy}+u_{yy}+u_x=0$$

Since $A=1$, $B=1$, $C=1$, and $B^2-4AC=-3<0$, the equation is elliptic. We have

$$\lambda_{1,2}=\frac{B\pm\sqrt{B^2-4AC}}{2A}=\frac{1}{2}\pm i\frac{\sqrt{3}}{2}$$

and hence,

$$\xi=y-\left(\frac{1}{2}+i\frac{\sqrt{3}}{2}\right)x$$

$$\eta=y-\left(\frac{1}{2}-i\frac{\sqrt{3}}{2}\right)x$$

Introducing the new variables

$$\alpha=\frac{1}{2}(\xi+\eta)=y-\frac{1}{2}x$$

$$\beta=\frac{1}{2i}(\xi-\eta)=-\frac{\sqrt{3}}{2}x$$

the given equation is then transformed into canonical form

$$u_{\alpha\alpha}+u_{\beta\beta}=\frac{2}{3}u_\alpha+\frac{2}{\sqrt{3}}u_\beta$$

3.4 General Solution

In general it is not so simple to determine the general solution of a given equation. Sometimes further simplification of the canonical form of an equation may yield the general solution. If the canonical form of the equation is simple, then the solution can be immediately ascertained.

EXAMPLE 3.4.1. Find the general solution of

$$x^2u_{xx}+2xyu_{xy}+y^2u_{yy}=0$$

In Example 3.2.2, using the transformation $\xi=y/x$, $\eta=y$, this equation was reduced to the canonical form

$$u_{\eta\eta}=0 \qquad \text{for } y\neq 0$$

Integrating twice with respect to η, we obtain

$$u(\xi,\eta)=\eta f(\xi)+g(\xi)$$

where $f(\xi)$ and $g(\xi)$ are arbitrary functions. In terms of the independent variables x and y, we have

$$u(x,y) = yf\left(\frac{y}{x}\right) + g\left(\frac{y}{x}\right)$$

EXAMPLE 3.4.2. Determine the general solution of

$$4u_{xx} + 5u_{xy} + u_{yy} + u_x + u_y = 2$$

Using the transformation $\xi = y - x$, $\eta = y - (x/4)$, the canonical form of this equation is (see Example 3.3.1)

$$u_{\xi\eta} = \tfrac{1}{3}u_\eta - \tfrac{8}{9}$$

By means of the substitution $v = u_\eta$, the preceding equation reduces to

$$v_\xi = \tfrac{1}{3}v - \tfrac{8}{9}$$

This can be easily integrated by separating the variables. Integrating with respect to ξ, we have

$$v = \tfrac{8}{3} + \tfrac{1}{3}e^{(\xi/3)}F(\eta)$$

Integrating again with respect to η, we obtain

$$u(\xi,\eta) = \tfrac{8}{3}\eta + \tfrac{1}{3}g(\eta)e^{\xi/3} + f(\xi)$$

where $f(\xi)$ and $g(\eta)$ are arbitrary functions. The general solution of the given equation is therefore

$$u(x,y) = \frac{8}{3}\left(y - \frac{x}{4}\right) + \frac{1}{3}g\left(y - \frac{x}{4}\right)e^{\frac{1}{3}(y-x)} + f(y-x)$$

EXAMPLE 3.4.3. Obtain the general solution of

$$3u_{xx} + 10u_{xy} + 3u_{yy} = 0$$

Since $B^2 - 4AC = 64 > 0$, the equation is hyperbolic. Thus, from Eqs. (3.3.2), the characteristics are

$$y = 3x + c_1$$

$$y = \tfrac{1}{3}x + c_2$$

Using the transformation

$$\xi = y - 3x$$

$$\eta = y - \tfrac{1}{3}x$$

the given equation is reduced to

$$\frac{64}{3}u_{\xi\eta} = 0$$

Hence, we obtain

$$u_{\xi\eta} = 0$$

Integration yields

$$u(\xi,\eta)=f(\xi)+g(\eta)$$

In terms of the original variables, the general solution is

$$u(x,y)=f(y-3x)+g\left(y-\frac{x}{3}\right)$$

3.5 Summary and Further Simplification

We summarize the classification of the linear second-order partial differential equations with constant coefficients in two independent variables.

hyperbolic: $$u_{rs}=a_1u_r+a_2u_s+a_3u+f_1 \tag{3.5.1}$$

$$u_{rr}-u_{ss}=a_1^*u_r+a_2^*u_s+a_3^*u+f_1^* \tag{3.5.2}$$

parabolic: $$u_{ss}=b_1u_r+b_2u_s+b_3u+f_2 \tag{3.5.3}$$

elliptic: $$u_{rr}+u_{ss}=c_1u_r+c_2u_s+c_3u+f_3 \tag{3.5.4}$$

where r and s represent the new independent variables in the linear transformation

$$\begin{aligned} r&=r(x,y)\\ s&=s(x,y) \end{aligned} \tag{3.5.5}$$

and the Jacobian $J\neq 0$.

To simplify Eq. (3.5.1) further, we introduce the new dependent variable

$$v=ue^{-(ar+bs)} \tag{3.5.6}$$

where a and b are undetermined coefficients. Finding the derivatives, we obtain

$$\begin{aligned} u_r&=(v_r+av)e^{ar+bs}\\ u_s&=(v_s+bv)e^{ar+bs}\\ u_{rr}&=\left(v_{rr}+2av_r+a^2v\right)e^{ar+bs}\\ u_{rs}&=(v_{rs}+av_s+bv_r+abv)e^{ar+bs}\\ u_{ss}&=\left(v_{ss}+2bv_s+b^2v\right)e^{ar+bs} \end{aligned}$$

Substitution of these in Eq. (3.5.1) yields

$$v_{rs}+(b-a_1)v_r+(a-a_2)v_s+(ab-a_1a-a_2b-a_3)v=f_1e^{-(ar+bs)}$$

In order that the first derivatives vanish, we set

$$a=a_2 \quad \text{and} \quad b=a_1$$

Thus, the above equation becomes

$$v_{rs}=(a_1a_2+a_3)v+g_1$$

where $g_1 = f_1 e^{-(a_2 r + a_1 s)}$. In a similar manner, we can transform Eqs. (3.5.2)–(3.5.4). Thus, we have the following transformed equations corresponding to Eqs. (3.5.1)–(3.5.4):

$$\begin{aligned} &\text{hyperbolic:} && v_{rs} = h_1 v + g_1 \\ & && v_{rr} - v_{ss} = h_1^* v + g_1^* \\ &\text{parabolic:} && v_{ss} = h_2 v + g_2 \\ &\text{elliptic:} && v_{rr} + v_{ss} = h_3 v + g_3 \end{aligned} \tag{3.5.7}$$

In the case of partial differential equations in several independent variables or in higher order, the classification is considerably more complex. For further reading, see Courant and Hilbert.

Exercises

1. Determine the region in which the given equation is hyperbolic, parabolic, or elliptic, and transform the equation in the respective region to the canonical form.

a. $xu_{xx} + u_{yy} = x^2$

b. $u_{xx} + y^2 u_{yy} = y$

c. $u_{xx} + xyu_{yy} = 0$

d. $x^2 u_{xx} - 2xyu_{xy} + y^2 u_{yy} = e^x$

e. $u_{xx} + u_{xy} - xu_{yy} = 0$

f. $e^x u_{xx} + e^y u_{yy} = u$

g. $\sin^2 x u_{xx} + \sin 2x u_{xy} + \cos^2 x u_{yy} = x$

h. $u_{xx} - yu_{xy} + xu_x + yu_y + u = 0$

2. Obtain the general solution of

i. $x^2 u_{xx} + 2xyu_{xy} + y^2 u_{yy} + xyu_x + y^2 u_y = 0$

ii. $ru_{tt} - c^2 ru_{rr} - 2c^2 u_r = 0$, $c = \text{constant}$

Check the solutions by substitution. [Hint: Let $v = ru$

3. Find the characteristics, the characteristic coordinates, and reduce the following equations to canonical form.

a. $u_{xx} + 2u_{xy} + 3u_{yy} + 4u_x + 5u_y + u = e^x$

b. $2u_{xx} - 4u_{xy} + 2u_{yy} + 3u = 0$

c. $u_{xx} + 5u_{xy} + 4u_{yy} + 7u_y = \sin x$

d. $u_{xx} + u_{yy} + 2u_x + 8u_y + u = 0$

e. $u_{xy}+2u_{yy}+9u_x+u_y=2$

f. $6u_{xx}-u_{xy}+u=y^2$

g. $u_{xy}+u_x+u_y=3x$

h. $u_{yy}-9u_x+7u_y=\cos y$

4. Determine the general solutions of

i. $u_{xx}-\dfrac{1}{c^2}u_{yy}=0, \qquad c=\text{constant}$

ii. $u_{xx}+u_{yy}=0$ [Hint: Let $c=i=\sqrt{-1}$ in (i)

iii. $u_{xxxx}+2u_{xxyy}+u_{yyyy}=0$ [Hint: Let $z=x+iy$

iv. $u_{xx}-3u_{xy}+2u_{yy}=0$

v. $u_{xx}+u_{xy}=0$

vi. $u_{xx}+10u_{xy}+9u_{yy}=y$

5. Transform the following equations to the form

$$v_{\xi\eta}=cv, \qquad c=\text{constant}$$

by introducing the new variables

$$v=ue^{-(a\xi+b\eta)}$$

where a and b are undetermined coefficients.

i. $u_{xx}-u_{yy}+3u_x-2u_y+u=0$

ii. $3u_{xx}+7u_{xy}+2u_{yy}+u_y+u=0$

6. Given the parabolic equation

$$u_{xx}=au_t+bu_x+cu+f$$

where the coefficients are constants. By the substitution

$$u=ve^{\frac{1}{2}bx}$$

for the case $c=-(b^2/4)$, show that the given equation is reduced to the heat equation

$$v_{xx}=av_t+g$$

where

$$g=fe^{-bx/2}$$

CHAPTER FOUR

The Cauchy Problem

4.1 The Cauchy Problem

In the theory of ordinary differential equations, by the initial value problem we mean the problem of finding the solution of a given differential equation with appropriate number of initial conditions prescribed at an initial point. For example, the second-order ordinary differential equation

$$\frac{d^2u}{dt^2}=f\left(t,u,\frac{du}{dt}\right)$$

and the initial conditions

$$u(t_0)=\alpha$$

$$\frac{du}{dt}(t_0)=\beta$$

constitute the initial value problem.

An analogous problem can be defined in the case of partial differential equations. Here we shall state the problem involving second-order partial differential equation in two independent variables.

We consider a second-order partial differential equation for the function u in the independent variables x and y, and suppose that this equation can be solved explicitly for u_{yy}, and hence can be represented in the form

$$u_{yy}=F(x,y,u,u_x,u_y,u_{xx},u_{xy}) \tag{4.1.1}$$

For some value $y=y_0$, we prescribe the initial values of the unknown function and of the derivative with respect to y

$$u(x,y_0)=f(x)$$

$$u_y(x,y_0)=g(x) \tag{4.1.2}$$

The problem of determining the solution of Eq. (4.1.1) satisfying the initial conditions (4.1.2) is known as the *initial-value problem*. For instance, the initial-value problem of a vibrating string is the problem of finding the solution of the equation

$$u_{tt}=c^2u_{xx}$$

satisfying the initial conditions

$$u(x,t_0)=u_0(x)$$
$$u_t(x,t_0)=v_0(x)$$

where $u_0(x)$ is the initial displacement and $v_0(x)$ is the initial velocity.

In initial-value problems, the initial values usually refer to the data assigned at $y=y_0$. It is not essential that these values be given along the line $y=y_0$, they may very well be prescribed along some curve L_0 in the xy plane. In such a context, the problem is called the *Cauchy problem* instead of the initial value problem, although the two names are actually synonymous.

Let us consider the equation

$$Au_{xx}+Bu_{xy}+Cu_{yy}=F(x,y,u,u_x,u_y) \tag{4.1.3}$$

where A,B,C are functions of x and y. Let (x_0,y_0) denote points on a smooth curve L_0 in the xy plane. Also let the parametric equations of this curve L_0 be

$$x_0=x_0(\lambda) \qquad y_0=y_0(\lambda) \tag{4.1.4}$$

where λ is a parameter.

We suppose that two functions $f(\lambda)$ and $g(\lambda)$ are prescribed along the curve L_0. The Cauchy problem is now one of determining the solution $u(x,y)$ of Eq. (4.1.3) in the neighborhood of the curve L_0 satisfying the Cauchy conditions

$$u=f(\lambda) \tag{4.1.5a}$$

$$\frac{\partial u}{\partial n}=g(\lambda) \tag{4.1.5b}$$

on the curve L_0. n is the direction of the normal to L_0 which lies to the left of L_0 in the counterclockwise direction of increasing arc length. The functions $f(\lambda)$ and $g(\lambda)$ are called the *Cauchy data*.

For every point on L_0, the value of u is specified by Eq. (4.1.5a). Thus, the curve L_0 represented by Eq. (4.1.4) with the condition (4.1.5a) yields a twisted curve L in (x,y,u) space whose projection on the xy plane is the curve L_0. Thus, the solution of the Cauchy problem is a surface, called an integral surface, in the (x,y,u) space passing through L and satisfying the condition (4.1.5b) which represents a tangent plane to the integral surface along L.

If the function $f(\lambda)$ is differentiable, then along the curve L_0, we have

$$\frac{du}{d\lambda}=\frac{\partial u}{\partial x}\frac{dx}{d\lambda}+\frac{\partial u}{\partial y}\frac{dy}{d\lambda}=\frac{df}{d\lambda} \tag{4.1.6}$$

and

$$\frac{\partial u}{\partial n}=\frac{\partial u}{\partial x}\frac{dx}{dn}+\frac{\partial u}{\partial y}\frac{dy}{dn}=g \tag{4.1.7}$$

But

$$\frac{dx}{dn}=-\frac{dy}{ds} \quad \text{and} \quad \frac{dy}{dn}=\frac{dx}{ds} \tag{4.1.8}$$

Eq. (4.1.7) may be written as

$$\frac{\partial u}{\partial n}=\frac{\partial u}{\partial x}\frac{dy}{ds}+\frac{\partial u}{\partial y}\frac{dx}{ds}=g \tag{4.1.9}$$

Since

$$\begin{vmatrix} \dfrac{dx}{d\lambda} & \dfrac{dy}{d\lambda} \\ \dfrac{-dy}{ds} & \dfrac{dx}{ds} \end{vmatrix}=\frac{(dx)^2+(dy)^2}{dsd\lambda}\neq 0, \tag{4.1.10}$$

it is possible to find u_x and u_y on L_0 from the system of Eqs. (4.1.6 and (4.1.9). Now that u_x and u_y are known on L_0, we find the higher derivatives by first differentiating u_x and u_y with respect to λ. Thus, we have

$$\frac{\partial^2 u}{\partial x^2}\frac{dx}{d\lambda}+\frac{\partial^2 u}{\partial x\partial y}\frac{dy}{d\lambda}=\frac{d}{d\lambda}\left(\frac{\partial u}{\partial x}\right) \tag{4.1.11}$$

$$\frac{\partial^2 u}{\partial x\partial y}\frac{dx}{d\lambda}+\frac{\partial^2 u}{\partial y^2}\frac{dy}{d\lambda}=\frac{d}{d\lambda}\left(\frac{\partial u}{\partial y}\right) \tag{4.1.12}$$

From Eq. (4.1.3), we have

$$A\frac{\partial^2 u}{\partial x^2}+B\frac{\partial^2 u}{\partial x\partial y}+C\frac{\partial^2 u}{\partial y^2}=F \tag{4.1.13}$$

where F is known since u_x and u_y have been found. The system of equations can be solved for u_{xx}, u_{xy}, and u_{yy}, if

$$\begin{vmatrix} \dfrac{dx}{d\lambda} & \dfrac{dy}{d\lambda} & 0 \\ 0 & \dfrac{dx}{d\lambda} & \dfrac{dy}{d\lambda} \\ A & B & C \end{vmatrix}=C\left(\frac{dx}{d\lambda}\right)^2-B\frac{dx}{d\lambda}\frac{dy}{d\lambda}+A\left(\frac{dy}{d\lambda}\right)^2\neq 0 \tag{4.1.14}$$

The equation

$$A\left(\frac{dy}{dx}\right)^2 - B\frac{dy}{dx} + C = 0 \tag{4.1.15}$$

is the characteristic equation. It is then evident that the necessary condition for obtaining the second derivatives is that the curve L_0 must not be a characteristic curve.

If the coefficients of Eq. (4.1.3) and the function (4.1.5) are analytic, then all the derivatives of higher orders can be computed by the above process. The solution can then be represented in the form of a Taylor series:

$$u(x,y) = \sum_{n=0}^{\infty} \sum_{k=0}^{n} \frac{1}{k!(n-k)!} \frac{\partial^n u_0}{\partial x_0^k \partial y_0^{n-k}} (x-x_0)^k (y-y_0)^{n-k} \tag{4.1.16}$$

which can be shown to converge in the neighborhood of the curve L_0. Thus we may state the famous Cauchy-Kowalewsky theorem.

4.2 Cauchy-Kowalewsky Theorem

Let the partial differential equation be given in the form

$$u_{yy} = F(y, x_1, x_2, \ldots, x_n, u, u_y, u_{x_1}, u_{x_2}, \ldots, u_{x_n},$$
$$u_{x_1 y}, u_{x_2 y}, \ldots, u_{x_n y}, u_{x_1 x_1}, u_{x_2 x_2}, \ldots, u_{x_n x_n}) \tag{4.2.1}$$

and let the initial conditions

$$u = f(x_1, x_2, \ldots, x_n) \tag{4.2.2}$$

$$u_y = g(x_1, x_2, \ldots, x_n) \tag{4.2.3}$$

be given on the noncharacteristic manifold $y = y_0$.

If the function F is analytic in some neighborhood of the point $(y^0, x_1^0, x_2^0, \ldots, x_n^0, u^0, u_y^0, \ldots)$ and if the functions f and g are analytic in some neighborhood of the point $(x_1^0, x_2^0, \ldots, x_n^0)$, then the Cauchy problem has a unique analytic solution in some neighborhood of the point $(y^0, x_1^0, x_2^0, \ldots, x_n^0)$.

For the proof, see Petrovsky [30].

The preceeding statement seems equally applicable to hyperbolic, parabolic or elliptic equations. However, we shall see that difficulties arise in formulating the Cauchy problem for nonhyperbolic equations. Consider, for instance, the Hadamard example.

The problem consists of the elliptic equation

$$u_{xx} + u_{yy} = 0$$

and the initial conditions

$$u(x,0)=0$$

$$u_y(x,0)=n^{-1}\sin nx$$

The solution of this problem is

$$u(x,y)=n^{-2}\sinh ny\sin nx,$$

which can be easily verified.

It can be seen that, when n tends to infinity, the function $n^{-1}\sin nx$ tends uniformly to zero. But, the solution $n^{-2}\sinh ny\sin nx$ does not become small, as n increases for any nonzero y. It is obvious then that the solution does not depend continuously on the data. Thus, it is not a properly posed problem.

4.3 Homogeneous Wave Equation

To study the Cauchy problems for hyperbolic partial differential equations, it is quite natural to begin investigating the simplest and yet most important equation, the one dimensional wave equation, by the method of characteristics. The essential characteristic of the solution of the general wave equation is preserved in this simplified case.

We shall consider the following Cauchy problem of an infinite string with the initial conditions:

$$u_{tt}-c^2u_{xx}=0 \tag{4.3.1}$$

$$u(x,0)=f(x) \tag{4.3.2}$$

$$u_t(x,0)=g(x) \tag{4.3.3}$$

By the method of characteristics described in Chapter 3, the characteristic equation according to Eq. (3.2.4) is

$$dx^2-c^2dt^2=0$$

which reduces to

$$dx+cdt=0$$

$$dx-cdt=0$$

The integrals are the straight lines

$$x+ct=c_1$$

$$x-ct=c_2$$

Introducing the characteristic coordinates

$$\xi=x+ct$$

$$\eta=x-ct$$

we obtain

$$u_{xx}=u_{\xi\xi}+2u_{\xi\eta}+u_{\eta\eta}$$

$$u_{tt}=c^2(u_{\xi\xi}-2u_{\xi\eta}+u_{\eta\eta})$$

Substitution of these in Eq. (4.3.1) yields

$$-4c^2u_{\xi\eta}=0$$

Since $c\neq 0$, we have

$$u_{\xi\eta}=0$$

Integrating with respect to ξ, we obtain

$$u_\eta=\psi^*(\eta)$$

where $\psi^*(\eta)$ is the arbitrary function of η. Integrating again with respect to η, we obtain

$$u(\xi,\eta)=\int\psi^*(\eta)\,d\eta+\phi(\xi)$$

If we set $\psi(\eta)=\int\psi^*(\eta)\,d\eta$, we have

$$u(\xi,\eta)=\phi(\xi)+\psi(\eta)$$

where ϕ and ψ are arbitrary functions. Transforming to the original variables x and t, we find the general solution of the wave equation

$$u(x,t)=\phi(x+ct)+\psi(x-ct) \tag{4.3.4}$$

provided ϕ and ψ are twice differential functions.

Now applying the initial conditions (4.3.2) and (4.3.3), we obtain

$$u(x,0)=f(x)=\phi(x)+\psi(x) \tag{4.3.5}$$

$$u_t(x,0)=g(x)=c\phi'(x)-c\psi(x) \tag{4.3.6}$$

Integration of Eq. (4.3.6) gives

$$\phi(x)-\psi(x)=\frac{1}{c}\int_{x_0}^{x}g(\tau)\,d\tau+K \tag{4.3.7}$$

where x_0 and K are arbitrary constants. Solving for ϕ and ψ from Eqs. (4.3.5) and (4.3.7), we obtain

$$\phi(x)=\frac{1}{2}f(x)+\frac{1}{2c}\int_{x_0}^{x}g(\tau)\,d\tau+\frac{K}{2}$$

$$\psi(x)=\frac{1}{2}f(x)-\frac{1}{2c}\int_{x_0}^{x}g(\tau)\,d\tau-\frac{K}{2}$$

The solution is thus given by

$$u(x,t)=\frac{1}{2}[f(x+ct)+f(x-ct)]+\frac{1}{2c}\left[\int_{x_0}^{x+ct}g(\tau)\,d\tau-\int_{x_0}^{x-ct}g(\tau)\,d\tau\right]$$

$$=\frac{1}{2}[f(x+ct)+f(x-ct)]+\frac{1}{2c}\int_{x-ct}^{x+ct}g(\tau)\,d\tau \tag{4.3.8}$$

This is called the *d'Alembert's solution* of the Cauchy problem for the one dimensional wave equation.

It is easy to verify that $u(x,t)$, represented by (4.3.8), is the solution of the wave equation (4.3.1). For this, $f(x)$ must be twice continuously differentiable and $g(x)$ continuously differentiable. This essentially proves the existence of the d'Alembert's solution. By direct substitution, it can also be shown that the solution (4.3.8) is uniquely determined by the initial conditions (4.3.2) and (4.3.3). As for the continuous dependence of the solution on the initial data, we state as follows.

For every $\varepsilon > 0$ and for each time interval $0 \leqslant t \leqslant t_0$, there exists a number $\delta(\varepsilon, t_0)$ such that

$$|u(x,t) - u^*(x,t)| < \varepsilon,$$

whenever

$$|f(x) - f^*(x)| < \delta, \qquad |g(x) - g^*(x)| < \delta.$$

The proof follows immediately from Eq. (4.3.8). We have

$$\begin{aligned} |u(x,t) - u^*(x,t)| &\leqslant \tfrac{1}{2}|f(x+ct) - f^*(x+ct)| \\ &\quad + \tfrac{1}{2}|f(x-ct) - f^*(x-ct)| \\ &\quad + \frac{1}{2c}\int_{x-ct}^{x+ct} |g(\tau) - g^*(\tau)|\, d\tau < \varepsilon \end{aligned}$$

where $\varepsilon = \delta(1 + t_0)$.

EXAMPLE 4.3.1. Find the solution of the initial value problem

$$\begin{aligned} u_{tt} &= c^2 u_{xx} \\ u(x,0) &= \sin x \\ u_t(x,0) &= \cos x \end{aligned}$$

From (4.3.8), we have

$$\begin{aligned} u(x,t) &= \frac{1}{2}\left[\sin(x+ct) + \sin(x-ct)\right] + \frac{1}{2c}\int_{x-ct}^{x+ct} \cos\tau\, d\tau \\ &= \sin x \cos ct + \frac{1}{2c}\left[\sin(x+ct) - \sin(x-ct)\right] \\ &= \sin x \cos ct + \frac{1}{c}\cos x \sin ct \end{aligned}$$

It follows from the d'Alembert solution that, if an initial displacement or an initial velocity is located in a small neighborhood of some point (x_0, t_0), it can influence only the area $t > t_0$ bounded by two characteristic $x - ct =$ constant and $x + ct =$ constant with slope $\pm 1/c$ passing through the point (x_0, t_0), as shown in Fig. 4.3.1. This means that the initial displacement

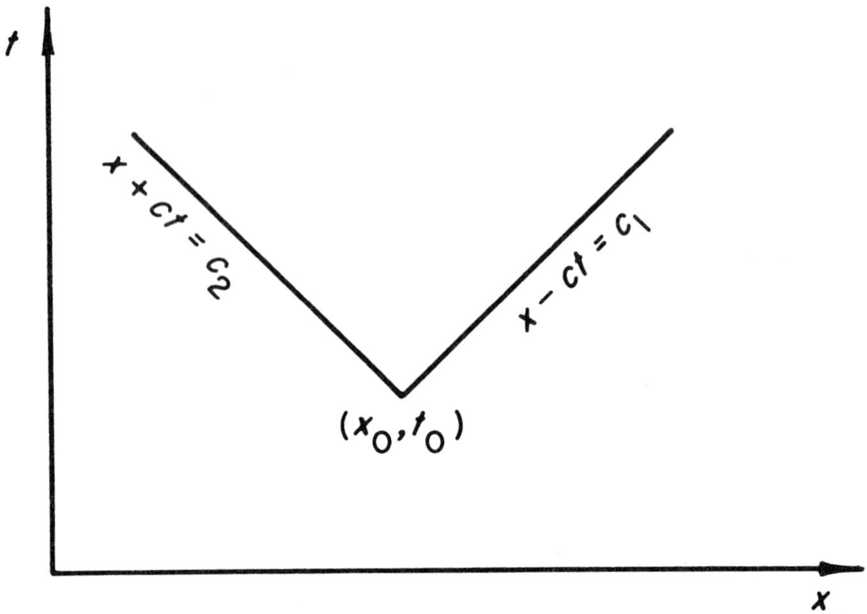

Figure 4.3.1

propagates at the speed c, whereas the effect of the initial velocity propagates at all speeds up to c. This infinite sector is called the *domain of influence* of the point (x_0, t_0).

According to Eq. (4.3.8), the value of $u(x_0, t_0)$ depends on the initial data f and g in the interval $[x_0 - ct_0, x_0 + ct_0]$ which is cut out of the initial line by the two characteristics $x - ct = \text{constant}$ and $x + ct = \text{constant}$ with slope $\pm 1/c$ passing through the point (x_0, t_0). The interval $[x_0 - ct_0, x_0 + ct_0]$ on the line $t = 0$ is called the *domain of dependence*, as shown in Fig. 4.3.2.

We will now examine the d'Alembert formula in greater detail.

$F(x \pm ct)$ is called the *progressing wave* or *wave*. By $F(x \pm ct)$ we mean $F(\zeta)$ evaluated at $\zeta = x \pm ct$. If we let $x' = ct$, then the transformation $\xi = x + ct = x + x'$ represents a translation of the coordinate system to the left by x'. Thus, $\phi(x + ct)$ may be interpreted as a wave which moves in the negative x-direction with speed c without change of shape. For instance,

$$u(x,t) = \sin(x + ct)$$

represents a sine wave travelling with speed c in the negative x-direction without changing its shape. In a similar manner $\psi(x - ct)$ may be interpreted as a wave which moves in the positive x-direction with speed c without change of shape.

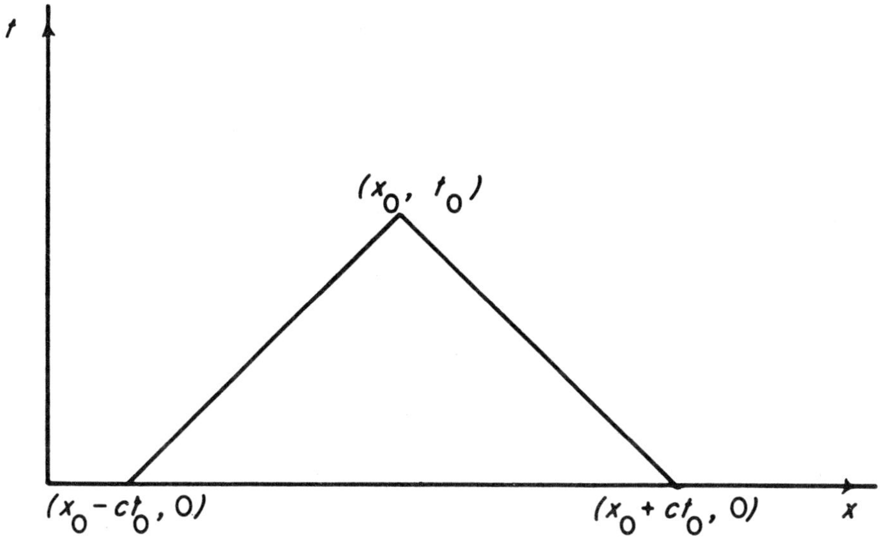

Figure 4.3.2

Obviously, the solution

$$u(x,t)=\phi(x+ct)+\psi(x-ct)$$

is the sum of two waves traveling in opposite directions, and the shape of $u(x,t)$ will change with time.

To interpret the d'Alembert formula we consider two cases:

Case 1. We first consider the case when the initial velocity is zero, that is,

$$g(x)=0$$

Then the d'Alembert solution has the form

$$u(x,t)=\tfrac{1}{2}[f(x+ct)+f(x-ct)]$$

Now suppose the initial displacement $f(x)$ is different from zero in an interval $(-b,b)$. Then in this case, the forward and the backward waves are represented by

$$u=\tfrac{1}{2}f(x)$$

The waves are first initially superimposed, and then they separate and travel in opposite directions.

We consider $f(x)$ which has the form of a triangle. We draw a triangle with the ordinate $x=0$ one-half that of the given function at that point, as shown in Fig. 4.3.3. If we displace these graphs and then take the sum of

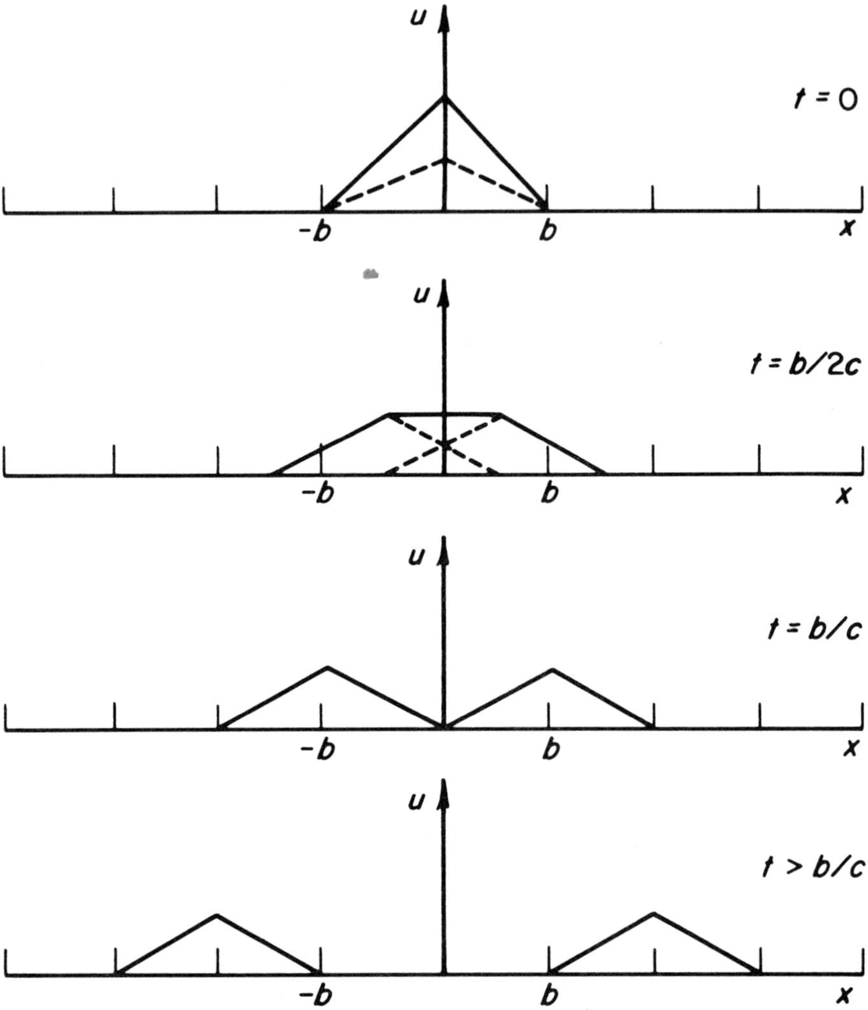

Figure 4.3.3

the ordinates of the displaced graphs, we obtain the shape of the string at any time t.

As can be seen from the figure, the waves travel in opposite directions away from each other. After both waves have passed the region of initial disturbance, the string returns to its rest position.

Case 2. We consider the case, when the initial displacement is zero, that is,

$$f(x)=0$$

and the d'Alembert formula assumes the form

$$u(x,t)=\frac{1}{2c}\int_{x-ct}^{x+ct}g(\tau)d\tau=\frac{1}{2}[G(x+ct)-G(x-ct)]$$

where

$$G(x)=\frac{1}{c}\int_{x_0}^{x}g(\tau)d\tau.$$

If we take for the initial velocity

$$g(x)=\begin{cases}0 & |x|>b\\ g_0 & |x|\leqslant b,\end{cases}$$

then the function $G(x)$ is equal to zero for values of x in the interval $x\leqslant -b$, and

$$G(x)=\begin{cases}\dfrac{1}{c}\displaystyle\int_{-b}^{x}g_0\,d\tau=\frac{g_0}{c}(x+b) & -b\leqslant x\leqslant b\\[2ex] \dfrac{1}{c}\displaystyle\int_{-b}^{x}g_0\,d\tau=\frac{2bg_0}{c} & x>b\end{cases}$$

As in the previous case, the two waves which differ in sign travel in opposite directions on the x-axis. After some time t, the two functions $(1/2)G(x)$ and $-(1/2)G(x)$ move a distance ct. Thus, the graph of u at time t is obtained by summing the ordinates of the displaced graphs as shown in Fig. 4.3.4. As t approaches infinity, the string will reach a state of rest, but it will not, in general, assume its original position. This displacement is known as the *residual displacement*.

In the preceding examples, we note that $f(x)$ is continuous but not continuously differentiable and $g(x)$ is discontinuous. To these initial data, there corresponds generalized solution. By generalized solution we mean the following.

Let us suppose that the function $u(x,t)$ satisfies the initial conditions (4.3.2) and (4.3.3). Let $u(x,t)$ be the limit of a uniformly convergent sequence of solutions $u_n(x,t)$ which satisfies the wave equation (4.3.1) and the initial conditions

$$u_n(x,0)=f_n(x)$$

$$\frac{\partial u_n}{\partial t}(x,0)=g_n(x)$$

Let $f_n(x)$ be a continuously differentiable function and converge uniformly to $f(x)$, and let $g_n(x)$ be a continuously differentiable function and $\int_{x_0}^{x}g_n(\tau)d\tau$ approach uniformly to $\int_{x_0}^{x}g(\tau)d\tau$. Then the function $u(x,t)$ is called the *generalized solution* of the problem (4.3.1–4.3.3).

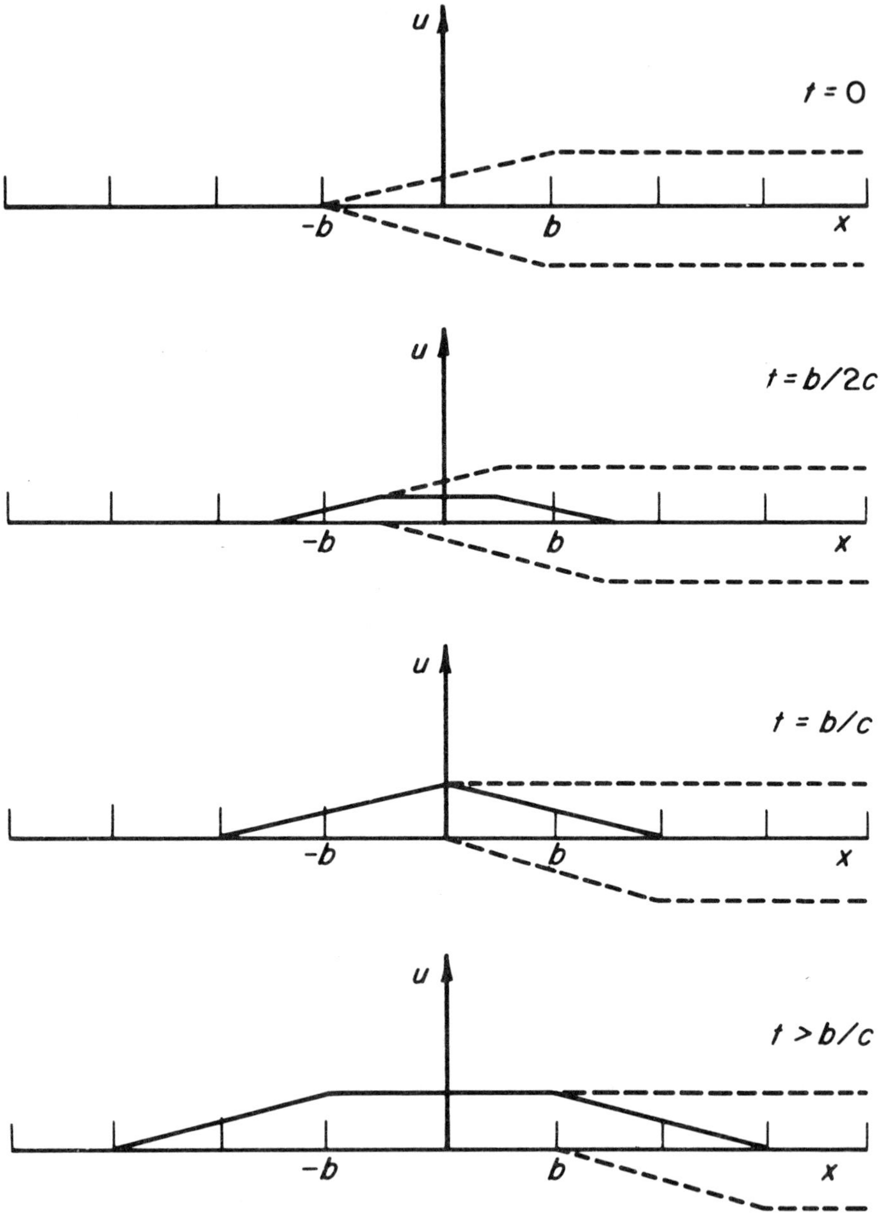

Figure 4.3.4

4.4 Initial-boundary Value Problems

We have just determined the solution of the initial value problem for the infinite vibrating string. We will now study the effect of a boundary on the solution.

(A) Semiinfinite String with a Fixed End

Let us first consider a semiinfinite vibrating string with a fixed end, that is,

$$\begin{aligned} u_{tt} &= c^2 u_{xx}, && 0<x<\infty, && t>0 \\ u(x,0) &= f(x), && 0 \leqslant x<\infty \\ u_t(x,0) &= g(x), && 0 \leqslant x<\infty \\ u(0,t) &= 0, && 0 \leqslant t<\infty \end{aligned} \tag{4.4.1}$$

It is evident here that the boundary condition at $x=0$ produces a wave moving to the right with the velocity c. Thus, for $x>ct$, the solution is the same as that of the infinite string, and the displacement is influenced only by the initial data on the interval $[x-ct, x+ct]$, as shown in Fig. 4.4.1.

When $x<ct$, the interval $[x-ct, x+ct]$ extends onto the negative x-axis where f and g are not prescribed.

Figure 4.4.1

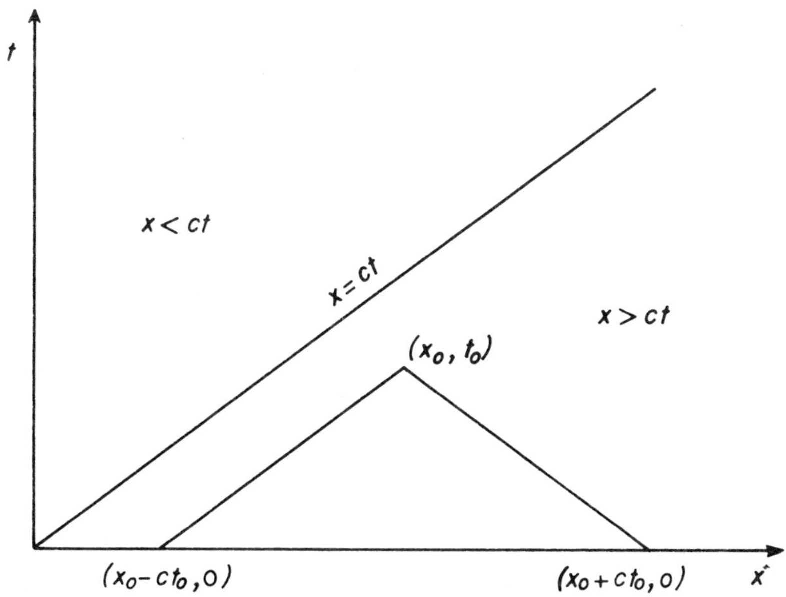

But from the d'Alembert formula

$$u(x,t)=\phi(x+ct)+\psi(x-ct) \tag{4.4.2}$$

where

$$\phi(\xi)=\frac{1}{2}f(\xi)+\frac{1}{2c}\int_0^{\xi}g(\tau)d\tau+\frac{K}{2} \tag{4.4.3}$$

$$\psi(\eta)=\frac{1}{2}f(\eta)-\frac{1}{2c}\int_0^{\eta}g(\tau)d\tau-\frac{K}{2} \tag{4.4.4}$$

we see that

$$u(0,t)=\phi(ct)+\psi(-ct)=0$$

Hence

$$\psi(-ct)=-\phi(ct)$$

If we let $\alpha=-ct$, then

$$\psi(\alpha)=-\phi(-\alpha)$$

Replacing α by $x-ct$, we obtain for $x<ct$

$$\psi(x-ct)=-\phi(ct-x),$$

and hence

$$\psi(x-ct)=-\frac{1}{2}f(ct-x)-\frac{1}{2c}\int_0^{ct-x}g(\tau)d\tau-\frac{K}{2}$$

The solution of the initial-boundary value problem, therefore, is given by

$$u(x,t)=\frac{1}{2}[f(x+ct)+f(x-ct)]+\frac{1}{2c}\int_{x-ct}^{x+ct}g(\tau)d\tau \text{ for } x>ct \tag{4.4.5}$$

$$u(x,t)=\frac{1}{2}[f(x+ct)-f(ct-x)]+\frac{1}{2c}\int_{ct-x}^{x+ct}g(\tau)d\tau \text{ for } x<ct \tag{4.4.6}$$

In order for this solution to exist, f must be twice continuously differentiable and g must be continuously differentiable, and in addition

$$f(0)=f''(0)=g(0)=0$$

Formula (4.4.6) has an interesting physical interpretation. If we draw the characteristics through the point (x_0,t_0) in the region $x>ct$, we see, as pointed out earlier, that the displacement at (x_0,t_0) is determined by the initial values on $[x_0-ct_0,x_0+ct_0]$.

If the point (x_0,t_0) lies in the region $x<ct$ as shown in Fig. 4.4.2, we see that the characteristic $x+ct=x_0+ct_0$ intersects the x-axis at $(x_0+ct_0,0)$. However, the characteristic $x-ct=x_0-ct_0$ intersects the t-axis at $(0,t_0-x_0/c)$, and the characteristic $x+ct=ct_0-x_0$ through this point intersects the x-axis at $(ct_0-x_0,0)$. Thus, the disturbance at $(ct_0-x_0,0)$ travels along the backward characteristic $x+ct=ct_0-x_0$, and is reflected at $(0,t_0-x_0/c)$ as a forward wave $-\phi(ct_0-x_0)$.

EXAMPLE 4.4.1. Determine the solution of the initial-boundary value problem

$$\begin{aligned} u_{tt} &= 4u_{xx}, && x>0, && t>0 \\ u(x,0) &= |\sin x|, && x>0 \\ u_t(x,0) &= 0, && x\geqslant 0 \\ u(0,t) &= 0, && && t\geqslant 0 \end{aligned}$$

For $x>2t$

$$\begin{aligned} u(x,t) &= \tfrac{1}{2}[f(x+2t)+f(x-2t)] \\ &= \tfrac{1}{2}[|\sin(x+2t)|+|\sin(x-2t)|] \end{aligned}$$

and $x<2t$

$$\begin{aligned} u(x,t) &= \tfrac{1}{2}[f(x+2t)-f(2t-x)] \\ &= \tfrac{1}{2}[|\sin(x+2t)|-|\sin(2t-x)|] \end{aligned}$$

Notice that $u(0,t)=0$ is satisfied by $u(x,t)$ for $x<2t$ (that is, $t>0$).

(B) Semiinfinite String with a Free End

Let us consider a semiinfinite string with a free end at $x=0$. Thus, we will determine the solution of

$$\begin{aligned} u_{tt} &= c^2u_{xx}, && 0<x<\infty, && t>0 \\ u(x,0) &= f(x), && 0\leqslant x<\infty \\ u_t(x,0) &= g(x), && 0\leqslant x<\infty \\ u_x(0,t) &= 0, && 0\leqslant t<\infty \end{aligned} \tag{4.4.7}$$

As in the case of the fixed end, for $x>ct$ the solution is the same as that of the infinite string. For $x<ct$, from the d'Alembert formula

$$u(x,t)=\phi(x+ct)+\psi(x-ct),$$

we have

$$u_x(x,t)=\phi'(x+ct)+\psi'(x-ct)$$

Thus

$$u_x(0,t)=\phi'(ct)+\psi'(-ct)=0$$

Integration yields

$$\phi(ct)-\psi(-ct)=K$$

where K is a constant. Now, if we let $\alpha=-ct$, we obtain

$$\psi(\alpha)=\phi(-\alpha)-K$$

Replacing α by $x-ct$, we have

$$\psi(x-ct)=\phi(ct-x)-K$$

and hence

$$\psi(x-ct)=\frac{1}{2}f(ct-x)+\frac{1}{2c}\int_0^{ct-x}g(\tau)d\tau-\frac{K}{2}$$

The solution of the initial-boundary value problem, therefore, is given by

$$u(x,t)=\frac{1}{2}[f(x+ct)+f(x-ct)]+\frac{1}{2c}\int_{x-ct}^{x+ct}g(\tau)d\tau \text{ for } x>ct \tag{4.4.8}$$

$$u(x,t)=\frac{1}{2}[f(x+ct)+f(ct-x)]+\frac{1}{2c}\left[\int_0^{x+ct}g(\tau)d\tau+\int_0^{ct-x}g(\tau)d\tau\right] \tag{4.4.9}$$

for $x<ct$

We note here that for this solution to exist, f must be twice continuously differentiable and g must be continuously differentiable, and in addition,

$$f'(0)=g'(0)=0.$$

EXAMPLE 4.4.2. Find the solution of the initial boundary value problem

$$\begin{aligned} &u_{tt}=u_{xx}, && 0<x<\infty, && t>0\\ &u(x,0)=\cos\frac{\pi x}{2}, && 0\leqslant x<\infty\\ &u_t(x,0)=0, && 0\leqslant x<\infty\\ &u_x(0,t)=0, && && t\geqslant 0 \end{aligned}$$

For $x>t$

$$\begin{aligned} u(x,t)&=\frac{1}{2}\left[\cos\frac{\pi}{2}(x+t)+\cos\frac{\pi}{2}(x-t)\right]\\ &=\cos\frac{\pi}{2}x\cos\frac{\pi}{2}t \end{aligned}$$

and for $x<t$

$$\begin{aligned} u(x,t)&=\frac{1}{2}\left[\cos\frac{\pi}{2}(x+t)+\cos\frac{\pi}{2}(t-x)\right]\\ &=\cos\frac{\pi}{2}x\cos\frac{\pi}{2}t \end{aligned}$$

4.5 Nonhomogeneous Boundary Conditions

In the case of initial-boundary value problems with nonhomogeneous boundary condition, such as,

$$\begin{aligned} &u_{tt}=c^2u_{xx}, && x>0, && t>0\\ &u(x,t)=f(x), && x\geqslant 0\\ &u_t(x,0)=g(x), && x\geqslant 0\\ &u(0,t)=p(t), && && t\geqslant 0 \end{aligned} \tag{4.5.1}$$

we proceed in a similar manner as in the case of homogeneous boundary conditions. Using Eq. (4.4.2), we apply the boundary condition to obtain

$$u(0,t)=\phi(ct)+\psi(-ct)=p(t)$$

If we let $\alpha=-ct$, we have

$$\psi(\alpha)=p\left(-\frac{\alpha}{c}\right)-\phi(-\alpha)$$

Replacing α by $x-ct$, the preceding relation becomes

$$\psi(x-ct)=p\left(t-\frac{x}{c}\right)-\phi(ct-x)$$

Thus, for $x<ct$

$$u(x,t)=p\left(t-\frac{x}{c}\right)+\tfrac{1}{2}\left[f(x+ct)-f(ct-x)\right] + \frac{1}{2c}\int_{ct-x}^{x+ct} g(\tau)\,d\tau$$

The solution for $x>ct$ is given by the solution of the infinite string (4.4.5).

In this case, in addition to the differentiability conditions satisfied by f and g, as in the case of the problem with the homogeneous boundary condition, p must be twice continuously differentiable in t and

$$p(0)=f(0),\ p'(0)=g(0),\ p''(0)=c^2f''(0)$$

Let us next consider the initial-boundary value problem

$$\begin{aligned} &u_{tt}=c^2u_{xx}, && x>0, \quad t>0\\ &u(x,0)=f(x), && x\geqslant 0\\ &u_t(x,0)=g(x), && x\geqslant 0\\ &u_x(0,t)=q(t), && t\geqslant 0 \end{aligned}$$

Using (4.4.2), we apply the boundary condition to obtain

$$u_x(0,t)=\phi'(ct)+\psi'(-ct)=q(t)$$

Then integration yields

$$\phi(ct)-\psi(-ct)=c\int_0^t q(\tau)\,d\tau+K$$

If we let $\alpha=-ct$, then

$$\psi(\alpha)=\phi(-\alpha)-c\int_0^{-\alpha/c} q(\tau)\,d\tau-K$$

Replacing α by $x-ct$, we obtain

$$\psi(x-ct)=\phi(ct-x)-c\int_0^{t-x/c} q(\tau)\,d\tau-K$$

The solution of the initial-boundary value problem for $x<ct$, therefore, is

given by

$$u(x,t)=\frac{1}{2}[f(x+ct)+f(ct-x)]+\frac{1}{2c}\left[\int_0^{x+ct}g(\tau)d\tau+\int_0^{ct-x}g(\tau)d\tau\right]$$
$$-c\int_0^{t-\frac{x}{c}}q(\tau)d\tau$$

Here f and g must satisfy the differentiability conditions, as in the case of the problem with the homogeneous boundary condition. In addition

$$f'(0)=q(0),\quad g'(0)=q'(0)$$

The solution for the initial-boundary value problem involving the elastic boundary condition

$$u_x(0,t)+hu(0,t)=0,\quad h=\text{constant}$$

can also be constructed in a similar manner from the d'Alembert formula.

4.6 Finite String With Fixed Ends

The problem of the finite string is more complicated than that of the infinite string due to the repeated reflection of waves from the boundaries.

Let us consider the vibration of the string of length l fixed at both ends. The problem is that of finding the solution of

$$\begin{aligned}
&u_{tt}=c^2u_{xx}, && 0<x<l, && t>0\\
&u(x,0)=f(x), && 0\leqslant x\leqslant l\\
&u_t(x,0)=g(x), && 0\leqslant x\leqslant l\\
&u(0,t)=0, && && t\geqslant 0\\
&u(l,t)=0, && && t\geqslant 0
\end{aligned}\tag{4.6.1}$$

From the previous results, we know that the solution of the wave equation is

$$u(x,t)=\phi(x+ct)+\psi(x-ct)$$

Applying the initial conditions, we have

$$u(x,0)=\phi(x)+\psi(x)=f(x),\qquad 0\leqslant x\leqslant l$$
$$u_t(x,0)=c\phi'(x)-c\psi'(x)=g(x),\quad 0\leqslant x\leqslant l$$

Solving for ϕ and ψ, we find

$$\phi(\xi)=\frac{1}{2}f(\xi)+\frac{1}{2c}\int_0^{\xi}g(\tau)d\tau+\frac{K}{2},\quad 0\leqslant\xi\leqslant l\tag{4.6.2}$$

$$\psi(\eta)=\frac{1}{2}f(\eta)-\frac{1}{2c}\int_0^{\eta}g(\tau)d\tau-\frac{K}{2},\quad 0\leqslant\eta\leqslant l\tag{4.6.3}$$

Hence

$$u(x,t)=\frac{1}{2}[f(x+ct)+f(x-ct)]+\frac{1}{2c}\int_{x-ct}^{x+ct}g(\tau)d\tau\tag{4.6.4}$$

for $0 \leqslant x+ct \leqslant l$ and $0 \leqslant x-ct \leqslant l$. The solution is thus uniquely determined by the initial data in the region

$$t \leqslant \frac{x}{c}, \quad t \leqslant \frac{l-x}{c}, \quad t \geqslant 0$$

For larger times, the solution depends on the boundary conditions. Applying the boundary conditions, we obtain

$$u(0,t)=\phi(ct)+\psi(-ct)=0, \quad t \geqslant 0 \tag{4.6.5}$$

$$u(l,t)=\phi(l+ct)+\psi(l-ct)=0, \quad t \geqslant 0 \tag{4.6.6}$$

If we set $\alpha = -ct$, Eq. (4.6.5) becomes

$$\psi(\alpha) = -\phi(-\alpha), \quad \alpha \leqslant 0 \tag{4.6.7}$$

and if we set $\alpha = l+ct$, Eq. (4.6.6) takes the form

$$\phi(\alpha) = -\psi(2l-\alpha), \quad \alpha \geqslant l \tag{4.6.8}$$

With $\xi = -\eta$ we may write Eq. (4.6.2) as

$$\phi(-\eta) = \frac{1}{2}f(-\eta) + \frac{1}{2c}\int_0^{-\eta} g(\tau)\,d\tau + \frac{K}{2}, \; 0 \leqslant -\eta \leqslant l \tag{4.6.9}$$

Thus from (4.6.7) and (4.6.9), we have

$$\psi(\eta) = -\frac{1}{2}f(-\eta) - \frac{1}{2c}\int_0^{-\eta} g(\tau)\,d\tau - \frac{K}{2}, \; -l \leqslant \eta \leqslant 0 \tag{4.6.10}$$

We see that the range of $\psi(\eta)$ is extended to $-l \leqslant \eta \leqslant l$.

If we put $\alpha = \xi$ in Eq. (4.6.8), we obtain

$$\phi(\xi) = -\psi(2l-\xi), \quad \xi \geqslant l \tag{4.6.11}$$

Then by putting $\eta = 2l-\xi$ in Eq. (4.6.3), we obtain

$$\psi(2l-\xi) = \frac{1}{2}f(2l-\varepsilon) - \frac{1}{2c}\int_0^{2l-\varepsilon} g(\tau)\,d\tau - \frac{K}{2}, \; 0 \leqslant 2l-\xi \leqslant l \tag{4.6.12}$$

Substitution of this in Eq. (4.6.11) yields

$$\phi(\xi) = -\frac{1}{2}f(2l-\xi) + \frac{1}{2c}\int_0^{2l-\xi} g(\tau)\,d\tau + \frac{K}{2}, l \leqslant \xi \leqslant 2l \tag{4.6.13}$$

The range of $\phi(\xi)$ is thus extended to $0 \leqslant \xi \leqslant 2l$. Continuing in this manner, we obtain $\phi(\xi)$ for all $\xi \geqslant 0$ and $\psi(\eta)$ for all $\eta \leqslant l$. Hence, the solution is determined for all $0 \leqslant x \leqslant l$ and $t \geqslant 0$.

In order to observe the effect of the boundaries on the propagation of waves, the characteristics are drawn through the end points until they meet the boundaries, and then continue inward as shown in Fig. 4.6.1. It can be seen from the diagram that only direct waves propagate in region 1. In regions 2 and 3, both direct and reflected waves propagate. In regions 4, 5, 6, ..., several waves propagate along the characteristics reflected from both of the boundaries $x=0$ and $x=l$.

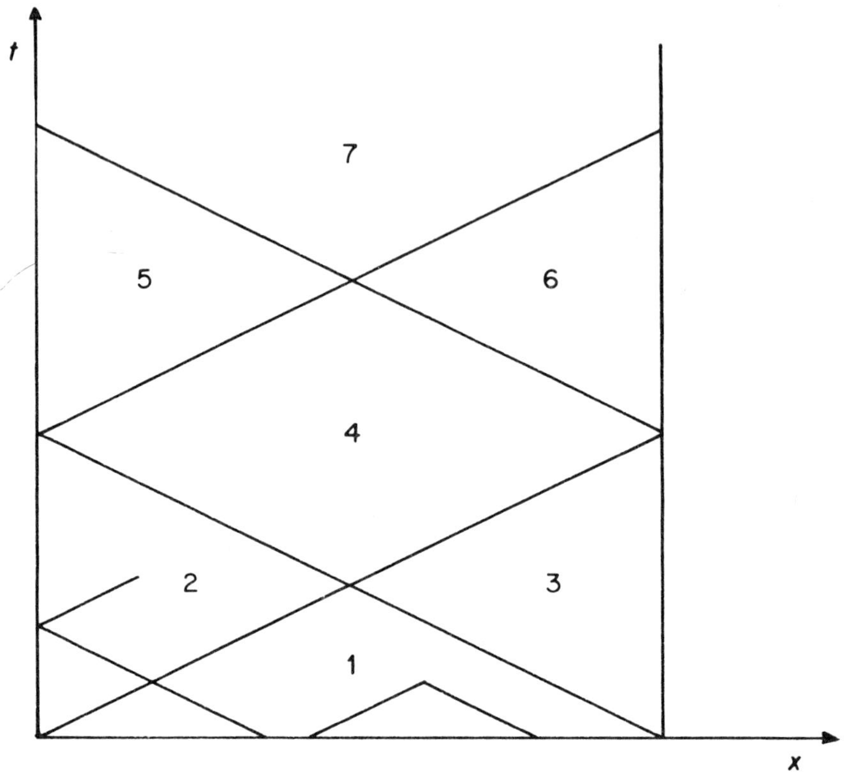

Figure 4.6.1

EXAMPLE 4.6.1. Determine the solution of

$$u_{tt} = c^2 u_{xx}, \qquad 0 < x < l, \qquad t > 0$$
$$u(x,0) = \sin \pi x / l, \qquad 0 \leqslant x \leqslant l$$
$$u_t(x,0) = 0, \qquad 0 \leqslant x \leqslant l$$
$$u(0,t) = 0, \qquad t \geqslant 0$$
$$u(l,t) = 0, \qquad t \geqslant 0$$

From Eqs. (4.6.2) and (4.6.3), we have

$$\phi(\xi) = \frac{1}{2} \sin \frac{\pi \xi}{l} + \frac{K}{2}, \qquad 0 \leqslant \xi \leqslant l$$
$$\psi(\eta) = \frac{1}{2} \sin \frac{\pi \eta}{l} - \frac{K}{2}, \qquad 0 \leqslant \eta \leqslant l$$

Using Eq. (4.6.10), we obtain

$$\begin{aligned}\psi(\eta) &= -\frac{1}{2}\sin\left(-\frac{\pi\eta}{l}\right)-\frac{K}{2}, \quad -l\leqslant\eta\leqslant 0\\ &= \frac{1}{2}\sin\frac{\pi\eta}{l}-\frac{K}{2}\end{aligned}$$

From Eq. (4.6.13), we find

$$\phi(\xi) = -\frac{1}{2}\sin\frac{\pi}{l}(2l-\xi)+\frac{K}{2}, \quad l\leqslant\xi\leqslant 2l$$

Again by Eq. (4.6.7) and from the preceding $\phi(\xi)$, we have

$$\psi(\eta) = \frac{1}{2}\sin\frac{\pi\eta}{l}-\frac{K}{2}, \quad -2l\leqslant\eta\leqslant -l$$

Proceeding in this manner, we determine the solution

$$\begin{aligned}u(x,t) &= \phi(\xi)+\psi(\eta)\\ &= \frac{1}{2}\left[\sin\frac{\pi}{l}(x+ct)+\sin\frac{\pi}{l}(x-ct)\right]\end{aligned}$$

for all x in $(0,l)$ and for all $t>0$.

The solution of the finite initial-boundary value problem

$$\begin{aligned}&u_{tt} = c^2u_{xx}, && 0<x<l, && t>0\\ &u(x,0) = f(x), && 0\leqslant x\leqslant l &&\\ &u_t(x,0) = g(x), && 0\leqslant x\leqslant l &&\\ &u(0,t) = p(t), && && t\geqslant 0\\ &u(l,t) = q(t), && && t\geqslant 0\end{aligned}$$

can be determined by the same method.

4.7 Nonhomogeneous Wave Equation

We shall consider next the Cauchy problem for the nonhomogeneous wave equation

$$u_{tt} = c^2u_{xx}+h^*(x,t) \tag{4.7.1}$$

with the initial conditions

$$\begin{aligned}u(x,0) &= f(x)\\ u_t(x,0) &= g^*(x)\end{aligned} \tag{4.7.2}$$

By the coordinate transformation

$$y = ct \tag{4.7.3}$$

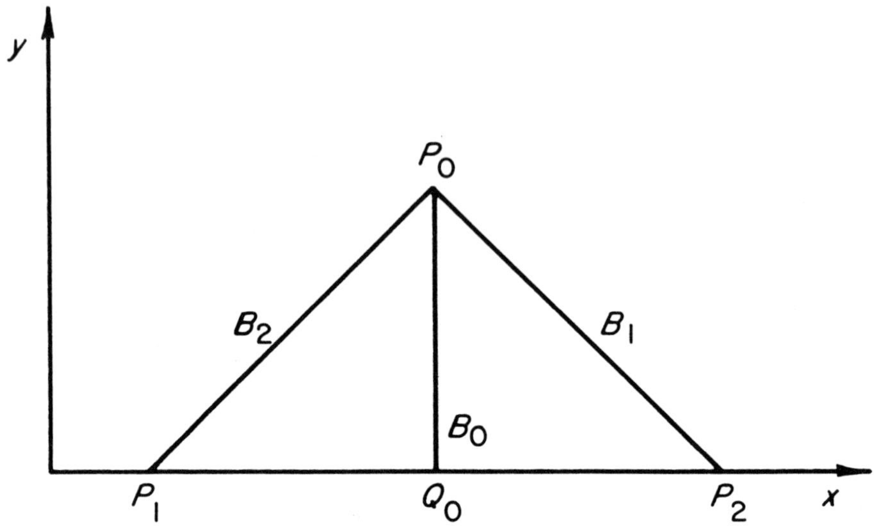

Figure 4.7.1

the problem is reduced to

$$u_{xx}-u_{yy}=h(x,y) \tag{4.7.4}$$

$$u(x,0)=f(x) \tag{4.7.5}$$

$$u_y(x,0)=g(x) \tag{4.7.6}$$

where $h(x,y)=-h^*/c^2$ and $g(x)=g^*/c$.

Let $P_0(x_0,y_0)$ be a point of the plane, and let Q_0 be the point $(x_0,0)$ on the initial line $y=0$. Then the characteristics $x\pm y=$constant of Eq. (4.7.4) are two straight lines drawn through the point P_0 with slopes ± 1. Obviously, they intersect the x-axis at the points $P_1(x_0-y_0,0)$ and $P_2(x_0+y_0,0)$, as shown in Fig. 4.7.1. Let the sides of the triangle $P_0P_1P_2$ be designated by B_0, B_1 and B_2, and let R be the region representing the interior of the triangle and its boundaries B. Integrating both sides of Eq. (4.7.4), we obtain

$$\int\int_R (u_{xx}-u_{yy})\,dR=\int\int_R h(x,y)\,dR \tag{4.7.7}$$

Now we apply Green's theorem, and obtain

$$\int\int_R (u_{xx}-u_{yy})\,dR=\oint_B (u_x\,dy+u_y\,dx) \tag{4.7.8}$$

Since B is composed of B_0, B_1, we note that

$$\int_{B_0} (u_x dy + u_y dx) = \int_{x_0 - y_0}^{x_0 + y_0} u_y dx$$

$$\int_{B_1} (u_x dy + u_y dx) = \int_{B_1} (-u_x dx - u_y dy)$$
$$= u(x_0 + y_0, 0) - u(x_0, y_0)$$

$$\int_{B_2} (u_x dy + u_y dx) = \int_{B_2} (u_x dx + u_y dy)$$
$$= u(x_0 - y_0, 0) - u(x_0, y_0)$$

Hence,

$$\oint_B (u_x dy + u_y dx) = -2u(x_0, y_0) + u(x_0 - y_0, 0) + u(x_0 + y_0, 0) + \int_{x_0 - y_0}^{x_0 + y_0} u_y dx \tag{4.7.9}$$

Combining Eqs. (4.7.7), (4.7.8) and (4.7.9), we obtain

$$u(x_0, y_0) = \frac{1}{2}[u(x_0 + y_0, 0) + u(x_0 - y_0, 0)] + \frac{1}{2}\int_{x_0 - y_0}^{x_0 + y_0} u_y dx - \frac{1}{2}\int\int_R h(x,y) dR \tag{4.7.10}$$

We have chosen x_0, y_0 arbitrarily, and as a consequence, we replace x_0 by x and y_0 by y. Equation (4.7.10) thus becomes

$$u(x,y) = \frac{1}{2}[f(x+y) + f(x-y)] + \frac{1}{2}\int_{x-y}^{x+y} g(\tau) d\tau - \frac{1}{2}\int\int_R h(x,y) dR$$

In terms of the original variables

$$u(x,t) = \frac{1}{2}[f(x+ct) + f(x-ct)] + \frac{1}{2c}\int_{x-ct}^{x+ct} g^*(\tau) d\tau - \frac{1}{2}\int\int_R h(x,t) dR \tag{4.7.11}$$

EXAMPLE 4.7.1. Determine the solution of

$$u_{xx} - u_{yy} = 1$$
$$u(x,0) = \sin x$$
$$u_y(x,0) = x$$

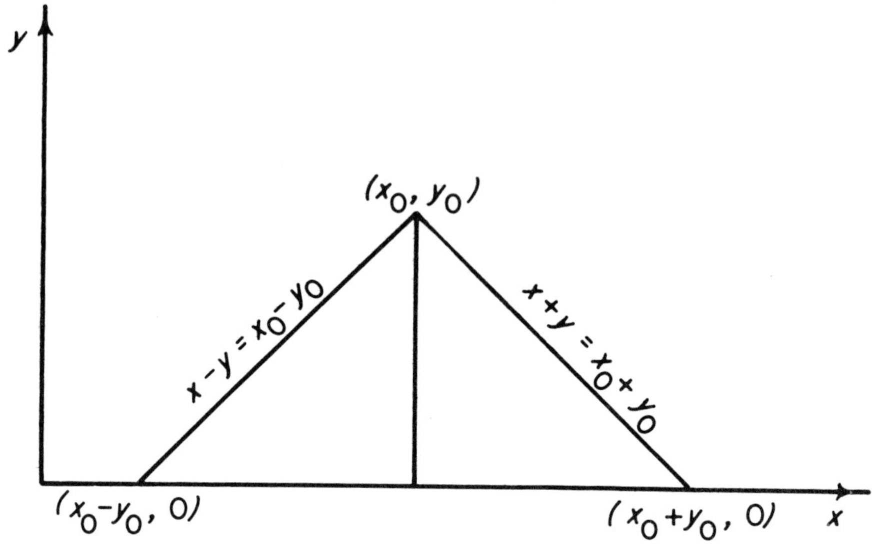

Figure 4.7.2

It is easy to see that the characteristics are $x+y=\text{constant}=x_0+y_0$ and $x-y=\text{constant}=x_0-y_0$, as shown in Fig. 4.7.2. Thus,

$$\begin{aligned} u(x_0,y_0) &= \frac{1}{2}\left[\sin(x_0+y_0)+\sin(x_0-y_0)\right] \\ &\quad +\frac{1}{2}\int_{x_0-y_0}^{x_0+y_0}\tau d\tau-\frac{1}{2}\int_0^{y_0}\int_{y+x_0-y_0}^{-y+x_0+y_0}dxdy \\ &= \frac{1}{2}\left[\sin(x_0+y_0)+\sin(x_0-y_0)\right]+x_0y_0-\frac{1}{2}y_0^2 \end{aligned}$$

Now dropping the subscript zero, we obtain the solution

$$u(x,y)=\frac{1}{2}\left[\sin(x+y)+\sin(x-y)\right]+xy-\frac{1}{2}y^2$$

4.8 Riemann Method

We shall discuss Riemann's method of integrating the linear hyperbolic equation

$$L[u]=u_{xy}+au_x+bu_y+cu=f(x,y). \tag{4.8.1}$$

L denotes the linear operator and $a(x,y), b(x,y), c(x,y)$, and $f(x,y)$ are differentiable functions in some domain D^*. The method consists essentially of the derivation of an integral formula which represents the solution of the Cauchy problem.

Let $v(x,y)$ be a function having continuous second-order partial derivatives. Then we may write

$$\begin{aligned} vu_{xy} - uv_{xy} &= (vu_x)_y - (uv_y)_x \\ vau_x &= (avu)_x - u(av)_x \\ vbu_y &= (bvu)_y - u(bv)_y \end{aligned} \tag{4.8.2}$$

so that

$$vL[u] - uM[v] = U_x + V_y \tag{4.8.3}$$

where M is the operator represented by

$$M[v] = v_{xy} - (av)_x - (bv)_y + cv \tag{4.8.4}$$

and

$$\begin{aligned} U &= auv - uv_y \\ V &= buv + vu_x \end{aligned} \tag{4.8.5}$$

The operator M is called the *adjoint operator* of L. If $M = L$, then the operator L is said to be *self-adjoint*. Now applying Green's theorem, we have

$$\int\int_D (U_x + V_y)\,dx\,dy = \oint_C U\,dy - V\,dx \tag{4.8.6}$$

where C is the closed curve bounding the region of integration D which is in D^*.

Let Λ be a smooth initial curve which is continuous, shown in Fig. 4.8.1. Since Eq. (4.8.1) is in first canonical form, x and y are the characteristic coordinates. We assume that the tangent to Λ is nowhere parallel to x or y axes. Let $P(\alpha,\beta)$ be a point at which the solution to the Cauchy problem is sought. Line PQ parallel to the x axis intersects the initial curve Λ at Q and line PR parallel to the y axis intersects the curve Λ at R. We suppose that u and u_x or u_y are prescribed along Λ.

Let C be the closed contour $PQRP$ bounding D. Since $dy = 0$ on PQ and $dx = 0$ on PR, it follows immediately from Eqs. (4.8.3) and (4.8.6) that

$$\int\int_D (vL[u] - uM[v])\,dx\,dy = \int_Q^R (U\,dy - V\,dx) + \int_R^P U\,dy - \int_P^Q V\,dx \tag{4.8.7}$$

From Eq. (4.8.5), we find

$$\int_P^Q V\,dx = \int_P^Q bvu\,dx + \int_P^Q vu_x\,dx$$

Integrating by parts, we obtain

$$\int_P^Q vu_x\,dx = [uv]_P^Q - \int_P^Q uv_x\,dx$$

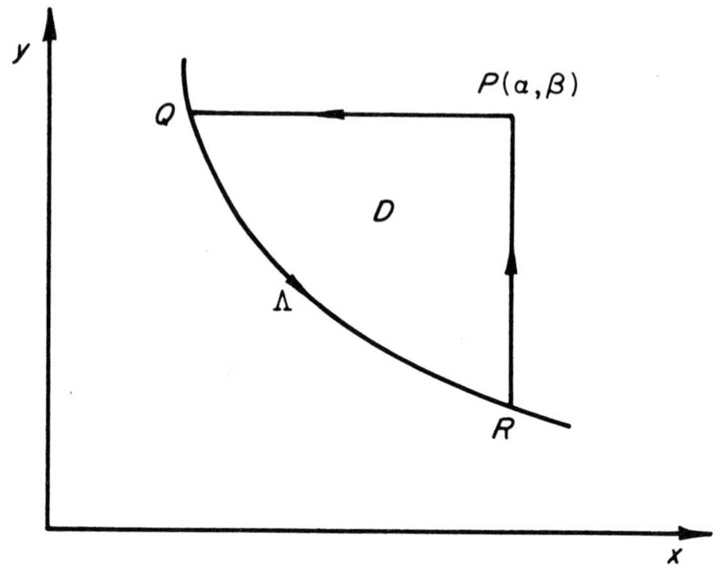

Figure 4.8.1

Hence, we may write

$$\int_P^Q V\,dx = [uv]_P^Q + \int_P^Q u(bv - v_x)\,dx$$

Substitution of this integral in Eq. (4.8.7) yields

$$[uv]_P = [uv]_Q + \int_P^Q u(bv - v_x)\,dx - \int_R^P u(av - v_y)\,dy - \int_Q^R (U\,dy - V\,dx)$$

$$+ \int\int_D (vL[u] - uM[v])\,dx\,dy \qquad (4.8.8)$$

Suppose we can choose the function $v(x,y;\alpha,\beta)$ to be the solution of the adjoint equation

$$M[v] = 0 \qquad (4.8.9)$$

and satisfying the conditions

$$\begin{aligned} v_x &= bv && \text{when} \quad y = \beta \\ v_y &= av && \text{when} \quad x = \alpha \\ v &= 1 && \text{when} \quad x = \alpha \quad y = \beta \end{aligned} \qquad (4.8.10)$$

The function $v(x,y;\alpha,\beta)$ is called the *Riemann function*. Since $L[u]=f$, Eq. (4.8.8) reduces to

$$[u]_P=[uv]_Q-\int_Q^R uv(a\,dy-b\,dx)+\int_Q^R(uv_y\,dy+vu_x\,dx)+\int\int_D vf\,dx\,dy$$

(4.8.11)

This gives us the value of u at the point P when u and u_x are prescribed along the curve Λ. When u and u_y are prescribed, the identity

$$[uv]_R-[uv]_Q=\int_Q^R[(uv)_x\,dx+(uv)_y\,dy]$$

may be used to put Eq. (4.8.8) in the form

$$[u]_P=[uv]_R-\int_Q^R uv(a\,dy-b\,dx)-\int_Q^R(uv_x\,dx+vu_y\,dy)+\int\int_D vf\,dx\,dy$$

(4.8.12)

Figure 4.8.2

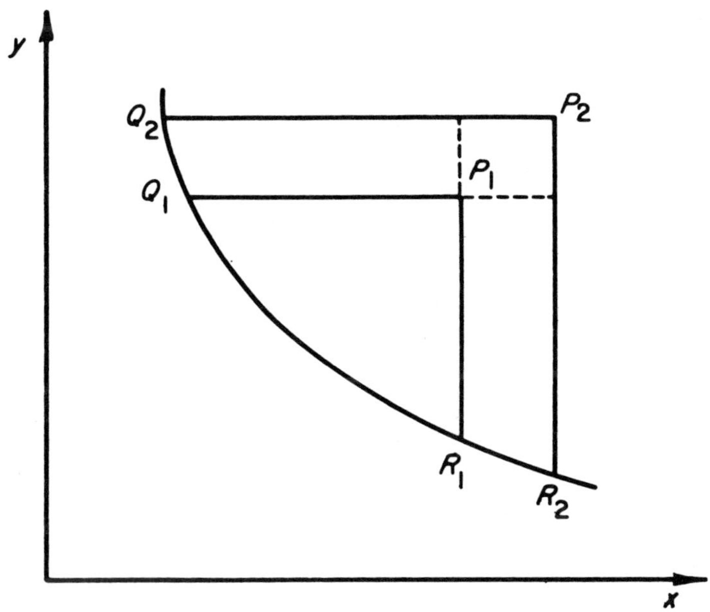

By adding Eqs. (4.8.11) and (4.8.12), the value of u at P is given by

$$[u]_P=\frac{1}{2}([uv]_Q+[uv]_R)-\int_Q^R uv(a\,dy-b\,dx)-\frac{1}{2}\int_Q^R u(v_x\,dx-v_y\,dy)$$
$$+\frac{1}{2}\int_Q^R v(u_x\,dx-u_y\,dy)+\iint_D vf\,dx\,dy \tag{4.8.13}$$

which is the solution of the Cauchy problem in terms of the Cauchy data given along the curve Λ. It is easy to see that the solution at the point (α,β) depends only on the Cauchy data along the arc QR on Λ. If the initial data were to change outside this arc QR, the solution would change only outside the triangle PQR. Thus, from Fig. 4.8.2, we can see that each characteristic separates the region in which the solution remains unchanged from the region in which it varies. Because of this fact, the unique continuation of the solution across any characteristic is not possible. This is evident from Fig. 4.8.2. The solution on the right of the characteristic P_1R_1 is determined by the initial data given on Q_1R_2, whereas the solution on the left is determined by the initial data given on Q_1R_1. If the initial data on R_1R_2 were changed, the solution on the right of P_1R_1 will only be affected.

It should be remarked here that the initial curve can intersect each characteristic at only one point. Suppose, for example, the initial curve Λ intersects the characteristic at two points as shown in Fig. 4.8.3. Then, the

Figure 4.8.3

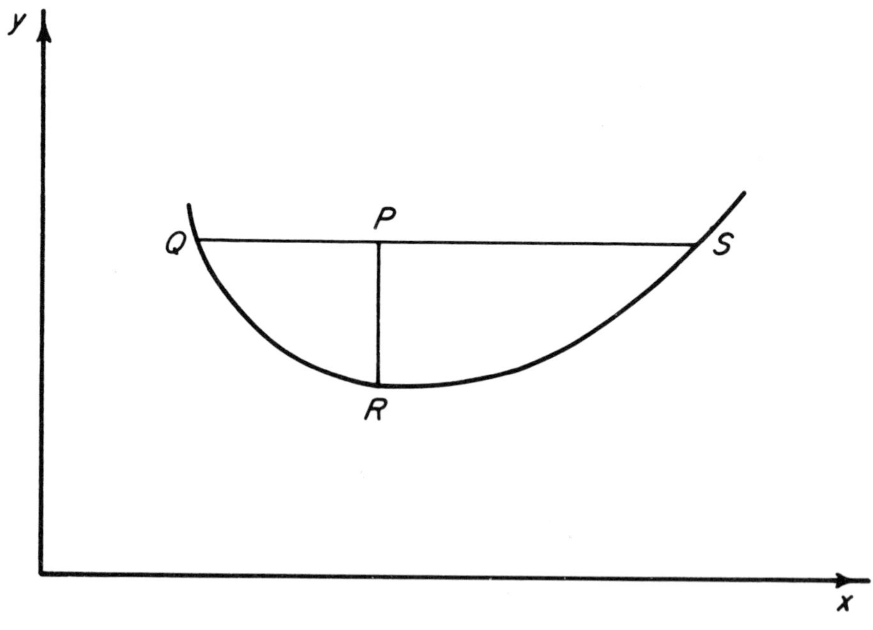

solution at P obtained from the initial data on QR will be different from the solution obtained from the initial data on RS. Hence, the Cauchy problem, in this case, is not solvable.

EXAMPLE 4.8.1. The telegraph equation

$$w_{tt}+a^*w_t+b^*w=c^2w_{xx}$$

may be transformed into canonical form

$$L[u]=u_{\xi\eta}+ku=0$$

by the successive transformations

$$w=ue^{-a^*t/2}$$

and

$$\xi=x+ct \qquad \eta=x-ct$$

where $k=(a^{*2}-4b^*)/16c^2$.

We apply Riemann's method to determine the solution satisfying the initial conditions

$$u(x,0)=f(x)$$
$$u_t(x,0)=g(x)$$

Since

$$t=\frac{1}{2c}(\xi-\eta),$$

the line $t=0$ corresponds to the straight line $\xi=\eta$ in the ξ-η plane. The initial conditions may thus be transformed into

$$[u]_{\xi=\eta}=f(\xi) \tag{4.8.14}$$

$$[u_\xi-u_\eta]_{\xi=\eta}=c^{-1}g(\xi) \tag{4.8.15}$$

We shall now determine the Riemann function $v(\xi,\eta;\alpha,\beta)$ which satisfies

$$v_{\xi\eta}+kv=0 \tag{4.8.16}$$

$$v_\xi(\xi,\beta;\alpha,\beta)=0 \tag{4.8.17}$$

$$v_\eta(\alpha,\eta;\alpha,\beta)=0 \tag{4.8.18}$$

$$v(\alpha,\beta;\alpha,\beta)=1 \tag{4.8.19}$$

The differential equation (4.8.16) is self-adjoint, that is,

$$L[v]=M[v]=v_{\xi\eta}+kv$$

We assume that the Riemann function is of the form

$$v(\xi,\eta;\alpha,\beta)=F(s)$$

with the argument $s=(\xi-\alpha)(\eta-\beta)$. Substituting this value in Eq. (4.8.16),

we obtain

$$sF_{ss}+F_s+kF=0$$

If we let $\lambda=\sqrt{4ks}$, the above equation becomes

$$F''(\lambda)+\frac{1}{\lambda}F'(\lambda)+F(\lambda)=0$$

This is the Bessel's equation of order zero, and the solution is

$$F(\lambda)=J_0(\lambda),$$

disregarding $Y_0(\lambda)$ which is unbounded at $\lambda=0$. Thus, the Riemann function is

$$v(\xi,\eta;\alpha,\beta)=J_0\left(\sqrt{4k(\xi-\alpha)(\eta-\beta)}\ \right)$$

which satisfies Eq. (4.8.16) and which is equal to one on the characteristics $\xi=\alpha$ and $\eta=\beta$. Since $J_0'(0)=0$, Eqs. (4.8.17) and (4.8.18) are satisfied. From this, it immediately follows that

$$[v_\xi]_{\xi=\eta}=\frac{\sqrt{k}\,(\xi-\beta)}{\sqrt{(\xi-\alpha)(\xi-\beta)}}[J_0'(\lambda)]_{\xi=\eta}$$

$$[v_\eta]_{\xi=\eta}=\frac{\sqrt{k}\,(\xi-\alpha)}{\sqrt{(\xi-\alpha)(\xi-\beta)}}[J_0'(\lambda)]_{\xi=\eta}$$

Thus, we have

$$[v_\xi-v_\eta]_{\xi=\eta}=\frac{\sqrt{k}\,(\alpha-\beta)}{\sqrt{(\xi-\alpha)(\xi-\beta)}}[J_0'(\lambda)]_{\xi=\eta} \tag{4.8.20}$$

From the initial condition

$$u(Q)=f(\beta)\quad\text{and}\quad u(R)=f(\alpha) \tag{4.8.21}$$

and substituting Eqs. (4.8.15), (4.8.19), and (4.8.20) into Eq. (4.8.13), we obtain

$$\begin{aligned}u(\alpha,\beta)&=\tfrac{1}{2}[f(\alpha)+f(\beta)]\\&\quad-\frac{1}{2}\int_\beta^\alpha\frac{\sqrt{k}\,(\alpha-\beta)}{\sqrt{(\tau-\alpha)(\tau-\beta)}}J_0'\left[\sqrt{4k(\tau-\alpha)(\tau-\beta)}\ \right]f(\tau)\,d\tau\\&\quad+\frac{1}{2c}\int_\beta^\alpha J_0\left[\sqrt{4k(\tau-\alpha)(\tau-\beta)}\ \right]g(\tau)\,d\tau\end{aligned} \tag{4.8.22}$$

Replacing α and β by ξ and η, and substituting the original variables x and t, we obtain

$$u(x,t)=\frac{1}{2}[f(x+ct)+f(x-ct)]+\frac{1}{2}\int_{x-ct}^{x+ct}G(x,t,\tau)\,d\tau \tag{4.8.23}$$

where

$$G(x,t,\tau)=\left\{-2\sqrt{k}\; ctf(\tau)J_0'\left[\sqrt{4k[(\tau-x)^2-c^2t^2]}\;\right]\right\}\Big/\sqrt{(\tau-x)^2-c^2t^2}$$
$$+c^{-1}g(\tau)J_0\left\{\sqrt{4k[(\tau-x)^2-c^2t^2]}\;\right\}$$

If we set $k=0$, we arrive at the d'Alembert formula for the wave equation

$$u(x,t)=\frac{1}{2}[f(x+ct)+f(x-ct)]+\frac{1}{2c}\int_{x-ct}^{x+ct}g(\tau)\,d\tau$$

4.9 Goursat Problem

The Goursat problem is that of finding the solution of a linear hyperbolic equation

$$u_{xy}=a_1(x,y)u_x+a_2(x,y)u_y+a_3(x,y)u+h(x,y) \tag{4.9.1}$$

satisfying the prescribed conditions

$$u(x,y)=f(x) \tag{4.9.2}$$

on a characteristic, say, $y=0$, and

$$u(x,y)=g(x) \tag{4.9.3}$$

on a monotonic increasing curve $y=y(x)$ which, for simplicity, is assumed to intersect the characteristic at the origin.

The solution in the region between the x-axis and the monotonic curve in the first quadrant can be determined by the method of successive approximations. The proof is given in Garabedian [15].

EXAMPLE 4.9.1. Determine the solution of the Goursat problem

$$u_{tt}=c^2u_{xx} \tag{4.9.4}$$
$$u(x,t)=f(x) \quad \text{on} \quad x-ct=0 \tag{4.9.5}$$
$$u(x,t)=g(x) \quad \text{on} \quad t=t(x) \tag{4.9.6}$$

where $f(0)=g(0)$.

The general solution of the wave equation is

$$u(x,t)=\phi(x+ct)+\psi(x-ct)$$

Applying the prescribed conditions, we obtain

$$f(x)=\phi(2x)+\psi(0) \tag{4.9.7}$$

$$g(x)=\phi(x+ct(x))+\psi(x-ct(x)) \tag{4.9.8}$$

It is evident that

$$f(0)=\phi(0)+\psi(0)=g(0)$$

Now, if $s=x-ct(x)$, the inverse of it is $x=\alpha(s)$. Thus, Eq. (4.9.8) may be written as

$$g(\alpha(s)) = \phi(x + ct(x)) + \psi(s) \tag{4.9.9}$$

Replacing x by $(x + ct(x))/2$ in Eq. (4.9.7), we obtain

$$f\left(\frac{x + ct(x)}{2}\right) = \phi(x + ct(x)) + \psi(0) \tag{4.9.10}$$

Thus, using (4.9.10), (4.9.9) becomes

$$\psi(s) = g(\alpha(s)) - f\left(\frac{\alpha(s) + ct(\alpha(s))}{2}\right) + \psi(0)$$

Replacing s by $x - ct$, we have

$$\psi(x - ct) = g(\alpha(x - ct)) - f\left(\frac{\alpha(x - ct) + ct(\alpha(x - ct))}{2}\right) + \psi(0)$$

Hence, by the solution is given by

$$u(x,t) = f\left(\frac{x + ct}{2}\right) - f\left(\frac{\alpha(x - ct) + ct(\alpha(x - ct))}{2}\right) + g(\alpha(x - ct)) \tag{4.9.11}$$

Let us consider a special case when the curve $t = t(x)$ is a straight line represented by $t - kx = 0$ with a constant $k > 0$. Then $s = x - ckx$ and hence $x = s/(1 - ck)$. Using these values in (4.9.11), we obtain

$$u(x,t) = f\left(\frac{x + ct}{2}\right) - f\left(\frac{(1 + ck)(x - ct)}{2(1 - ck)}\right) + g\left(\frac{x - ct}{1 - ck}\right) \tag{4.9.12}$$

When the values of u are prescribed on both the characteristics, the problem of finding u of a linear hyperbolic equation is called a *characteristic initial value problem*. This is the degenerate case of the Goursat problem.

Consider the characteristic initial value problem

$$u_{xy} = h(x,y) \tag{4.9.13}$$

$$u(x,0) = f(x) \tag{4.9.14}$$

$$u(0,y) = g(y) \tag{4.9.15}$$

where f and g are continuously differentiable and $f(0) = g(0)$.

Integrating Eq. (4.9.13), we obtain

$$u(x,y) = \int_0^x \int_0^y h(\xi,\eta)\, d\eta\, d\xi + \phi(x) + \psi(y) \tag{4.9.16}$$

where ϕ and ψ are arbitrary functions. Applying the prescribed conditions (4.9.14) and (4.9.15), we have

$$u(x,0) = \phi(x) + \psi(0) = f(x) \tag{4.9.17}$$

$$u(0,y) = \phi(0) + \psi(y) = g(x) \tag{4.9.18}$$

Thus

$$\phi(x)+\psi(y)=f(x)+g(x)-\phi(0)-\psi(0) \tag{4.9.19}$$

But from (4.9.17), we have

$$\phi(0)+\psi(0)=f(0) \tag{4.9.20}$$

Hence, from (4.9.16), (4.9.19), and (4.9.20), we obtain

$$u(x,y)=f(x)+g(y)-f(0)+\int_0^x\int_0^y h(\xi,\eta)\,d\eta\,d\xi \tag{4.9.21}$$

EXAMPLE 4.9.2. Determine the solution of the characteristic initial value problem

$$u_{tt}=c^2u_{xx}$$
$$u(x,t)=f(x) \quad \text{on} \quad x+ct=0$$
$$u(x,t)=g(x) \quad \text{on} \quad x-ct=0$$

where $f(0)=g(0)$

Here it is not necessary to reduce the given equation into canonical form. The general solution of the wave equation is

$$u(x,t)=\phi(x+ct)+\psi(x-ct)$$

The characteristics are

$$x+ct=0$$
$$x-ct=0$$

Applying the prescribed conditions, we have

$$u(x,t)=\phi(2x)+\psi(0)=f(x) \quad \text{on} \quad x+ct=0 \tag{4.9.22}$$

$$u(x,t)=\phi(0)+\psi(2x)=g(x) \quad \text{on} \quad x-ct=0 \tag{4.9.23}$$

We observe that these equations are compatible, since $f(0)=g(0)$.

Now replacing x by $(x+ct)/2$ in Eq. (4.9.22) and replacing x by $(x-ct)/2$ in Eq. (4.9.23), we have

$$\phi(x+ct)=f\left(\frac{x+ct}{2}\right)-\psi(0)$$

$$\psi(x-ct)=g\left(\frac{x-ct}{2}\right)-\phi(0)$$

Hence the solution is given by

$$u(x,t)=f\left(\frac{x+ct}{2}\right)+g\left(\frac{x-ct}{2}\right)-f(0) \tag{4.9.24}$$

We note that this solution can be obtained by substituting $k=-1/c$ into (4.9.12).

EXAMPLE 4.9.3. Find the solution of the characteristic initial value problem

$$y^3 u_{xx} - y u_{yy} + u_y = 0 \tag{4.9.25}$$

$$u(x,y) = f(x) \quad \text{on} \quad x + \frac{y^2}{2} = 4 \quad \text{for} \quad 2 \leqslant x \leqslant 4$$

$$u(x,y) = g(x) \quad \text{on} \quad x - \frac{y^2}{2} = 0 \quad \text{for} \quad 0 \leqslant x \leqslant 2$$

with $f(2) = g(2)$.

Since the equation is hyperbolic except for $y = 0$, we reduce it to the canonical form

$$u_{\xi\eta} = 0$$

where $\xi = x + y^2/2$ and $\eta = x - y^2/2$. Thus, the general solution is

$$u(x,y) = \phi\left(x + \frac{y^2}{2}\right) + \psi\left(x - \frac{y^2}{2}\right) \tag{4.9.26}$$

Applying the prescribed conditions, we have

$$f(x) = \phi(4) + \psi(2x - 4) \tag{4.9.27}$$

$$g(x) = \phi(2x) + \psi(0) \tag{4.9.28}$$

Now, if we replace $(2x-4)$ by $(x - y^2/2)$ in (4.9.27) and $(2x)$ by $(x + y^2/2)$ in (4.9.28), we obtain

$$\psi\left(x - \frac{y^2}{2}\right) = f\left(\frac{x}{2} - \frac{y^2}{4} + 2\right) - \phi(4)$$

$$\phi\left(x + \frac{y^2}{2}\right) = g\left(\frac{x}{2} + \frac{y^2}{4}\right) - \psi(0)$$

Thus

$$u(x,y) = f\left(\frac{x}{2} - \frac{y^2}{4} + 2\right) + g\left(\frac{x}{2} + \frac{y^2}{4}\right) - \phi(4) - \psi(0)$$

But from (4.9.27) and (4.9.28), we see that

$$f(2) = \phi(4) + \psi(0) = g(2)$$

Hence

$$u(x,y) = f\left(\frac{x}{2} - \frac{y^2}{4} + 2\right) + g\left(\frac{x}{2} + \frac{y^2}{4}\right) - f(2)$$

Exercises

1. Determine the solution of each of the following initial-value problems:

a.	$u_{tt}-c^2u_{xx}=0$	$u(x,0)=0$	$u_t(x,0)=1$
b.	$u_{tt}-c^2u_{xx}=0$	$u(x,0)=\sin x$	$u_t(x,0)=x^2$
c.	$u_{tt}-c^2u_{xx}=0$	$u(x,0)=x^3$	$u_t(x,0)=x$
d.	$u_{tt}-c^2u_{xx}=0$	$u(x,0)=\cos x$	$u_t(x,0)=e^{-1}$
e.	$u_{tt}-c^2u_{xx}=0$	$u(x,0)=\log(1+x^2)$	$u_t(x,0)=2$
f.	$u_{tt}-c^2u_{xx}=0$	$u(x,0)=x$	$u_t(x,0)=\sin x$

2. Determine the solution of each of the following initial-value problems:

a.	$u_{tt}-c^2u_{xx}=x$	$u(x,0)=0$	$u_t(x,0)=3$
b.	$u_{tt}-c^2u_{xx}=x+ct$	$u(x,0)=x$	$u_t(x,0)=\sin x$
c.	$u_{tt}-c^2u_{xx}=e^x$	$u(x,0)=5$	$u_t(x,0)=x^2$
d.	$u_{tt}-c^2u_{xx}=\sin x$	$u(x,0)=\cos x$	$u_t(x,0)=1+x$
e.	$u_{tt}-c^2u_{xx}=xe^t$	$u(x,0)=\sin x$	$u_t(x,0)=0$
f.	$u_{tt}-c^2u_{xx}=2$	$u(x,0)=x^2$	$u_t(x,0)=\cos x$

3. A gas which is contained in a sphere of radius R is at rest initially, and the initial condensation is given by s_0 inside the sphere and zero outside the sphere. The condensation is related to the velocity potential by

$$s=(1/c^2)u_t$$

at all times, and the velocity potential satisfies

$$u_{tt}=\nabla^2 u$$

Determine the condensation s for all $t>0$.

4. Solve the initial value problem

$$u_{xx}+2u_{xy}-3u_{yy}=0$$
$$u(x,0)=\sin x$$
$$u_y(x,0)=x$$

5. Find the longitudinal oscillation of a rod subject to the initial conditions

$$u(x,0)=\sin x$$
$$u_t(x,0)=x$$

6. By the Riemann method, solve

a. $\sin^2\mu\phi_{xx}-\cos^2\mu\phi_{yy}-(\lambda^2\sin^2\mu\cos^2\mu)\phi=0$

$$\phi(0,y)=f_1(y) \qquad \phi(x,0)=g_1(x)$$
$$\phi_x(0,y)=f_2(y) \qquad \phi_y(x,0)=g_2(x)$$

b. $x^2u_{xx}-t^2u_{tt}=0$

$$u(x,t_1)=f(x)$$
$$u_t(x,t_2)=g(x)$$

7. Determine the solution of the initial-boundary value problem

$$\begin{aligned} u_{tt}&=4u_{xx}, && 0<x<\infty, && t>0\\ u(x,0)&=x^4, && 0\leqslant x<\infty\\ u_t(x,0)&=0, && 0\leqslant x<\infty\\ u(0,t)&=0, && && t\geqslant 0 \end{aligned}$$

8. Determine the solution of the initial-boundary value problem

$$\begin{aligned} u_{tt}&=9u_{xx}, && 0<x<\infty, && t>0\\ u(x,0)&=0, && 0\leqslant x<\infty\\ u_t(x,0)&=x^3, && 0\leqslant x<\infty\\ u_x(x,0)&=0, && && t\geqslant 0 \end{aligned}$$

9. Determine the solution of the initial-boundary value problem

$$\begin{aligned} u_{tt}&=16u_{xx}, && 0<x<\infty, && t>0\\ u(x,0)&=\sin x, && 0\leqslant x<\infty\\ u_t(x,0)&=x^2, && 0\leqslant x<\infty\\ u(0,t)&=0, && && t\geqslant 0 \end{aligned}$$

10. In the initial-boundary value problem

$$\begin{aligned} u_{tt}&=c^2u_{xx}, && 0<x<l, && t>0\\ u(x,0)&=f(x), && 0\leqslant x\leqslant l\\ u_t(x,0)&=q(x), && 0\leqslant x\leqslant l\\ u(0,t)&=0, && && t\geqslant 0 \end{aligned}$$

if f and g are extended as the odd functions, show that $u(x,t)$ is given by (4.4.5) for $x>ct$ and (4.4.6) for $x<ct$.

11. In the initial-boundary value problem

$$\begin{aligned} u_{tt}&=c^2u_{xx}, && 0<x<l, && t>0\\ u(x,0)&=f(x), && 0\leqslant x\leqslant l\\ u_t(x,0)&=g(x), && 0\leqslant x\leqslant l\\ u_x(0,t)&=0, && && t\geqslant 0 \end{aligned}$$

if f and g are extended as the even functions, show that $u(x,t)$ is given by (4.4.8) for $x>ct$ and (4.4.9) for $x<ct$.

12. Determine the solution of the initial-boundary value problem

$$\begin{aligned} &u_{tt}=c^2u_{xx}, && 0<x<\infty, && t>0\\ &u(x,0)=f(x), && 0\leqslant x<\infty && \\ &u_t(x,0)=0, && 0\leqslant x<\infty && \\ &u_x(0,t)+hu(0,t)=0, && && t\geqslant 0, h=\text{constant} \end{aligned}$$

State the compatibility condition of f.

13. Find the solution of the problem

$$\begin{aligned} &u_{tt}=c^2u_{xx}, && at<x<\infty, && t>0\\ &u(x,0)=f(x), && 0<x<\infty && \\ &u_t(x,0)=0, && 0<x<\infty && \\ &u(at,t)=0, && && t>0 \end{aligned}$$

where $f(0)=0$ and a is a constant.

14. Find the solution of the initial-boundary value problem

$$\begin{aligned} &u_{tt}=u_{xx}, && 0<x<2, && t>0\\ &u(x,0)=\sin\pi x/2, && 0\leqslant x\leqslant 2 && \\ &u_t(x,0)=0, && 0\leqslant x\leqslant 2 && \\ &u(0,t)=0, && && t\geqslant 0\\ &u(2,t)=0, && && t\geqslant 0. \end{aligned}$$

15. Find the solution of the initial-boundary value problem

$$\begin{aligned} &u_{tt}=4u_{xx}, && 0<x<1, && t>0\\ &u(x,0)=0, && 0\leqslant x\leqslant 1 && \\ &u_t(x,0)=x(1-x), && 0\leqslant x\leqslant 1 && \\ &u(0,t)=0, && && t\geqslant 0\\ &u(1,t)=0, && && t\geqslant 0 \end{aligned}$$

16. Determine the solution of the initial-boundary value problem

$$\begin{aligned} &u_{tt}=c^2u_{xx}, && 0<x<l, && t>0\\ &u(x,0)=f(x), && 0\leqslant x\leqslant l && \\ &u_t(x,0)=g(x), && 0\leqslant x\leqslant l && \\ &u_x(0,t)=0, && && t\geqslant 0\\ &u_x(l,t)=0, && && t\geqslant 0 \end{aligned}$$

by extending f and g as the even functions about $x=0$ and $x=l$.

17. Determine the solution of the initial-boundary value problem

$$\begin{aligned} &u_{tt}=c^2u_{xx}, && 0<x<l, && t>0\\ &u(x,0)=f(x), && 0\leqslant x\leqslant l && \\ &u_t(x,0)=g(x), && 0\leqslant x\leqslant l && \\ &u(0,t)=p(t), && && t\geqslant 0\\ &u(l,t)=q(t), && && t\geqslant 0 \end{aligned}$$

18. Determine the solution of the initial-boundary value problem

$$\begin{array}{lll} u_{tt}=c^2u_{xx}, & 0<x<l, & t>0 \\ u(x,0)=f(x), & 0\leqslant x\leqslant l & \\ u_t(x,0)=g(x), & 0\leqslant x\leqslant l & \\ u_x(0,t)=p(t), & & t\geqslant 0 \\ u_x(l,t)=q(t), & & t\geqslant 0 \end{array}$$

19. Solve the characteristic initial value problem

$$xy^3u_{xx}-x^3yu_{yy}-y^3u_x+x^3u_y=0$$

$$u(x,y)=f(x) \quad \text{on} \quad y^2-x^2=8 \quad \text{for } 0\leqslant x\leqslant 2$$

$$u(x,y)=g(x) \quad \text{on} \quad y^2+x^2=16 \quad \text{for } 2\leqslant x\leqslant 4$$

with $f(2)=g(2)$.

20. Solve the Goursat problem

$$xy^3u_{xx}-x^3yu_{yy}-y^3u_x+x^3u_y=0$$

$$u(x,y)=f(x) \quad \text{on} \quad y^2+x^2=16 \quad \text{for } 0\leqslant x\leqslant 4$$

$$u(x,y)=f(y) \quad \text{on} \quad x=0 \quad \text{for } 0\leqslant y\leqslant 4$$

where $f(0)=g(4)$.

21. Solve

$$\begin{array}{lll} u_{tt}=c^2u_{xx} & & \\ u(x,t)=f(x) & \text{on} & t=t(x) \\ u(x,t)=g(x) & \text{on} & x+ct=0 \end{array}$$

where $f(0)=g(0)$.

22. Solve the characteristic initial value problem

$$xu_{xx}-x^3u_{yy}-u_x=0, \quad x\neq 0$$

$$u(x,y)=f(y) \quad \text{on} \quad y-\frac{x^2}{2}=0 \quad \text{for} \quad 0\leqslant y\leqslant 2$$

$$u(x,y)=g(y) \quad \text{on} \quad y+\frac{x^2}{2}=4 \quad \text{for} \quad 2\leqslant y\leqslant 4$$

where $f(2)=g(2)$.

23. Solve

$$u_{xx}+10u_{xy}+9u_{yy}=0$$

$$u(x,0)=f(x)$$

$$u_y(x,0)=g(x)$$

24. Solve

$$4u_{xx} + 5u_{xy} + u_{yy} + u_x + u_y = 2$$
$$u(x,0) = f(x)$$
$$u_y(x,0) = g(x)$$

25. Solve

$$3u_{xx} + 10u_{xy} + 3u_{yy} = 0$$
$$u(x,0) = f(x)$$
$$u_y(x,0) = g(x)$$

26. Solve

$$u_{xx} - 3u_{xy} + 2u_{yy} = 0$$
$$u(x,0) = f(x)$$
$$u_y(x,0) = g(x)$$

27. Solve

$$x^2u_{xx} - t^2u_{tt} = 0 \quad x > 0, \quad t > 0$$
$$u(x,1) = f(x)$$
$$u_t(x,1) = g(x)$$

CHAPTER FIVE

Fourier Series

This chapter is devoted to the theory of Fourier series. Although the treatment can be extensive, the exposition of the theory here will be concise, but sufficient for its application to many problems of mathematical physics.

The reason that the Fourier theory of trigonometric series is of great practical importance is because certain types of discontinuous functions which cannot be expanded in power series can be expanded in Fourier series. More important, a wide class of problems in physics and engineering possesses periodic phenomena, and as a consequence, Fourier's trigonometric series become an indispensable tool in the analyses of these problems.

We shall begin our study with the basic concepts and definitions of some properties of real-valued functions.

5.1 Piecewise Continuous Functions

We define the *left hand limit* of a single-valued function $f(x)$ at x_0 as the finite limit of $f(x)$ if it exists, as x approaches x_0 from the left, and is denoted by $f(x_0-)$. That is, if h is positive,

$$f(x_0-)=\lim_{h\to 0} f(x_0-h) \tag{5.1.1}$$

Similarly, the *right hand limit* is defined as

$$f(x_0+)=\lim_{h\to 0} f(x_0+h) \tag{5.1.2}$$

These limits are shown in Fig. 5.1.1.

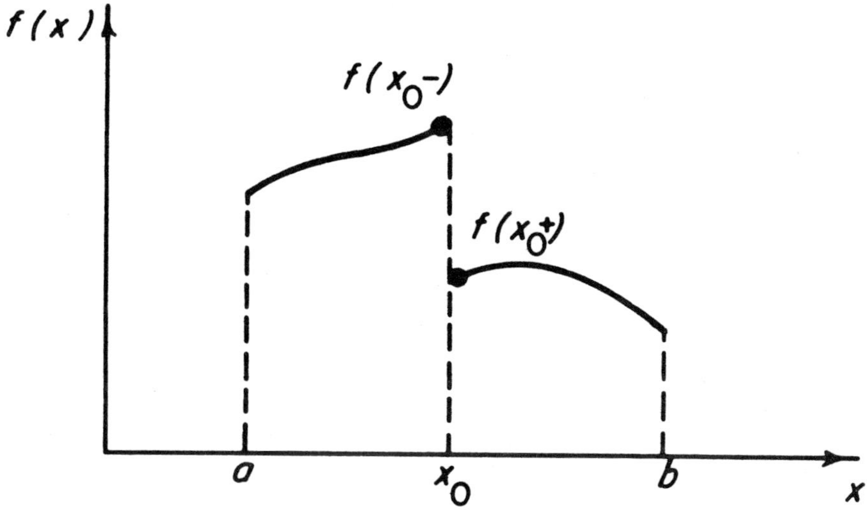

Figure 5.1.1

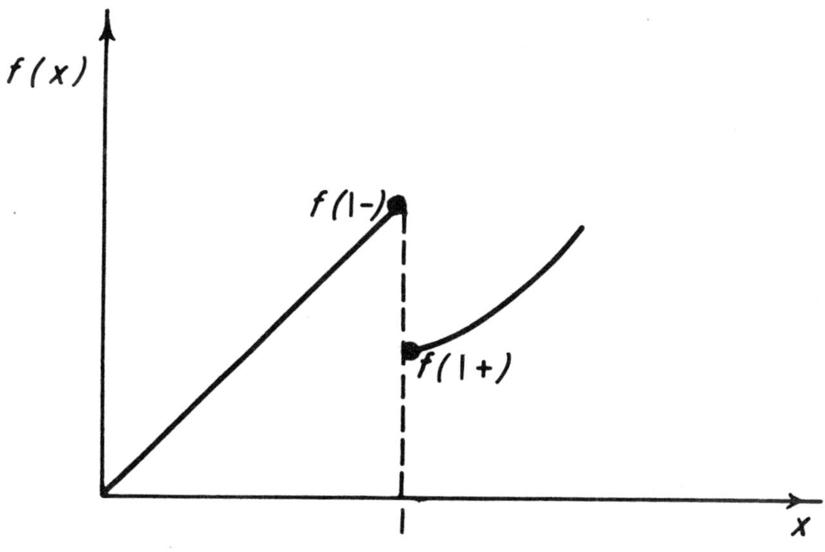

Figure 5.1.2

We note that if $f(x)$ is continuous at x_0, then

$$f(x_0-)=f(x_0+)=f(x_0)$$

Thus, for example, the function shown in Fig. 5.1.2

$$f(x)=\begin{cases} x & \text{when} \quad 0\leqslant x<1 \\ x^2-\frac{1}{2} & \text{when} \quad 1<x\leqslant 2 \end{cases}$$

has the left hand limit $f(1-)=1$ and has the right hand limit $f(1+)=\frac{1}{2}$. The magnitude of the *jump discontinuity* which occurs at $x=1$ is

$$f(1-)-f(1+)=\tfrac{1}{2}$$

A single-valued function f is said to be *piecewise continuous* in an interval $[a,b]$, if there exist finitely many points $a=x_1<x_2<\ldots<x_n=b$, such that f is continuous in the intervals $x_j<x<x_{j+1}$ and the one-sided limits $f(x_j+)$ and $f(x_{j+1}-)$ exist for all $j=1,2,3,\ldots,n-1$.

A piecewise continuous function is shown in Fig. 5.1.3. The functions such as $1/x$ and $\sin 1/x$ fail to be piecewise continuous in the closed interval $[0,1]$ because the one-sided limit $f(0+)$ does not exist in both cases.

If f is piecewise continuous in an interval $[a,b]$, then it is necessarily bounded and integrable over that interval. Also, it follows immediately that the product of two piecewise continuous functions is piecewise continuous on the common interval.

Figure 5.1.3

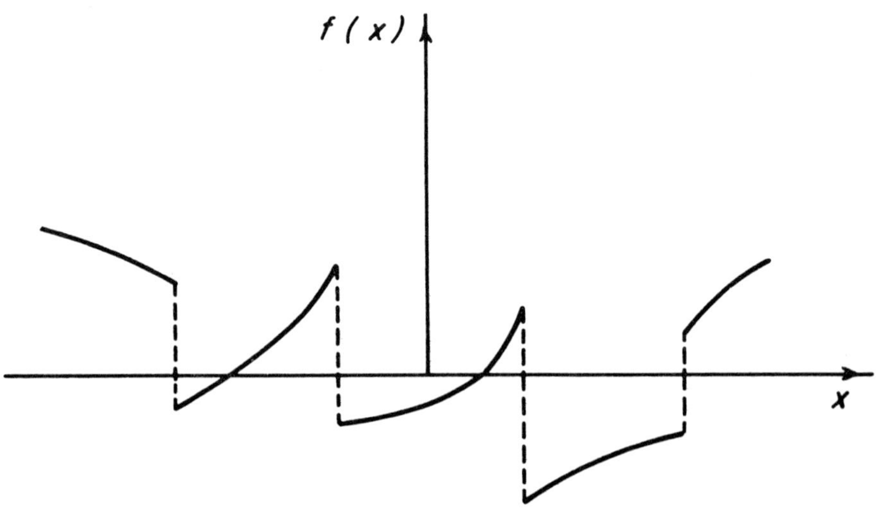

We define the *left hand derivative* of the function $f(x)$ at x_0 as

$$f'(x_0-)=\lim_{h\to 0}\frac{f(x_0-)-f(x_0-h)}{h} \tag{5.1.3}$$

and the *right hand derivative* of the function $f(x)$ at x_0 as

$$f'(x_0+)=\lim_{h\to 0}\frac{f(x_0+h)-f(x_0+)}{h} \tag{5.1.4}$$

where h is a positive increment.

It is clear that if f has a derivative f' at x_0, then the left hand and right hand derivatives exist and have the value $f'(x_0)$. However, the converse is not true. For example, the function

$$f(x)=\begin{cases} x & x\leqslant 0 \\ \cos x & x\geqslant 0\end{cases}$$

possesses the left and right hand derivatives at $x=0$, the magnitudes of which are 1 and 0 respectively, although $f'(0)$ does not exist.

If f is piecewise continuous in an interval $[a,b]$ and, if, in addition, the first derivative f' is continuous in each of the intervals $x_j<x<x_{j+1}$, and the limits $f'(x_j+)$ and $f'(x_j-)$ exist, then f is said to be *piecewise smooth*; if, in addition, the second derivative f'' is continuous in each of the intervals $x_j<x<x_{j+1}$, and the limits $f''(x_j+)$ and $f''(x_j-)$ exist, then f is said to be *piecewise very smooth.*

5.2 Even and Odd Functions

In analyses, certain properties of a function can often simplify the results appreciably and reduce the amount of work considerably. In this section, we formally introduce the even and odd behavior of a function.

A function $f(x)$ is said to be even if, for all x,

$$f(-x)=f(x) \tag{5.2.1}$$

and odd if, for all x,

$$f(-x)=-f(x) \tag{5.2.2}$$

In other words, an even function is one whose graph is symmetric about the y-axis and an odd function has graph which is symmetric about the origin as in Fig. 5.2.1. For example, x^2, $\cos x$ are even functions whereas x, $\sin x$ are odd functions.

We should note that not all functions are either even or odd. Some are neither, for instance e^x. However, any function can be written as the sum of an even and odd function. Thus

$$\begin{aligned} f(x)&=\tfrac{1}{2}[f(x)+f(-x)]+\tfrac{1}{2}[f(x)-f(-x)] \\ &=f_e(x)+f_o(x) \end{aligned} \tag{5.2.3}$$

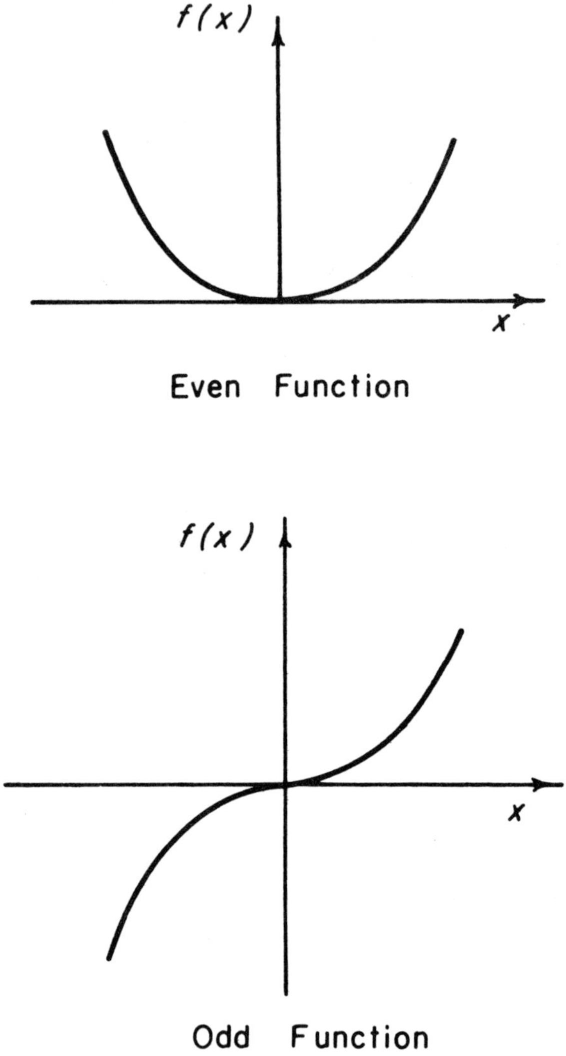

Figure 5.2.1

where f_e and f_o denote even and odd functions respectively. For example, the function e^x may be written as

$$\begin{aligned} e^x &= \frac{e^x + e^{-x}}{2} + \frac{e^x - e^{-x}}{2} \\ &= \cosh x + \sinh x \end{aligned}$$

where $\cosh x$ and $\sinh x$ are the even and odd parts of the exponential function.

EXAMPLE 5.2.1. Given $f(x)=1+2x^2+3x^4$. We test this function by the relations (5.2.1) and (5.2.2).

$$\begin{aligned}f(-x)&=1+2(-x)^2+3(-x)^4\\&=1+2x^2+3x^4\\&=f(x)\end{aligned}$$

Hence $f(x)$ is an even function.

EXAMPLE 5.2.2. Consider $f(x)=x+\sin x$. As before we observe that

$$\begin{aligned}f(-x)&=(-x)+\sin(-x)\\&=-x-\sin x\\&=-f(x)\end{aligned}$$

which shows that $f(x)$ is an odd function.

EXAMPLE 5.2.3. Consider the function $f(x)=e^x$. Since

$$f(-x)=e^{-x}$$

$f(x)$ is neither even nor odd. This is due to the fact that the negative sign cannot be factored out of the exponential function to put the resulting function in the original form.

One application where the symmetric property of even and odd functions is significant is in the evaluation of integrals. If $f(x)$ is either even or odd functions, we have

$$\int_{-a}^{a} f(x)\,dx=2\int_0^a f(x)\,dx \qquad \text{when } f(x) \text{ is even} \tag{5.2.4}$$

$$\int_{-a}^{a} f(x)\,dx=0 \qquad \text{when } f(x) \text{ is odd} \tag{5.2.5}$$

provided $f(x)$ is integrable. Graphically, as shown in Fig. 5.2.2, the integral represents the area under the curve, and thus for even function, the entire

Figure 5.2.2

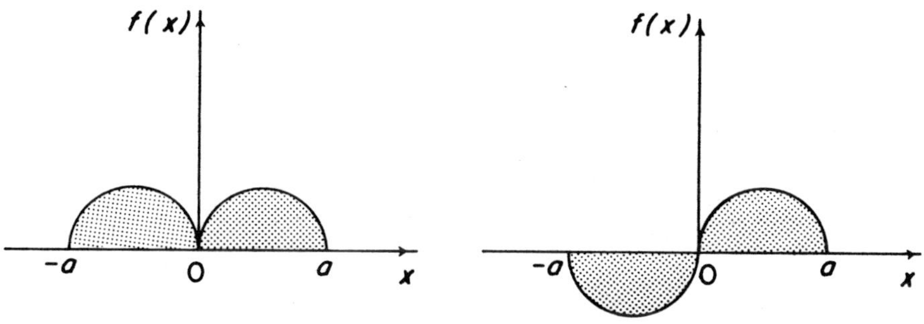

area is twice that of area under the curve from 0 to a whereas in the case of odd function the negative area under the curve from $-a$ to 0 cancels with the area under the curve from 0 to a so that the net area is zero.

It is easy to show (Exercise 4) that an even function times an even function or an odd function times an odd function is an even function, and an even function times an odd function yields an odd function.

5.3 Periodic Functions

A piecewise continuous function $f(x)$ in an interval $[a,b]$ is said to be periodic if there exists a real positive number p such that

$$f(x+p)=f(x) \tag{5.3.1}$$

for all x. p is called the *period* of f, and the smallest value of p is termed the *fundamental period*. A sample graph of a periodic function is given in Fig. 5.3.1.

If f is periodic with period p, then

$$\begin{aligned}
&f(x+p)=f(x)\\
&f(x+2p)=f(x+p+p)=f(x+p)\\
&f(x+3p)=f(x+2p+p)=f(x+2p)\\
&f(x+np)=f(x+(n-1)p+p)=f(x+(n-1)p)=f(x)
\end{aligned}$$

for any integer n. Hence for all integral values of n

$$f(x+np)=f(x) \tag{5.3.2}$$

It can be readily shown that if $f_1, f_2, \ldots, f_k$ have the period p and c_k are the constants, then

$$f=c_1 f_1+c_2 f_2+\ldots+c_k f_k \tag{5.3.3}$$

has the period p.

Well known examples of periodic functions are the sine and cosine functions. As a special case, a constant function is also a periodic function

Figure 5.3.1

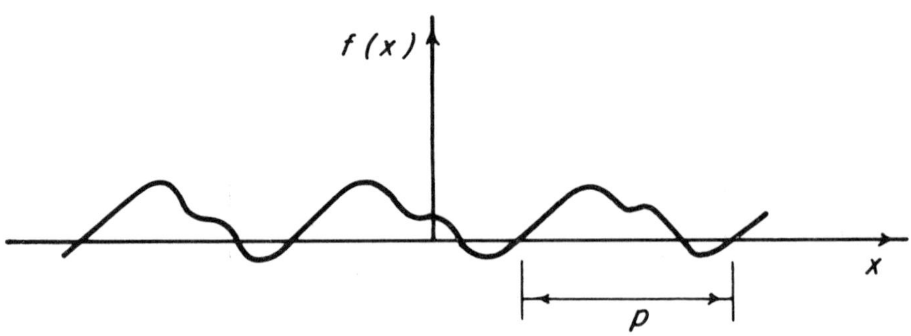

with arbitrary period p. Thus, by the relation (5.3.3), the series

$$a_0 + a_1\cos x + a_2\cos 2x + \ldots + b_1\sin x + b_2\sin 2x + \ldots$$

if it converges, obviously has the period 2π. Such type of series which occurs frequently in problems of mathematical physics, will be treated later.

5.4 Orthogonality

The functions of a sequence $\{\phi_n(x)\}$ are said to be *orthogonal* with respect to the weight function $q(x)$ on the interval $[a,b]$ if

$$\int_a^b \phi_m(x)\phi_n(x)q(x)\,dx = 0, \quad m \neq n \tag{5.4.1}$$

If $m=n$, then we have

$$\|\phi_n\| = \left[\int_a^b \phi_n^2 q\,dx\right]^{1/2} \tag{5.4.2}$$

which is called the *norm* of the orthogonal system $\{\phi_n\}$.

EXAMPLE 5.4.1. The functions $\sin mx, m=1,2,\ldots,$ form an orthogonal system on the interval $[-\pi,\pi]$, because

$$\int_{-\pi}^{\pi} \sin mx \sin nx\,dx = \begin{cases} 0, & m \neq n \\ \pi, & m=n \end{cases}$$

In this example we notice that the weight function is equal to unity, and the value of the norm is $\sqrt{\pi}$.

An orthogonal system $\phi_1, \phi_2, \ldots, \phi_n$ where n may be finite or infinite, which satisfies the relations

$$\int_a^b \phi_m(x)\phi_n(x)q(x)\,dx = \begin{cases} 0, & m \neq n \\ 1, & n=n \end{cases} \tag{5.4.3}$$

is called the *orthonormal* system of functions on $[a,b]$. It is evident that an orthonormal system can be obtained from an orthogonal system by dividing each function by its norm on $[a,b]$.

EXAMPLE 5.4.2. The sequence of functions

$$1, \cos x, \sin x, \ldots, \cos nx, \sin nx$$

form an orthogonal system on $[-\pi,\pi]$ since

$$\int_{-\pi}^{\pi} \sin mx \sin nx\,dx = \begin{cases} 0, & m \neq n \\ \pi, & m=n \end{cases}$$

$$\int_{-\pi}^{\pi} \sin mx \cos nx\,dx = 0 \qquad \text{for all } m,n \tag{5.4.4}$$

$$\int_{-\pi}^{\pi} \cos mx \cos nx\,dx = \begin{cases} 0, & m \neq n \\ \pi, & m=n \end{cases}$$

for positive integers m and n. To normalize this system, we divide the elements of the original orthogonal system by its norm. Hence

$$\frac{1}{\sqrt{2\pi}}, \frac{\cos x}{\sqrt{\pi}}, \frac{\sin x}{\sqrt{\pi}}, \ldots, \frac{\cos nx}{\sqrt{\pi}}, \frac{\sin nx}{\sqrt{\pi}}$$

form an orthonormal system.

5.5 Fourier Series

The functions

$$1, \cos x, \sin x, \cos 2x, \sin 2x \ldots$$

are mutually orthogonal to each other in the interval $[-\pi, \pi]$ and are linearly independent. Thus, we form a formal series representing $f(x)$. We write

$$f(x) \sim \frac{a_0}{2} + \sum_{k=1}^{\infty} (a_k \cos kx + b_k \sin kx) \tag{5.5.1}$$

where the symbol $\sim$ indicates an association of a_0, a_k, and b_k to f in some unique manner. The series may or may not be convergent. The coefficient $a_0/2$ instead of a_0 is used for convenience in formulation.

Let $f(x)$ be a Riemann integrable function defined on the interval $[-\pi, \pi]$. Suppose we define the nth partial sum

$$s_n(x) = \frac{a_0}{2} + \sum_{k=1}^{n} (a_k \cos kx + b_k \sin kx) \tag{5.5.2}$$

that is to represent $f(x)$ on $[-\pi, \pi]$. We shall seek the coefficients a_0, a_k, and b_k such that $s_n(x)$ represents the best approximation to $f(x)$ in the sense of least squares, that is, we seek to minimize the integral

$$I(a_0, a_k, b_k) = \int_{-\pi}^{\pi} [f(x) - s_n(x)]^2 dx \tag{5.5.3}$$

This is an extremal problem. A necessary condition for a_0, a_k, b_k, so that I be minimum, is that the first partial derivatives of I with respect to these coefficients vanish. Thus, substituting Eq. (5.5.2) into (5.5.3) and differentiating with respect to a_0, a_k, and b_k, we obtain

$$\frac{\partial I}{\partial a_0} = -\int_{-\pi}^{\pi} \left[f(x) - \frac{a_0}{2} - \sum_{j=1}^{n} (a_j \cos jx + b_j \sin jx) \right] dx \tag{5.5.4}$$

$$\frac{\partial I}{\partial a_k} = -2\int_{-\pi}^{\pi} \left[f(x) - \frac{a_0}{2} - \sum_{j=1}^{n} (a_j \cos jx + b_j \sin jx) \right] \cos kx \, dx \tag{5.5.5}$$

$$\frac{\partial I}{\partial b_k} = -2\int_{-\pi}^{\pi} \left[f(x) - \frac{a_0}{2} - \sum_{j=1}^{n} (a_j \cos jx + b_j \sin jx) \right] \sin kx \, dx \tag{5.5.6}$$

Using the orthogonality relations of the trigonometric functions (5.4.4) and noting that

$$\int_{-\pi}^{\pi} \cos mx\,dx = \int_{-\pi}^{\pi} \sin mx\,dx = 0 \tag{5.5.7}$$

where m and n are positive integers, Eqs. (5.5.4), (5.5.5) and (5.5.6) become

$$\frac{\partial I}{\partial a_0} = \pi a_0 - \int_{-\pi}^{\pi} f(x)\,dx \tag{5.5.8}$$

$$\frac{\partial I}{\partial a_k} = 2\pi a_k - 2\int_{-\pi}^{\pi} f(x)\cos kx\,dx \tag{5.5.9}$$

$$\frac{\partial I}{\partial b_k} = 2\pi b_k - 2\int_{-\pi}^{\pi} f(x)\sin kx\,dx \tag{5.5.10}$$

which must vanish for I to have an extremal value. Thus, we have

$$a_0 = \frac{1}{\pi}\int_{-\pi}^{\pi} f(x)\,dx \tag{5.5.11}$$

$$a_k = \frac{1}{\pi}\int_{-\pi}^{\pi} f(x)\cos kx\,dx \tag{5.5.12}$$

$$b_k = \frac{1}{\pi}\int_{-\pi}^{\pi} f(x)\sin kx\,dx \tag{5.5.13}$$

Note that a_0 is the special case of a_k which is the reason for writing $a_0/2$ rather than a_0 in Eq. (5.5.1). It immediately follows from Eqs. (5.5.8), (5.5.9) and (5.5.10) that

$$\frac{\partial^2 I}{\partial a_0^2} = \pi \tag{5.5.14}$$

$$\frac{\partial^2 I}{\partial a_k^2} = \frac{\partial^2 I}{\partial b_k^2} = 2\pi \tag{5.5.15}$$

and all mixed second order and all remaining higher order derivatives vanish. Now if we expand I in a Taylor series about $(a_0, a_1, \ldots, a_n, b_1, \ldots, b_n)$ we have

$$I(a_0+\Delta a_0, \ldots, b_n+\Delta b_n) = I(a_0, \ldots, b_n) + \Delta I \tag{5.5.16}$$

where ΔI stands for the remaining terms. Since the first derivatives, all mixed second derivatives and all remaining higher derivatives vanish, we obtain

$$\Delta I = \frac{1}{2!}\left[\frac{\partial^2 I}{\partial a_0^2}\Delta a_0^2 + \sum_{k=1}^{n}\left(\frac{\partial^2 I}{\partial a_k^2}\Delta a_k^2 + \frac{\partial^2 I}{\partial b_k^2}\Delta b_k^2\right)\right] \tag{5.5.17}$$

By virtue of Eqs. (5.5.14) and (5.5.15), ΔI is positive. Hence, for I to have a minimum value, the coefficients a_0, a_k, b_k must be given by Eqs. (5.5.11),

(5.5.12), and (5.5.13) respectively. These coefficients are called the *Fourier coefficients* of $f(x)$ and the series in Eq. (5.5.1) is said to be the *Fourier series* corresponding to $f(x)$.

We remark that the possibility of representing the given function $f(x)$ by a Fourier series does not imply that the Fourier series converges to the function $f(x)$. As a matter of fact, there exist Fourier series which diverge. A convergent trigonometric series need not be a Fourier series. For instance, the trigonometric series

$$\sum_{n=2}^{\infty} \frac{\sin nx}{\log n}$$

which is convergent for all values of x, is not a Fourier series, for there exist no integrable functions corresponding to this series.

5.6 Convergence in the Mean

Let $f(x)$ be piecewise continuous and periodic with period 2π. It is obvious that

$$\int_{-\pi}^{\pi} [f(x) - s_n(x)]^2 dx \geqslant 0 \tag{5.6.1}$$

where

$$s_n(x) = \frac{a_0}{2} + \sum_{k=1}^{n} a_k \cos kx + b_k \sin kx$$

Expanding

$$\int_{-\pi}^{\pi} [f(x) - s_n(x)]^2 dx = \int_{-\pi}^{\pi} [f(x)]^2 dx - 2\int_{-\pi}^{\pi} f(x) s_n(x)\, dx + \int_{-\pi}^{\pi} [s_n(x)]^2 dx$$

But, by the definitions of the Fourier coefficients (5.5.11), (5.5.12), and (5.5.13) and by the orthogonal relations for the trigonometric series (5.4.4), we have

$$\begin{aligned}\int_{-\pi}^{\pi} f(x) s_n(x)\, dx &= \int_{-\pi}^{\pi} f(x) \left[\frac{a_0}{2} + \sum_{k=1}^{n} (a_k \cos kx + b_k \sin kx) \right] dx \\ &= \frac{\pi a_0^2}{2} + \pi \sum_{k=1}^{n} (a_k^2 + b_k^2) \end{aligned} \tag{5.6.2}$$

and

$$\begin{aligned}\int_{-\pi}^{\pi} s_n^2(x)\, dx &= \int_{-\pi}^{\pi} \left[\frac{a_0}{2} + \sum_{k=1}^{n} (a_k \cos kx + b_k \sin kx) \right]^2 dx \\ &= \frac{\pi a_0^2}{2} + \pi \sum_{k=1}^{n} (a_k^2 + b_k^2) \end{aligned} \tag{5.6.3}$$

Consequently

$$\int_{-\pi}^{\pi}[f(x)-s_n(x)]^2\,dx=\int_{-\pi}^{\pi}f^2(x)\,dx-\left[\frac{\pi a_0^2}{2}+\pi\sum_{k=1}^{n}(a_k^2+b_k^2)\right]\geqslant 0 \tag{5.6.4}$$

It follows that

$$\frac{a_0^2}{2}+\sum_{k=1}^{n}(a_k^2+b_k^2)\leqslant\frac{1}{\pi}\int_{-\pi}^{\pi}f^2(x)\,dx \tag{5.6.5}$$

for all values of n. Since the right hand side of Eq. (5.6.5) is independent of n, we obtain

$$\frac{a_0^2}{2}+\sum_{k=1}^{\infty}(a_k^2+b_k^2)\leqslant\frac{1}{\pi}\int_{-\pi}^{\pi}f^2(x)\,dx \tag{5.6.6}$$

which is known as *Bessel's inequality*.

We see that the left hand side is nondecreasing and is bounded above, and therefore the series

$$\frac{a_0^2}{2}+\sum_{k=1}^{\infty}(a_k^2+b_k^2) \tag{5.6.7}$$

converges. Thus, the necessary condition for the convergence of Eq (5.6.7) is that

$$\lim_{k\to\infty}a_k=0 \qquad \lim_{k\to\infty}b_k=0 \tag{5.6.8}$$

The Fourier series is said to *converge in the mean* to $f(x)$ when

$$\lim_{n\to\infty}\int_{-\pi}^{\pi}\left[f(x)-\left(\frac{a_0}{2}+\sum_{k=1}^{n}a_k\cos kx+b_k\sin kx\right)\right]^2dx=0 \tag{5.6.9}$$

If the Fourier series converges in the mean to $f(x)$, then

$$\frac{a_0^2}{2}+\sum_{k=1}^{\infty}(a_k^2+b_k^2)=\frac{1}{\pi}\int_{-\pi}^{\pi}f^2(x)\,dx \tag{5.6.10}$$

which is called *Parseval's relation*. Furthermore, if the relation (5.6.9) holds true, the set of trigonometric functions 1, $\cos x$, $\sin x$, $\cos 2x$, $\sin 2x,\ldots$ is said to be *complete*.

5.7 Examples of Fourier Series

The Fourier coefficients (5.5.11), (5.5.12), and (5.5.13) of Sec. 5.5 may be obtained in a different way. Suppose the function $f(x)$ of period 2π has the expansion

$$f(x)=\frac{a_0}{2}+\sum_{k=1}^{\infty}(a_k\cos kx+b_k\sin kx) \tag{5.7.1}$$

If we assume that the infinite series is term-by-term integrable (we will see later that uniform convergence of the series is a sufficient condition for this), then

$$\int_{-\pi}^{\pi} f(x)\,dx = \int_{-\pi}^{\pi}\left[\frac{a_0}{2} + \sum_{k=1}^{\infty}(a_k\cos kx + b_k\sin kx)\right]dx$$
$$= \pi a_0$$

Hence

$$a_0 = \frac{1}{\pi}\int_{-\pi}^{\pi} f(x)\,dx \tag{5.7.2}$$

Again, we multiply both sides of Eq. (5.7.1) by $\cos nx$ and integrate from $-\pi$ to π. We obtain

$$\int_{-\pi}^{\pi} f(x)\cos nx\,dx = \int_{-\pi}^{\pi}\left[\frac{a_0}{2} + \sum_{k=1}^{\infty}(a_k\cos kx + b_k\sin kx)\right]\cos nx\,dx$$
$$= \pi a_k$$

Thus,

$$a_k = \frac{1}{\pi}\int_{-\pi}^{\pi} f(x)\cos kx\,dx \tag{5.7.3}$$

In a similar manner, we find that

$$b_k = \frac{1}{\pi}\int_{-\pi}^{\pi} f(x)\sin kx\,dx \tag{5.7.4}$$

The coefficients a_0, a_k, b_k just found are exactly the same as those obtained in Sec. 5.5.

EXAMPLE 5.7.1. Find the Fourier series expansion for the function, shown in Fig. 5.7.1.

$$f(x) = x + x^2 \qquad -\pi < x < \pi$$

Figure 5.7.1

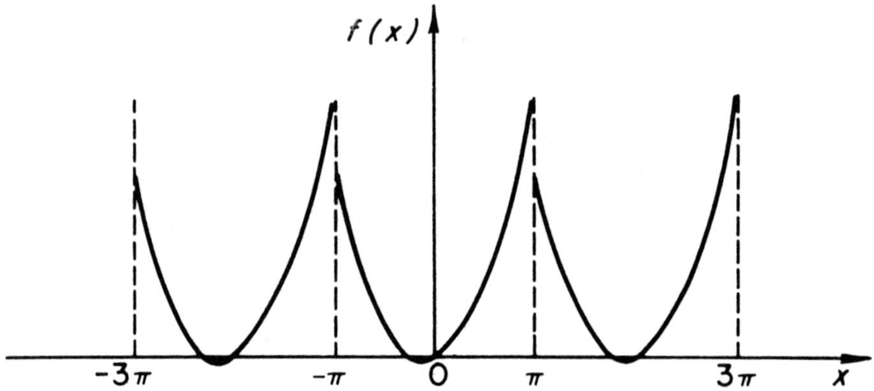

Here

$$\begin{aligned} a_0 &= \frac{1}{\pi}\int_{-\pi}^{\pi} f(x)\,dx \\ &= \frac{1}{\pi}\int_{-\pi}^{\pi} (x+x^2)\,dx \\ &= 2\frac{\pi^2}{3} \end{aligned}$$

and

$$\begin{aligned} a_k &= \frac{1}{\pi}\int_{-\pi}^{\pi} f(x)\cos kx\,dx \\ &= \frac{1}{\pi}\int_{-\pi}^{\pi} (x+x^2)\cos kx\,dx \\ &= \frac{1}{\pi}\left[\frac{x\sin kx}{k}\bigg|_{-\pi}^{\pi} - \int_{-\pi}^{\pi}\frac{\sin kx}{k}\,dx\right] \\ &\quad + \frac{1}{\pi}\left[\frac{x^2\sin kx}{k}\bigg|_{-\pi}^{\pi} - \int_{-\pi}^{\pi}\frac{2x\sin kx}{k}\,dx\right] \\ &= -\frac{2}{k\pi}\left[-\frac{x\cos kx}{k}\bigg|_{-\pi}^{\pi} + \int_{-\pi}^{\pi}\frac{\cos kx}{k}\,dx\right] \\ &= \frac{4}{k^2}\cos k\pi \\ &= \frac{4}{k^2}(-1)^5 \qquad \text{for } k=1,2,3,\ldots \end{aligned}$$

Also

$$\begin{aligned} b_k &= \frac{1}{\pi}\int_{-\pi}^{\pi} f(x)\sin kx\,dx \\ &= \frac{1}{\pi}\int_{-\pi}^{\pi} (x+x^2)\sin kx\,dx \\ &= \frac{1}{\pi}\left[-\frac{x\cos kx}{k}\bigg|_{-\pi}^{\pi} + \int_{-\pi}^{\pi}\frac{\cos kx}{k}\,dx\right] \\ &\quad + \frac{1}{\pi}\left[-x^2\frac{\cos kx}{k}\bigg|_{-\pi}^{\pi} + \int_{-\pi}^{\pi}\frac{2x\cos kx}{k}\,dx\right] \\ &= -\frac{2}{k}\cos k\pi + \frac{2}{k\pi}\left[\frac{x\sin kx}{k}\bigg|_{-\pi}^{\pi} - \int_{-\pi}^{\pi}\frac{\sin kx}{k}\,dx\right] \\ &= -\frac{2}{k}\cos k\pi \\ &= -\frac{2}{k}(-1)^k \qquad \text{for } k=1,2,3,\ldots \end{aligned}$$

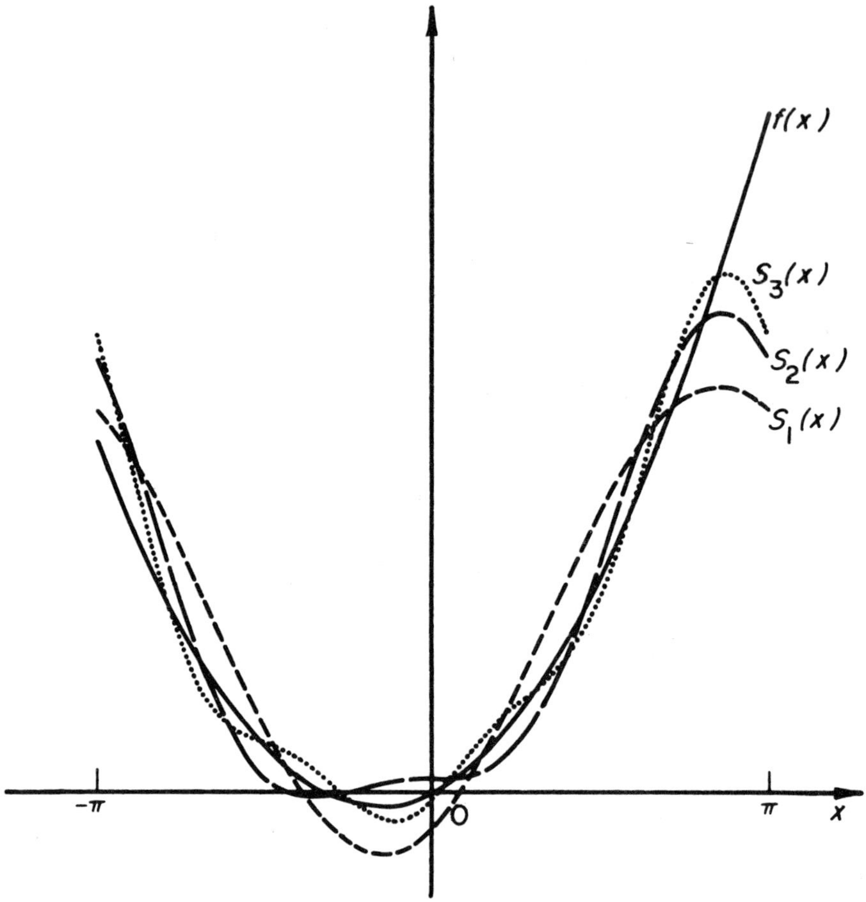

Figure 5.7.2

Therefore the Fourier series expansion for f is

$$f(x)=\frac{\pi^2}{3}+\sum_{k=1}^{\infty}\frac{4}{k^2}(-1)^k\cos kx-\frac{2}{k}(-1)^k\sin kx$$

$$=\frac{\pi^2}{3}-4\cos x+2\sin x+\cos 2x-\sin 2x-\ldots$$

The first two partial sums are

$$s_1=\frac{\pi^2}{3}-4\cos x+2\sin x$$

$$s_2=\frac{\pi^2}{3}-4\cos x+2\sin x+\cos 2x-\sin 2x$$

These are plotted as shown in Fig. 5.7.2. It can be seen that the first few terms already furnish a fairly good approximation to f in the interval $-\pi<x<\pi$. The approximation improves with the number of terms taken for each fixed x on $-\pi<x<\pi$ but not for $x=\pm\pi$. The behavior of the approximations at the discontinuous points will be discussed later.

EXAMPLE 5.7.2. Consider the periodic function, shown in Fig. 5.7.3,

$$f(x)=\begin{cases} -\pi & -\pi<x<0 \\ x & 0<x<\pi \end{cases}$$

In this case,

$$\begin{aligned} a_0 &= \frac{1}{\pi}\int_{-\pi}^{\pi} f(x)\,dx \\ &= \frac{1}{\pi}\left[\int_{-\pi}^{0} -\pi\,dx + \int_0^{\pi} x\,dx\right] \\ &= -\frac{\pi}{2} \end{aligned}$$

and

$$\begin{aligned} a_k &= \frac{1}{\pi}\int_{-\pi}^{\pi} f(x)\cos kx\,dx \\ &= \frac{1}{\pi}\left[\int_{-\pi}^{0} -\pi\cos kx\,dx + \int_0^{\pi} x\cos kx\,dx\right] \\ &= \frac{1}{k^2\pi}(\cos k\pi - 1) \\ &= \frac{1}{k^2\pi}\left[(-1)^k - 1\right] \end{aligned}$$

Figure 5.7.3

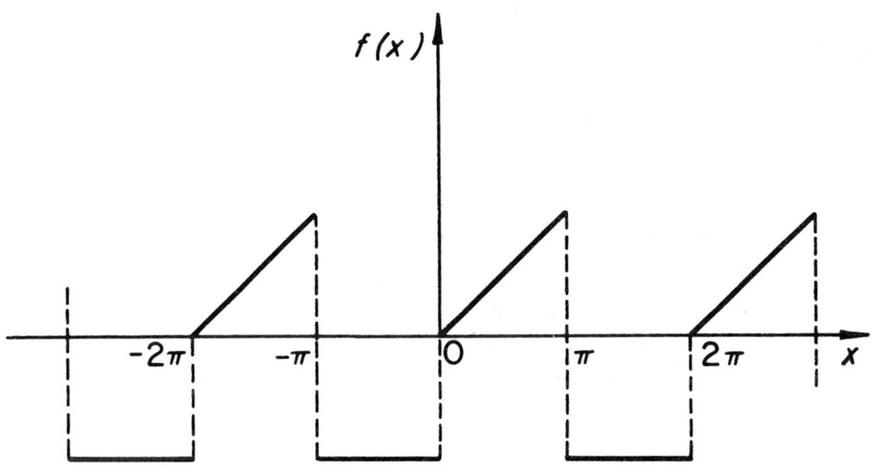

Also

$$b_k = \frac{1}{\pi}\int_{-\pi}^{\pi} f(x)\sin kx\,dx$$

$$= \frac{1}{\pi}\left[\int_{-i\pi}^{0} -\pi\sin kx\,dx + \int_{0}^{\pi} x\sin kx\,dx\right]$$

$$= \frac{1}{k}(1-2\cos k\pi)$$

$$= \frac{1}{k}\left[1-2(-1)^k\right]$$

Hence the Fourier series is

$$f(x) = -\frac{\pi}{4} + \sum_{k=1}^{\infty} \frac{1}{k^2\pi}\left[(-1)^k - 1\right]\cos kx + \frac{1}{k}\left[1-2(-1)^k\right]\sin kx$$

EXAMPLE 5.7.3. Consider the function $f(x)=x$ in the interval $-\pi<x<\pi$. We can readily find that

$$a_k = 0, \quad k=0,1,2,\ldots$$

The coefficients b_k are given by

$$b_k = \frac{1}{\pi}\int_{-\pi}^{\pi} f(x)\sin kx dx$$

$$= \frac{1}{\pi}\int_{-\pi}^{\pi} x\sin kx dx$$

$$= \frac{2}{k}(-1)^{k+1}$$

Hence

$$f(x) = 2\sum_{k=1}^{\infty}(-1)^{k+1}\frac{\sin kx}{k}$$

5.8 Cosine and Sine Series

Let $f(x)$ be an even function defined on the interval $[-\pi,\pi]$. Since cos kx is an even function, and $\sin kx$ an odd function, the function $f(x)\cos kx$ is an even function and the function $f(x)\sin kx$ an odd function. Thus, employing Eqs. (5.2.4) and (5.2.5), we find that the Fourier coefficients of

$f(x)$ are

$$a_k = \frac{1}{\pi}\int_{-\pi}^{\pi} f(x)\cos kx dx = \frac{2}{\pi}\int_0^{\pi} f(x)\cos kx dx, \quad k=0,1,2,\ldots \tag{5.8.1}$$

$$b_k = \frac{1}{\pi}\int_{-\pi}^{\pi} f(x)\sin kx dx = 0, \quad k=1,2,3,\ldots$$

Hence the Fourier series of an even function can be written as

$$f(x) \sim \frac{a_0}{2} + \sum_{k=1}^{\infty} a_k \cos kx \tag{5.8.2}$$

where the coefficients a_k are given by formula (5.8.1).

In a similar manner, if $f(x)$ is an odd function, the function $f(x)\cos kx$ is an odd function and the function $f(x)\sin kx$ is an even function. As a consequence, the Fourier coefficients of $f(x)$, in this case, are

$$a_k = \frac{1}{\pi}\int_{-\pi}^{\pi} f(x)\cos kx dx = 0, \quad k=0,1,2,\ldots$$

$$b_k = \frac{1}{\pi}\int_{-\pi}^{\pi} f(x)\sin kx dx = \frac{2}{\pi}\int_0^{\pi} f(x)\sin kx dx, \quad k=1,2,\ldots \tag{5.8.3}$$

Therefore, the Fourier series of an odd function can be written as

$$f(x) = \sum_{k=1}^{\infty} b_k \sin kx \tag{5.8.4}$$

where the coefficients b_k are given by formula (5.8.3).

EXAMPLE 5.8.1. Obtain the Fourier series of the function

$$f(x) = \begin{cases} -1, & -\pi < x < 0 \\ +1, & 0 < x < \pi \end{cases}$$

In this case, f is an odd function so that $a_k = 0$ for $k=0,1,2,3,\ldots$ and

$$\begin{aligned} b_k &= \frac{2}{\pi}\int_0^{\pi} f(x)\sin kx dx \\ &= \frac{2}{\pi}\int_0^{\pi} \sin kx dx \\ &= \frac{2}{k\pi}\left[1-(-1)^k\right] \end{aligned}$$

Thus $b_{2k}=0$ and $b_{2k-1}=[4/\pi(2k-1)]$. Therefore, the Fourier series of the function $f(x)$ is

$$f(x) = \frac{4}{\pi}\sum_{k=1}^{\infty} \frac{\sin(2k-1)x}{(2k-1)}$$

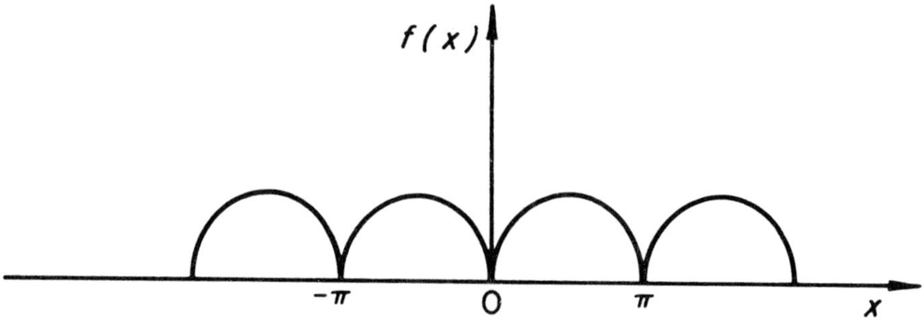

Figure 5.8.1

EXAMPLE 5.8.2. Expand $|\sin x|$ in Fourier series, Since $|\sin x|$ is an even function as shown in Fig. 5.8.1, $b_k = 0$ for $k = 1, 2, \ldots$ and

$$\begin{aligned} a_k &= \frac{2}{\pi}\int_0^\pi f(x)\cos kx\,dx \\ &= \frac{2}{\pi}\int_0^\pi \sin x\cos kx\,dx \\ &= \frac{1}{\pi}\int_0^\pi [\sin(1+k)x + \sin(1-k)x]\,dx \\ &= \frac{2[1+(-1)^k]}{\pi(1-k^2)} \quad \text{for } k = 0, 2, 3, \ldots \end{aligned}$$

For $k = 1, a_1 = \frac{2}{\pi}\int_0^\pi \sin x\cos x\,dx = 0.$

Hence the Fourier series of $f(x)$ is

$$f(x) = \frac{2}{\pi} + \frac{4}{\pi}\sum_{k=1}^{\infty} \frac{\cos 2kx}{(1-4k^2)}$$

In the preceding sections, we have prescribed the function $f(x)$ in the interval $(-\pi, \pi)$ and assumed $f(x)$ to be periodic with period 2π in the entire interval $(-\infty, \infty)$. In practice, we frequently encounter problems in which a function is defined only in the interval $(-\pi, \pi)$. In such a case, we simply extend the function periodically with period 2π, as in Fig. 5.8.2. In this way, we are able to represent the function $f(x)$ by the Fourier series expansion, although we are interested only in the expansion on $(-\pi, \pi)$.

If a function f is defined only in the interval $(0, \pi)$, we may extend f in two ways. The first is the *even extension* of f, denoted and defined by (see Fig. 5.8.3)

$$F_e(x) = \begin{cases} f(x) & 0 < x < \pi \\ f(-x) & -\pi < x < 0 \end{cases}$$

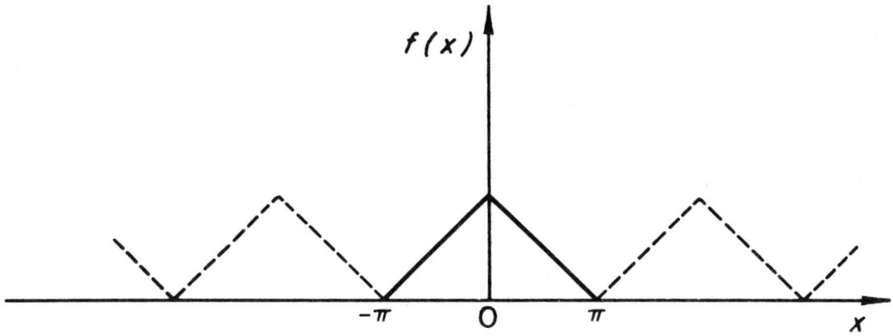

Figure 5.8.2

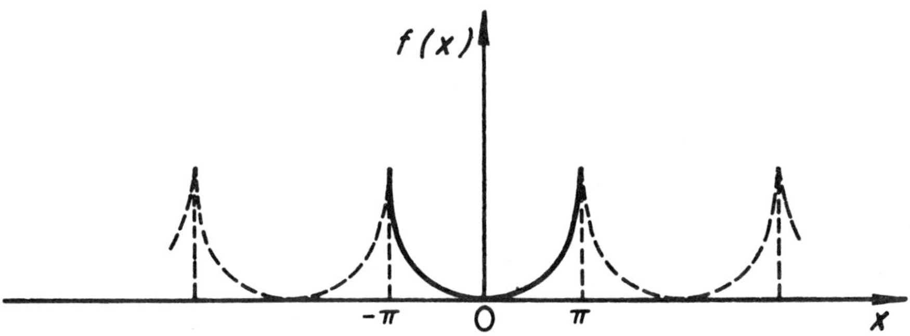

Figure 5.8.3

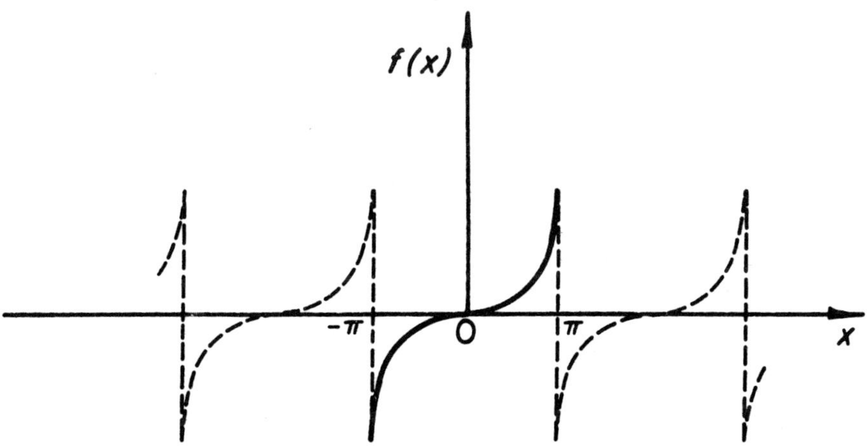

Figure 5.8.4

while the second is the *odd extension* of f, denoted and defined by (see Fig. 5.8.4)

$$F_o(x)=\begin{cases} f(x) & 0<x<\pi \\ -f(-x) & -\pi<x<0 \end{cases}$$

Since $F_e(x)$ and $F_o(x)$ are the even and odd functions with period 2π respectively, the Fourier series expansions of $F_e(x)$ and $F_o(x)$ are

$$F_e(x)=\frac{a_0}{2}+\sum_{k=1}^{\infty} a_k \cos kx$$

where

$$a_k=\frac{2}{\pi}\int_0^{\pi} f(x)\cos kx dx$$

and

$$F_o(x)=\sum_{k=1}^{\infty} b_k \sin kx$$

where

$$b_k=\frac{2}{\pi}\int_0^{\pi} f(x)\sin kx dx$$

5.9 Complex Fourier Series

It is sometimes convenient to represent a function by an expansion in complex form. This expansion can easily be derived from the Fourier series

$$f(x)=\frac{a_0}{2}+\sum_{k=1}^{\infty}(a_k\cos kx+b_k\sin kx)$$

Noting that

$$\sin x=\frac{e^{ix}-e^{-ix}}{2i} \qquad \cos x=\frac{e^{ix}+e^{-ix}}{2}$$

we write

$$\begin{aligned} f(x)&=\frac{a_0}{2}+\sum_{k=1}^{\infty}\left[a_k\left(\frac{e^{ikx}+e^{-ikx}}{2}\right)+b_k\left(\frac{e^{ikx}-e^{-ikx}}{2i}\right)\right] \\ &=\frac{a_0}{2}+\sum_{k=1}^{\infty}\left[\left(\frac{a_k-ib_k}{2}\right)e^{ikx}+\left(\frac{a_k+ib_k}{2}\right)e^{-ikx}\right] \\ &=c_0+\sum_{k=1}^{\infty}(c_k e^{ikx}+c_{-k}e^{-ikx}) \end{aligned}$$

where

$$c_0 = \frac{a_0}{2} = \frac{1}{2\pi}\int_{-\pi}^{\pi} f(x)\,dx$$

$$c_k = \frac{a_k - ib_k}{2} = \frac{1}{2\pi}\int_{-\pi}^{\pi} f(x)(\cos kx - i\sin kx)\,dx$$

$$= \frac{1}{2\pi}\int_{-\pi}^{\pi} f(x)e^{-ikx}\,dx$$

$$c_{-k} = \frac{a_k + ib_k}{2} = \frac{1}{2\pi}\int_{-\pi}^{\pi} f(x)(\cos kx + i\sin kx)\,dx$$

$$= \frac{1}{2\pi}\int_{-\pi}^{\pi} f(x)e^{ikx}\,dx$$

Thus, we obtain the Fourier series expansion for $f(x)$ in complex form, namely,

$$f(x) = \sum_{k=-\infty}^{\infty} c_k e^{ikx} \qquad -\pi < x < \pi \tag{5.9.1}$$

where

$$c_k = \frac{1}{2\pi}\int_{-\pi}^{\pi} f(x)e^{-ikx}\,dx \tag{5.9.2}$$

EXAMPLE 5.9.1. Obtain the complex Fourier series expansion for the function

$$f(x) = e^x \qquad -\pi < x < \pi$$

We find

$$c_k = \frac{1}{2\pi}\int_{-\pi}^{\pi} f(x)e^{-ikx}\,dx$$

$$= \frac{1}{2\pi}\int_{-\pi}^{\pi} e^x e^{-ikx}\,dx$$

$$= \frac{(1+ik)(-1)^k}{\pi(1+k^2)}\sinh\pi$$

and hence the Fourier series is

$$f(x) = \sum_{k=-\infty}^{\infty} \frac{(1+ik)(-1)^k}{\pi(1+k^2)}\sinh\pi e^{ikx}$$

5.10 Change of Interval

So far we have been concerned with functions defined on the interval $[-\pi,\pi]$. In many applications, however, this interval is restrictive, and the interval of interest may be arbitrary, say $[a,b]$.

If we introduce the new variable t by the transformation

$$x=\frac{1}{2}(b+a)+\frac{(b-a)}{2\pi}t \tag{5.10.1}$$

then the interval $a\leqslant x\leqslant b$ becomes $-\pi\leqslant t\leqslant\pi$. Thus, the function $f[(b+a)/2+((b-a)/2\pi)t]=F(t)$ obviously has period 2π. Expanding this function in a Fourier series, we obtain

$$F(t)=\frac{a_0}{2}+\sum_{k=1}^{\infty}(a_k\cos kt+b_k\sin kt) \tag{5.10.2}$$

where

$$a_k=\frac{1}{\pi}\int_{-\pi}^{\pi}F(t)\cos kt\,dt \quad k=0,1,2,\ldots$$

$$b_k=\frac{1}{\pi}\int_{-\pi}^{\pi}F(t)\sin kt\,dt \quad k=1,2,3,\ldots$$

On changing t into x, we find the expansion for $f(x)$ in $[a,b]$

$$f(x)=\frac{a_0}{2}+\sum_{k=1}^{\infty}\left[a_k\cos\frac{k\pi(2x-b-a)}{(b-a)}+b_k\sin\frac{k\pi(2x-b-a)}{(b-a)}\right] \tag{5.10.3}$$

where

$$a_k=\frac{2}{b-a}\int_a^b f(x)\cos\frac{k\pi(2x-b-a)}{(b-a)}\,dx, \quad k=0,1,2,\ldots \tag{5.10.4}$$

$$b_k=\frac{2}{b-a}\int_a^b f(x)\sin\frac{k\pi(2x-b-a)}{(b-a)}\,dx, \quad k=1,2,3,\ldots \tag{5.10.5}$$

It is sometimes convenient to take the interval in which the function f is defined as $[-l,l]$. It follows at once from the result just obtained that by letting $a=-l$ and $b=l$, the expansion for f in $[-l,l]$ takes the form

$$f(x)=\frac{a_0}{2}+\sum_{k=1}^{\infty}\left(a_k\cos\frac{k\pi x}{l}+b_k\sin\frac{k\pi x}{l}\right) \tag{5.10.6}$$

where

$$a_k=\frac{1}{l}\int_{-l}^{l}f(x)\cos\frac{k\pi x}{l}\,dx, \quad k=0,1,2,\ldots \tag{5.10.7}$$

$$b_k=\frac{1}{l}\int_{-l}^{l}f(x)\sin\frac{k\pi x}{l}\,dx, \quad k=1,2,3,\ldots \tag{5.10.8}$$

If f is an even function of period $2l$, then by Eq. (5.10.6), we can readily determine that

$$f(x)=\frac{a_0}{2}+\sum_{k=1}^{\infty} a_k\cos\frac{k\pi x}{l} \tag{5.10.9}$$

where

$$a_k=\frac{2}{l}\int_0^l f(x)\cos\frac{k\pi x}{l}\,dx, \quad k=0,1,2,\ldots \tag{5.10.10}$$

If f is an odd function of period $2l$, then by Eq. (5.10.6), the expansion for f is

$$f(x)=\sum_{k=1}^{\infty} b_k\sin\frac{k\pi x}{l} \tag{5.10.11}$$

where

$$b_k=\frac{2}{l}\int_0^l f(x)\sin\frac{k\pi x}{l}\,dx \tag{5.10.12}$$

EXAMPLE 5.10.1. Consider the odd periodic function f

$$f(x)=x \quad -2<x<2$$

as shown in Fig. 5.10.1. Here $l=2$. Since f is odd, $a_k=0$, and

$$\begin{aligned} b_k&=\frac{2}{l}\int_0^l f(x)\sin\frac{k\pi x}{l}\,dx\\ &=\frac{2}{2}\int_0^2 x\sin\frac{k\pi x}{2}\,dx\\ &=-\frac{4}{k\pi}(-1)^k \quad \text{for } k=1,2,3,\ldots \end{aligned}$$

Figure 5.10.1

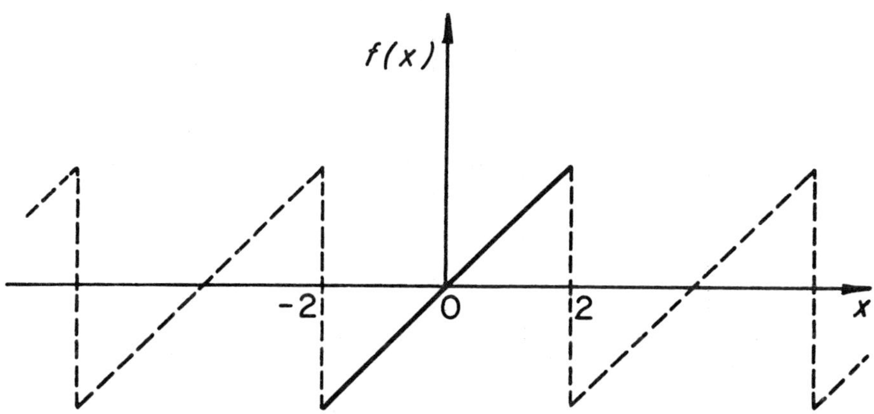

Therefore, the Fourier series of f is

$$f(x)=\sum_{k=1}^{\infty}\frac{4}{k\pi}(-1)^{k+1}\sin\frac{k\pi x}{2}$$

EXAMPLE 5.10.2. Given the function

$$f(x)=\begin{cases}1 & 0<x<\frac{1}{2}\\ 0 & \frac{1}{2}<x<1\end{cases}$$

In this case the period is $2l=2$ or $l=1$. Extend f as shown in Fig. 5.10.2. Since the extension is even, we have $b_k=0$ and

$$\begin{aligned}a_0&=\frac{2}{l}\int_0^l f(x)dx\\&=\frac{2}{1}\int_0^1 dx\\&=1\\a_k&=\frac{2}{l}\int_0^l f(x)\cos\frac{k\pi x}{l}dx\\&=\frac{2}{1}\int_0^1\cos k\pi x dx\\&=\frac{2}{k\pi}\sin\frac{k\pi}{2}\end{aligned}$$

Hence

$$f(x)=\frac{1}{2}+\sum_{k=1}^{\infty}\frac{2}{(2k-1)\pi}(-1)^{k-1}\cos(2k-1)\pi x$$

Figure 5.10.2

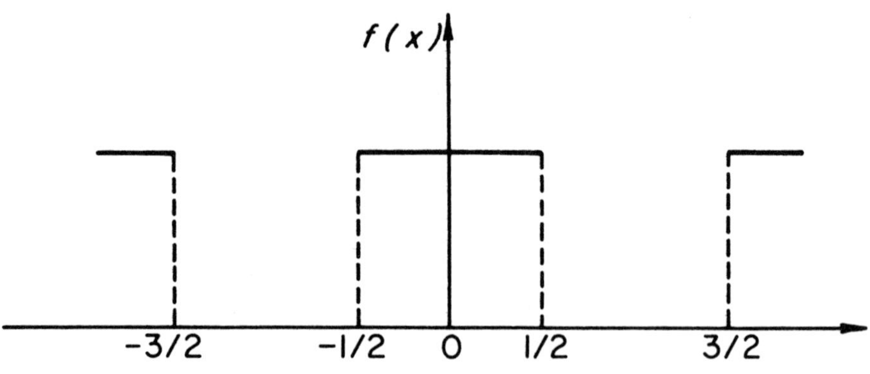

5.11 Pointwise Convergence

We have stated earlier that if $f(x)$ is piecewise continuous on the interval $[-\pi,\pi]$, then there exists a Fourier series expansion which converges in the mean to $f(x)$.

In this section, we shall investigate the pointwise convergence of the Fourier series of a given function. To begin with, we consider

$$f(x)\sim\frac{a_0}{2}+\sum_{k=1}^{\infty}(a_k\cos kx+b_k\sin kx)$$

where

$$a_k=\frac{1}{\pi}\int_{-\pi}^{\pi}f(x)\cos kx\,dx,\quad k=0,1,2,\ldots$$

$$b_k=\frac{1}{\pi}\int_{-\pi}^{\pi}f(x)\sin kx\,dx,\quad k=1,2,3,\ldots$$

Let $s_n(x)$ be the nth partial sum of the Fourier series of $f(x)$, that is,

$$s_n(x)=\frac{a_0}{2}+\sum_{k=1}^{n}(a_k\cos kx+b_k\sin kx)$$

Substituting for a_k and b_k in $s_n(x)$, we obtain

$$\begin{aligned}s_n(x)&=\frac{1}{2\pi}\int_{-\pi}^{\pi}f(t)\,dt+\frac{1}{\pi}\sum_{k=1}^{n}\left\{\left[\int_{-\pi}^{\pi}f(t)\cos kt\,dt\right]\cos kx\right.\\&\qquad\left.+\left[\int_{-\pi}^{\pi}f(t)\sin kt\,dt\right]\sin kx\right\}\\&=\frac{1}{\pi}\int_{-\pi}^{\pi}f(t)\left[\frac{1}{2}+\sum_{k=1}^{n}(\cos kt\cos kx+\sin kt\sin kx)\right]dt\\&=\frac{1}{\pi}\int_{-\pi}^{\pi}f(t)\left[\frac{1}{2}+\sum_{k=1}^{n}\cos k(t-x)dt\right]\end{aligned}\tag{5.11.1}$$

Summing the trigonometric identity

$$2\sin\alpha/2\cos k\alpha=\sin\left(k+\tfrac{1}{2}\right)\alpha-\sin\left(k-\tfrac{1}{2}\right)\alpha$$

from $k=1$ to $k=n$, we obtain

$$\begin{aligned}2\sin\frac{\alpha}{2}\left[\frac{1}{2}+\sum_{k=1}^{n}\cos k\alpha\right]&=\sin\frac{\alpha}{2}+\left[\sin\frac{3\alpha}{2}-\sin\frac{\alpha}{2}\right]\\&\qquad+\ldots+\left[\sin\left(n+\frac{1}{2}\right)\alpha-\sin\left(n-\frac{1}{2}\right)\alpha\right]\\&=\sin\left(n+\frac{1}{2}\right)\alpha\end{aligned}\tag{5.11.2}$$

Then using (5.11.2) in (5.11.1) we obtain

$$s_n(x)=\frac{1}{\pi}\int_{-\pi}^{\pi} f(t)\frac{\sin(n+1/2)(t-x)}{2\sin\left(\frac{t-x}{2}\right)}dt \tag{5.11.3}$$

Introducing the new variable $s=t-x$, we have

$$s_n(x)=\frac{1}{\pi}\int_{-\pi-x}^{\pi-x} f(s+x)\frac{\sin(n+1/2)s}{2\sin(s/2)}ds \tag{5.11.4}$$

Now, if $f(x)$ is piecewise continuous and periodic with period 2π, then $s_n(x)$ is also periodic with period 2π (Exercise 28). Thus,

$$s_n(x)=\frac{1}{\pi}\int_{-\pi}^{\pi} f(s+x)\frac{\sin(n+1/2)s}{2\sin(s/2)}ds \tag{5.11.5}$$

which is known as the *Dirichlet formula* for s_n. The kernel

$$\frac{\sin(n+1/2)s}{2\sin(s/2)} \tag{5.11.6}$$

called the *Dirichlet kernel*, is periodic with period 2π, and

$$\frac{1}{\pi}\int_{-\pi}^{\pi}\frac{\sin(n+1/2)s}{2\sin(s/2)}ds=\frac{1}{\pi}\int_{-\pi}^{\pi}\left[\frac{1}{2}+\sum_{k=1}^{n}\cos ks\right]ds=1 \tag{5.11.7}$$

Lemma 5.11.1 (Riemann-Lebesque Lemma) *If $g(x)$ is piecewise continuous on the interval $[a,b]$, then*

$$\lim_{\lambda\to\infty}\int_a^b g(x)\sin\lambda x\,dx=0 \tag{5.11.8}$$

PROOF. Consider the integral

$$I(\lambda)=\int_a^b g(x)\sin\lambda x\,dx \tag{5.11.9}$$

With the change of variable

$$x=t+\pi/\lambda,$$

we have

$$\sin\lambda x=\sin\lambda(t+\pi/\lambda)=-\sin\lambda t$$

and

$$I(\lambda)=-\int_{a-\pi/\lambda}^{b-\pi/\lambda} g(t+\pi/\lambda)\sin\lambda t\,dt \tag{5.11.10}$$

Since t is a dummy variable, we write the above integral as

$$I(\lambda)=-\int_{a-\pi/\lambda}^{b-\pi/\lambda} g(x+\pi/\lambda)\sin\lambda x\,dx \tag{5.11.11}$$

Addition of Eqs. (5.11.9) and (5.11.11) yields

$$\begin{aligned}2I(\lambda) &= \int_a^b g(x)\sin\lambda x dx - \int_{a-\pi/\lambda}^{b-\pi/\lambda} g(x+\pi/\lambda)\sin\lambda x dx \\ &= -\int_{a-\pi/\lambda}^{a} g(x+\pi/\lambda)\sin\lambda x dx + \int_{b-\pi/\lambda}^{b} g(x)\sin\lambda x dx \\ &\quad + \int_a^{b-\pi/\lambda} [g(x) - g(x+\pi/\lambda)]\sin\lambda x dx \qquad (5.11.12)\end{aligned}$$

First, let $g(x)$ be a continuous function in $[a,b]$. Then $g(x)$ is necessarily bounded, that is, there exists an M such that $|g(x)| \leqslant M$. Hence

$$\left|\int_{a-\pi/\lambda}^{a} g(x+\pi/\lambda)\sin\lambda x dx\right| = \left|\int_a^{a+\pi/\lambda} g(x)\sin\lambda x dx\right| \leqslant \frac{\pi M}{\lambda}$$

and

$$\left|\int_{b-\pi/\lambda}^{b} g(x)\sin\lambda x dx\right| \leqslant \frac{\pi M}{\lambda}$$

Consequently,

$$|I(\lambda)| \leqslant \frac{\pi M}{\lambda} + \int_a^{b-\pi/\lambda} |g(x) - g(x+\pi/\lambda)| dx \qquad (5.11.13)$$

Since $g(x)$ is a continuous function on a closed interval $[a,b]$, it is uniformly continuous on $[a,b]$ so that

$$|g(x) - g(x+\pi/\lambda)| < \varepsilon/(b-a) \qquad (5.11.14)$$

for all $\lambda > \Lambda$ and all x in $[a,b]$. We now choose Λ such that $\pi M/\lambda < \varepsilon/2$, whenever $\lambda > \Lambda$. Then

$$|I(\lambda)| < \frac{\varepsilon}{2} + \frac{\varepsilon}{2} = \varepsilon$$

If $g(x)$ is piecewise continuous in $[a,b]$, then the proof consists of a repeated application of the preceding argument to every subinterval of $[a,b]$ in which $g(x)$ is continuous. □

Theorem 5.11.1 (Pointwise Convergence Theorem) *If $f(x)$ is piecewise smooth and periodic with period 2π in $[-\pi,\pi]$, then for any x*

$$\frac{a_0}{2} + \sum_{k=1}^{\infty} (a_k \cos kx + b_k \sin kx) = \frac{1}{2}[f(x+) + f(x-)] \qquad (5.11.15)$$

where

$$a_k = \frac{1}{\pi}\int_{-\pi}^{\pi} f(x)\cos kx dx, \quad k = 0, 1, 2, \ldots$$

$$b_k = \frac{1}{\pi}\int_{-\pi}^{\pi} f(x)\sin kx dx, \quad k = 1, 2, 3, \ldots$$

PROOF. We have

$$s_n(x)=\frac{1}{\pi}\int_{-\pi}^{\pi} f(x+s)\frac{\sin(n+1/2)s}{2\sin(s/2)}\,ds$$
$$=\frac{1}{\pi}\int_{-\pi}^{0} f(x+s)\frac{\sin(n+1/2)s}{2\sin(s/2)}\,ds$$
$$+\frac{1}{\pi}\int_{0}^{\pi} f(x+s)\frac{\sin(n+1/2)s}{2\sin(s/2)}\,ds$$

Denoting the first integral by I_1, we write

$$I_1=\frac{1}{\pi}\int_{-\pi}^{0}[f(x+s)-f(x-)+f(x-)]\frac{\sin(n+1/2)s}{2\sin(s/2)}\,ds \tag{5.11.16}$$

By virtue of Eq. (5.11.7), we have

$$\frac{1}{\pi}\int_{-\pi}^{0} f(x-)\frac{\sin(n+1/2)s}{2\sin(s/2)}\,ds=\frac{f(x-)}{\pi}\int_{0}^{\pi}\frac{\sin(n+1/2)s}{2\sin(s/2)}\,ds=\frac{f(x-)}{2}$$

Thus,

$$I_1=\frac{f(x-)}{2}+\frac{1}{\pi}\int_{-\pi}^{0}\frac{f(x+s)-f(x-)}{2\sin(s/2)}\sin(n+1/2)s\,ds$$

But

$$\lim_{s\to 0}\frac{f(x+s)-f(x-)}{2\sin(s/2)}=\lim_{s\to 0}\left[\frac{f(x+s)-f(x-)}{s}\right]\frac{s}{2\sin(s/2)}$$
$$=\lim_{s\to 0}\left[\frac{f(x+s)-f(x-)}{s}\right]$$
$$=f'(x-)$$

which exists since $f(x)$ is piecewise smooth. Hence, the function

$$\frac{f(x+s)-f(x-)}{2\sin(s/2)}$$

is piecewise continuous. By the Riemann-Lebesque lemma,

$$\lim_{n\to\infty}\int_{-\pi}^{0}\frac{f(x+s)-f(x-)}{2\sin(s/2)}\sin(n+1/2)s\,ds=0$$

Therefore,

$$\lim_{n\to\infty} I_1=\frac{f(x-)}{2}$$

In a similar manner, if we denote

$$I_2=\frac{1}{\pi}\int_{0}^{\pi} f(x+s)\frac{\sin(n+1/2)s}{2\sin(s/2)}\,ds,$$

we obtain

$$\lim_{n\to\infty} I_2 = \frac{f(x+)}{2}$$

Finally, we have

$$\lim_{n\to\infty} s_n(x) = \lim_{n\to\infty} (I_1 + I_2) = \tfrac{1}{2}[f(x+) + f(x-)]$$

Hence

$$\frac{a_0}{2} + \sum_{k=1}^{\infty} (a_k \cos kx + b_k \sin kx) = \frac{1}{2}[f(x+) + f(x-)]$$

At the point where $f(x)$ is continuous, $f(x+) = f(x-) = f(x)$. In that case,

$$\frac{a_0}{2} + \sum_{k=1}^{\infty} (a_k \cos kx + b_k \sin kx) = f(x)$$

□

EXAMPLE 5.11.1. In Example 5.7.1, we find that the Fourier series expansion for $x + x^2$ in $[-\pi, \pi]$, as shown in Fig. 5.11.1 is

$$f(x) \sim \frac{\pi^2}{3} + \sum_{k=1}^{\infty} \frac{4}{k^2}(-1)^k \cos kx - \frac{2}{k}(-1)^k \sin kx$$

Since $f(x) = x + x^2$ is piecewise smooth, the series converges, and hence we write

$$x + x^2 = \frac{\pi^2}{3} + \sum_{k=1}^{\infty} \frac{4}{k^2}(-1)^k \cos kx - \frac{2}{k}(-1)^k \sin kx$$

Figure 5.11.1

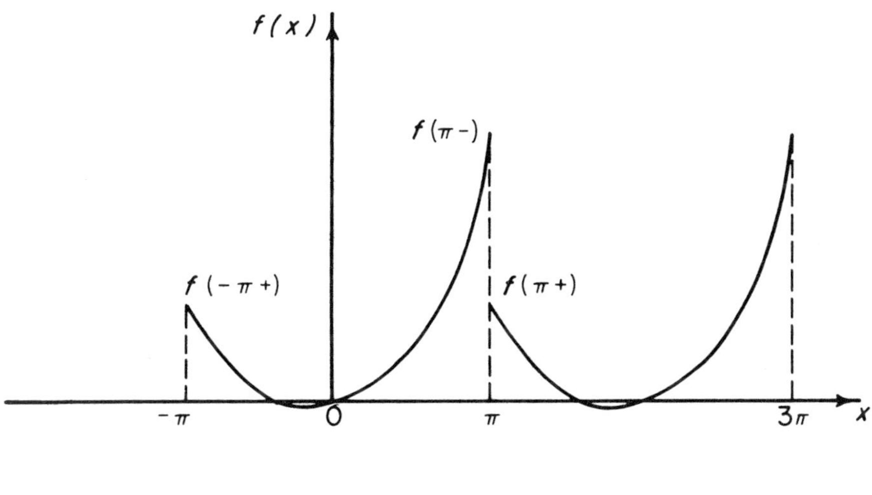

at points of continuity. At points of discontinuity, such as $x=\pi$, by virtue of the Pointwise Convergence Theorem,

$$\frac{1}{2}\left[(\pi+\pi^2)+(-\pi+\pi^2)\right]=\frac{\pi^2}{3}+\sum_{k=1}^{\infty}\frac{4}{k^2}(-1)^k\cos k\pi \quad (5.11.17)$$

since

$$f(\pi-)=\pi+\pi^2 \quad \text{and} \quad f(\pi+)=f(-\pi+)=-\pi+\pi^2.$$

Simplification of Eq. (5.11.17) gives

$$\pi^2=\frac{\pi^2}{3}+\sum_{k=1}^{\infty}\frac{4}{k^2}(-1)^{2k}$$

or

$$\frac{\pi^2}{6}=\sum_{k=1}^{\infty}1/k^2$$

5.12 Uniform Convergence

In the preceding section, we have proved the pointwise convergence of the Fourier series for a piecewise smooth function. Here, we shall consider another important theorem of uniform convergence.

Theorem 5.12.1 (Uniform and Absolute Convergence) *Let $f(x)$ be a continuous function with period 2π, and let $f'(x)$ be piecewise continuous in the interval $[-\pi,\pi]$. If, in addition, $f(-\pi)=f(\pi)$, then the Fourier series expansion for $f(x)$ is uniformly and absolutely convergent.*

PROOF. Let the Fourier series of $f(x)$ and $f'(x)$ be

$$\frac{a_0}{2}+\sum_{k=1}^{\infty}(a_k\cos kx+b_k\sin kx) \quad (5.12.1)$$

and

$$\frac{A_0}{2}+\sum_{k=1}^{\infty}(A_k\cos kx+B_k\sin kx) \quad (5.12.2)$$

respectively. Since f' is piecewise continuous and $f(-\pi)=f(\pi)$, we have

$$A_0=\frac{1}{\pi}\int_{-\pi}^{\pi}f'(x)dx=\frac{1}{\pi}\left[f(\pi)-f(-\pi)\right]=0$$

$$\begin{aligned}A_k&=\frac{1}{\pi}\int_{-\pi}^{\pi}f'(x)\cos kx\,dx\\&=\frac{1}{\pi}\left[f(x)\cos kx\right]_{-\pi}^{\pi}+\frac{k}{\pi}\int_{-\pi}^{\pi}f(x)\sin kx\,dx\\&=kb_k,\quad k=1,2,3,\ldots\end{aligned} \quad (5.12.3)$$

$$\begin{aligned}B_k &= \frac{1}{\pi}\int_{-\pi}^{\pi} f'(x)\sin kx\,dx \\ &= \frac{1}{\pi}\left[f(x)\sin kx\right]_{-\pi}^{\pi} - \frac{k}{\pi}\int_{-\pi}^{\pi} f(x)\cos kx\,dx \\ &= -ka_k, \quad k=1,2,3\ldots\end{aligned} \tag{5.12.4}$$

Moreover, A_k and B_k satisfy Bessel's inequality, that is,

$$\sum_{k=1}^{\infty}\left(A_k^2+B_k^2\right) \leqslant \frac{1}{\pi}\int_{-\pi}^{\pi}\left[f'(x)\right]^2 dx < \infty \tag{5.12.5}$$

Now, from (5.12.3) and (5.12.4), we write

$$\sum_{k=1}^{n}\left(a_k^2+b_k^2\right)^{1/2} = \sum_{k=1}^{n}\left[k^{-2}\left(A_k^2+B_k^2\right)\right]^{1/2} \tag{5.12.6}$$

which, after applying the Schwarz inequality (Exercise 27) to the right side, becomes

$$\sum_{k=1}^{n}\left(a_k^2+b_k^2\right)^{1/2} \leqslant \left[\sum_{k=1}^{n} k^{-2}\right]^{1/2}\left[\sum_{k=1}^{n}\left(A_k^2+B_k^2\right)\right]^{1/2} \tag{5.12.7}$$

From (5.12.5), it is evident that the series

$$\sum_{k=1}^{n}\left(A_k^2+B_k^2\right)$$

tends to a finite limit as $n\to\infty$, and so also does the series

$$\sum_{k=1}^{n} k^{-2}$$

which is the harmonic series of order two. Hence, there exists a constant M such that

$$\sum_{k=1}^{n}\left(a_k^2+b_k^2\right)^{1/2} \leqslant M$$

Since the left side is positive and bounded above by M, it must tend to a finite limit as $n\to\infty$. Thus, the series

$$\sum_{k=1}^{\infty}\left(a_k^2+b_k^2\right)^{1/2}$$

converges. Next, consider

$$a_k\cos kx + b_k\sin kx = \left(a_k^2+b_k^2\right)^{1/2}\cos(kx-\theta)$$

with $\theta = \arc\tan(b_k/a_k)$. Noting that $|\cos(kx-\theta)| \leqslant 1$, we have

$|a_k \cos kx + b_k \sin kx| \leqslant (a_k^2 + b_k^2)^{1/2}$ for all x. Since

$$\left|\frac{a_0}{2} + \sum_{k=1}^{\infty} a_k \cos kx + b_k \sin kx\right| \leqslant \left|\frac{a_0}{2}\right| + \sum_{k=1}^{\infty} |a_k \cos kx + b_k \sin kx|$$

$$\leqslant \left|\frac{a_0}{2}\right| + \sum_{k=1}^{\infty} (a_k^2 + b_k^2)^{1/2},$$

by the application of the Weierstrass M-test, the series

$$\frac{a_0}{2} + \sum_{k=1}^{\infty} (a_k \cos kx + b_k \sin kx)$$

converges uniformly and absolutely, and indeed, by the pointwise convergence theorem, converges to $f(x)$. □

In the preceding theorem, we have assumed that $f(x)$ is continuous and $f'(x)$ is piecewise continuous. With less stringent conditions on f, the following theorem can be proved.

Theorem 5.12.2 *Let $f(x)$ be piecewise smooth in the interval $[-\pi,\pi]$. If $f(x)$ is periodic with period 2π, then the Fourier series for f converges uniformly to f in every closed interval containing no discontinuity.*

We note that the partial sums $s_n(x)$ of a Fourier series cannot approach the function $f(x)$ uniformly over any interval containing a point of discontinuity of f. The behavior of the deviation of $s_n(x)$ from $f(x)$ in such an interval is known as the *Gibbs phenomenon*. For instance, in the Example 5.8.1, the Fourier series of the function

$$f(x) = \begin{cases} -1 & -\pi < x < 0 \\ 1 & 0 < x < \pi \end{cases}$$

was given by

$$f(x) = \frac{4}{\pi} \sum_{k=1}^{\infty} \frac{\sin(2k-1)x}{(2k-1)}$$

If we plot the partial sums $s_n(x)$ against the x-axis, as shown in Fig. 5.12.1, we find that s_n oscillate above and below the value of f. It can be observed that, near the discontinuous points $x=0$ and $x=\pi$, s_n deviate from the function rather significantly. Although the magnitude of oscillation decreases at all points in the interval for large n, but very near the points of discontinuity, the amplitude remains practically independent of n as n increases. This illustrates the fact that the Fourier series of a function f does not converge uniformly on any interval which contains a discontinuity.

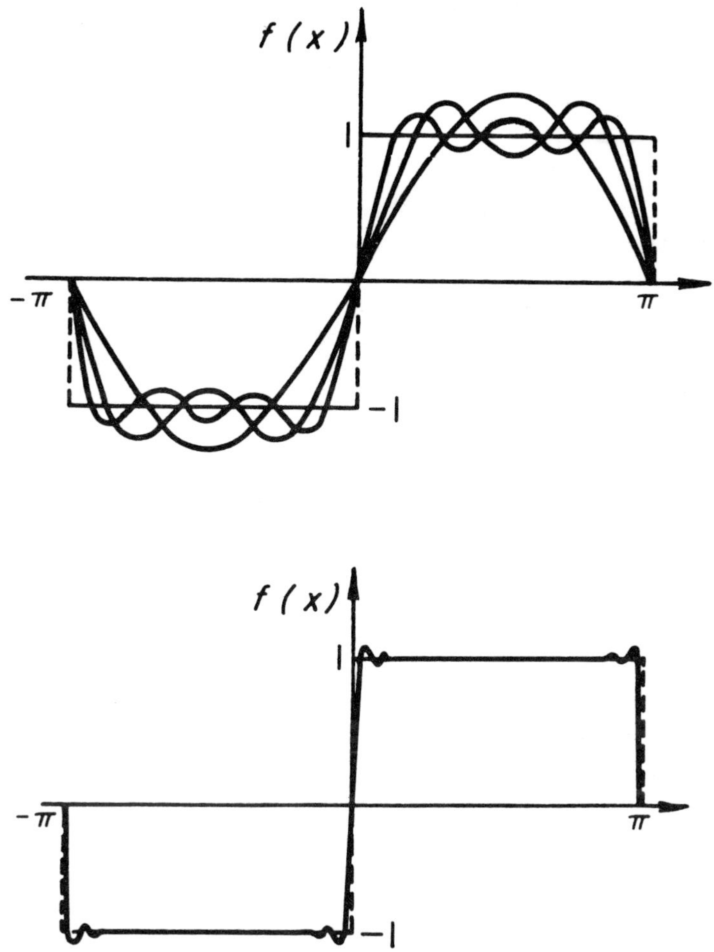

Figure 5.12.1

5.13 Differentiation and Integration

Termwise differentiation of Fourier series is, in general, not permissible. For example, the Fourier series for $f(x)=x$ (Example 5.7.3) is

$$x=2\left[\sin x-\frac{\sin 2x}{2}+\frac{\sin 3x}{3}-\dots\right]$$

which converges for all x, whereas the series after formal term by term differentiation,

$$2[\cos x-\cos 2x+\cos 3x-\dots]$$

diverges for all x. The difficulty arises from the fact that the given function $f(x)=x$ in $[-\pi,\pi]$ when extended periodically is discontinuous at the points $\pm\pi, \pm 3\pi, \ldots$ We shall see below that the continuity of the periodic function is one of the conditions that must be met for the termwise differentiation of a Fourier series.

Theorem 5.13.1 (Differentiation Theorem) *Let $f(x)$ be a continuous function in the interval $[-\pi,\pi]$ with $f(-\pi)=f(\pi)$, and let $f'(x)$ be piecewise smooth in that interval. Then the Fourier series for f' can be obtained by termwise differentiation of the series for f, and the differentiated series converges pointwise to f' at points of continuity and to $[f'(x)+f'(-x)]/2$ at discontinuous points.*

PROOF. Let the Fourier series of f and f' be

$$\frac{a_0}{2}+\sum_{k=1}^{\infty}(a_k\cos kx+b_k\sin kx)$$

and

$$\frac{A_0}{2}+\sum_{k=1}^{\infty}(A_k\cos kx+B_k\sin kx)$$

respectively. It has been shown in Theorem 5.12.1, that if $f(-\pi)=f(\pi)$, we have $A_0=0, A_k=kb_k$, and $B_k=-ka_k$. Thus,

$$f'(x)=\frac{A_0}{2}+\sum_{k=1}^{\infty}(A_k\cos kx+B_k\sin kx)$$

becomes

$$f'(x)=\sum_{k=1}^{\infty}(-ka_k\sin kx+kb_k\cos kx)$$

This is precisely the result which is obtained by termwise differentiation of

$$f(x)=\frac{a_0}{2}+\sum_{k=1}^{\infty}(a_k\cos kx+b_k\sin kx)$$

At a point of discontinuity, termwise differentiation is still valid in the sense that

$$\frac{1}{2}[f'(x+)+f'(x-)]=\sum_{k=1}^{\infty}(-ka_k\sin kx+kb_k\cos kx)$$

Since f' is piecewise smooth, according to Theorem 5.12.2, the derived series converges pointwise to $f'(x)$ at points of continuity and to $[f'(x+)+f'(x-)]/2$ at discontinuous points. □

The termwise integration of Fourier series is possible under more general conditions than termwise differentiation. We recall that in calculus, the series of functions to be integrated must converge uniformly in order to assure the convergence of a termwise integrated series. However, in the case of Fourier series, this condition is not necessary.

Theorem 5.13.2 (Integration Theorem) *Let $f(x)$ be piecewise continuous in $[-\pi,\pi]$, and periodic with period 2π. Then the Fourier series of $f(x)$*

$$\frac{a_0}{2}+\sum_{k=1}^{\infty}(a_k\cos kx+b_k\sin kx)$$

whether convergent or not, can be integrated term by term between any limits.

PROOF. We wish to prove

$$\int_a^b f(x)\,dx=\int_a^b \frac{a_0}{2}\,dx+\sum_{k=1}^{\infty}\left[\int_a^b(a_k\cos kx+b_k\sin kx)\,dx\right], \qquad (5.13.1)$$

so it suffices to show that

$$\int_a^b\left[f(x)-\frac{a_0}{2}\right]dx=\sum_{k=1}^{\infty}\frac{1}{k}\left[a_k(\sin kb-\sin ka)-b_k(\cos kb-\cos ka)\right]$$

First, we define $F(x)$ to be

$$F(x)=\int_0^x\left[f(t)-\frac{a_0}{2}\right]dt$$

Since f is piecewise continuous, F is continuous. Furthermore,

$$F'(x)=f(x)-\frac{a_0}{2}$$

is piecewise continuous. In addition

$$\begin{aligned}F(x+2\pi)&=\int_0^x\left[f(t)-\frac{a_0}{2}\right]dt+\int_x^{x+2\pi}\left[f(t)-\frac{a_0}{2}\right]dt\\&=F(x)+\int_{-\pi}^{\pi}\left[f(t)-\frac{a_0}{2}\right]dt\\&=F(x),\end{aligned}$$

since f is periodic with period 2π and

$$a_0=\frac{1}{\pi}\int_{-\pi}^{\pi}f(x)\,dx$$

As F was just shown to be a continuous periodic function with a piecewise continuous first derivative, we can expand F in a Fourier series

$$F(x)=\frac{A_0}{2}+\sum_{k=1}^{\infty}(A_k\cos kx+B_k\sin kx)$$

which converges uniformly and absolutely, according to Theorem 5.12.1. In the same theorem, it was shown that

$$A_k=-b_k/k \quad \text{and} \quad B_k=a_k/k \quad \text{for} \quad k=1,2,3\ldots$$

Hence

$$F(x)=\frac{A_0}{2}+\sum_{k=1}^{\infty}\left[\frac{1}{k}(a_k\sin kx-b_k\cos kx)\right]$$

and from the definition of $F(x)$, it follows that

$$\int_0^x f(t)\,dt=\frac{a_0x}{2}+\frac{A_0}{2}+\sum_{k=1}^{\infty}\frac{1}{k}(a_k\sin kx-b_k\cos kx)$$

Since

$$\int_a^b f(t)\,dt=\int_0^b f(t)\,dt-\int_0^a f(t)\,dt$$

we have

$$\int_a^b f(x)\,dx=\frac{a_0}{2}(b-a)+\sum_{k=1}^{\infty}\frac{1}{k}(a_k\sin kb-b_k\cos kb)$$
$$-\sum_{k=1}^{\infty}\frac{1}{k}(a_k\sin ka-b_k\cos ka)$$

Because these series are absolutely convergent, we rearrange terms and obtain

$$\int_a^b f(x)\,dx=\frac{a_0}{2}(b-a)$$
$$+\sum_{k=1}^{\infty}\frac{1}{k}\left[a_k(\sin kb-\sin ka)-b_k(\cos kb-\cos ka)\right]$$

which is precisely the result obtained by integrating the Fourier series for $f(x)$ formally term by term. □

EXAMPLE 5.13.1. In Example 5.8.2, we have found that $f(x)=|\sin x|$ is represented by the Fourier series

$$\sin x=\frac{2}{\pi}+\frac{4}{\pi}\sum_{k=1}^{\infty}\frac{\cos 2kx}{(1-4k^2)} \qquad 0<x<\pi$$

Since $f(x)=|\sin x|$ is continuous in the interval $[-\pi,\pi]$ and $f(-\pi)=f(\pi)$, we differentiate term by term, obtaining

$$\cos x = -\frac{8}{\pi}\sum_{k=1}^{\infty}\frac{k\sin 2kx}{(1-4k^2)}$$

by use of Theorem 5.13.1, since $f'(x)$ is piecewise smooth in $[-\pi,\pi]$. In this way, we obtain the Fourier sine series expansion of the cosine function in $(0,\pi)$. Note that the reverse process is not permissible.

EXAMPLE 5.13.2. Consider the function $f(x)=x$ in the interval $-\pi<x<\pi$. As shown in Example 5.7.3,

$$x=2\left[\sin x-\frac{\sin 2x}{2}+\frac{\sin 3x}{3}-\dots\right]$$

By Theorem 5.13.2, we can integrate the series term by term from a to x to obtain

$$\frac{1}{2}(x^2-a^2)=2\left[-\left(\cos x-\frac{\cos 2x}{2^2}+\frac{\cos 3x}{3^2}-\dots\right)\right.$$
$$\left.+\left(\cos a-\frac{\cos 2a}{2^2}+\frac{\cos 3a}{3^2}-\dots\right)\right]$$

To determine the sum of the series of constants, we write

$$\frac{x^2}{4}=C-\sum_{k=1}^{\infty}(-1)^{k+1}\frac{\cos kx}{k^2}$$

where C is a constant. Since the series on the right is the Fourier series which converges uniformly, we can integrate the series term by term from $-\pi$ to π to obtain

$$\int_{-\pi}^{\pi}\frac{x^2}{2}\,dx=2\left[\int_{-\pi}^{\pi}C\,dx-\sum_{k=1}^{\infty}\frac{(-1)^{k+1}}{k^2}\int_{-\pi}^{\pi}\cos kx\,dx\right]$$

$$\frac{\pi^3}{3}=2(2\pi C)$$

Hence

$$C=\frac{\pi^2}{12}$$

Therefore, by integrating the Fourier series of $f(x)=x$ in $(-\pi,\pi)$, we obtain the Fourier series expansion for the function $f(x)=x^2$ as

$$x^2=4\left[\frac{\pi^2}{12}-\sum_{k=1}^{\infty}(-1)^{k+1}\frac{\cos kx}{k^2}\right]$$

5.14 Double Fourier Series

The theory of series expansions for functions of two variables is analogous to that of series expansions for functions of one variable. Here we shall present a short description of double Fourier series.

We have seen earlier that, if $f(x)$ is piecewise continuous and periodic with period 2π, then the Fourier series

$$f(x) \sim \frac{a_0}{2} + \sum_{m=1}^{\infty} (a_m \cos mx + b_m \sin mx)$$

converges in the mean to $f(x)$. If f is continuously differentiable, then its Fourier series converges uniformly.

For the sake of simplicity and convenience, let us consider the function $f(x,y)$ which is continuously differentiable (a stronger condition than necessary). Let $f(x,y)$ be periodic with period 2π, that is,

$$f(x+2\pi,y) = f(x,y+2\pi) = f(x,y)$$

Then, if we hold y fixed, we can expand $f(x,y)$ into a uniformly convergent Fourier series

$$f(x,y) = \frac{a_0(y)}{2} + \sum_{m=1}^{\infty} \left[a_m(y) \cos mx + b_m(y) \sin mx \right] \tag{5.14.1}$$

in which the coefficients are functions of y, namely,

$$a_m(y) = \frac{1}{\pi} \int_{-\pi}^{\pi} f(x,y) \cos mx dx$$

$$b_m(y) = \frac{1}{\pi} \int_{-\pi}^{\pi} f(x,y) \sin mx dx$$

These coefficients are continuously differentiable in y, and therefore, we can expand them in uniformly convergent series

$$a_m(y) = \frac{a_{m0}}{2} + \sum_{n=1}^{\infty} a_{mn} \cos ny + b_{mn} \sin ny$$

$$b_m(y) = \frac{c_{m0}}{2} + \sum_{n=1}^{\infty} c_{mn} \cos ny + d_{mn} \sin ny \tag{5.14.2}$$

where

$$a_{mn} = \frac{1}{\pi^2} \int_{-\pi}^{\pi} \int_{-\pi}^{\pi} f(x,y) \cos mx \cos ny \, dx \, dy$$

$$b_{mn} = \frac{1}{\pi^2} \int_{-\pi}^{\pi} \int_{-\pi}^{\pi} f(x,y) \cos mx \sin ny \, dx \, dy$$

$$c_{mn} = \frac{1}{\pi^2} \int_{-\pi}^{\pi} \int_{-\pi}^{\pi} f(x,y) \sin mx \cos ny \, dx \, dy$$

$$d_{mn} = \frac{1}{\pi^2} \int_{-\pi}^{\pi} \int_{-\pi}^{\pi} f(x,y) \sin mx \sin ny \, dx \, dy \tag{5.14.3}$$

Substitution of a_m and b_m into Eq. (5.14.1) yields

$$f(x,y)=\frac{a_{00}}{4}+\frac{1}{2}\sum_{n=1}^{\infty}\left[a_{0n}\cos ny+b_{0n}\sin ny\right]$$
$$+\frac{1}{2}\sum_{m=1}^{\infty}\left[a_{m0}\cos mx+c_{m0}\sin mx\right]$$
$$+\sum_{m=1}^{\infty}\sum_{n=1}^{\infty}\left[a_{mn}\cos mx\cos ny+b_{mn}\cos mx\sin ny\right.$$
$$\left.+c_{mn}\sin mx\cos ny+d_{mn}\sin mx\sin ny\right] \tag{5.14.4}$$

which is called the *double Fourier series*.

a. When $f(-x,y)=f(x,y)$ and $f(x,-y)=f(x,y)$, all the coefficients vanish except a_{mn}, and the double Fourier series reduces to

$$f(x,y)=\sum_{m=1}^{\infty}\sum_{n=1}^{\infty}a_{mn}\cos mx\cos ny \tag{5.14.5}$$

where

$$a_{mn}=\frac{4}{\pi^2}\int_0^{\pi}\int_0^{\pi}f(x,y)\cos mx\cos ny\,dx\,dy$$

b. When $f(-x,y)=f(x,y)$ and $f(x,-y)=-f(x,y)$, we have

$$f(x,y)=\frac{1}{2}\sum_{n=1}^{\infty}b_{0n}\sin ny+\sum_{m=1}^{\infty}\sum_{n=1}^{\infty}b_{mn}\cos mx\sin ny \tag{5.14.6}$$

where

$$b_{mn}=\frac{4}{\pi^2}\int_0^{\pi}\int_0^{\pi}f(x,y)\cos mx\sin ny\,dx\,dy$$

c. When $f(-x,y)=-f(x,y)$ and $f(x,-y)=f(x,y)$, we have

$$f(x,y)=\frac{1}{2}\sum_{m=1}^{\infty}c_{m0}\sin mx+\sum_{m=1}^{\infty}\sum_{n=1}^{\infty}c_{mn}\sin mx\cos ny \tag{5.14.7}$$

where

$$c_{mn}=\frac{4}{\pi^2}\int_0^{\pi}\int_0^{\pi}f(x,y)\sin mx\cos ny\,dx\,dy$$

d. When $f(-x,y)=-f(x,y)$ and $f(x,-y)=-f(x,y)$, we have

$$f(x,y)=\sum_{m=1}^{\infty}\sum_{n=1}^{\infty}d_{mn}\sin mx\sin ny \tag{5.14.8}$$

where

$$d_{mn}=\frac{4}{\pi^2}\int_0^{\pi}\int_0^{\pi}f(x,y)\sin mx\sin ny\,dx\,dy$$

EXAMPLE 5.14.1. Expand the function $f(x,y)=xy$ into double Fourier series in the interval $-\pi<x<\pi, -\pi<y<\pi$.

Since $f(-x,y)=-xy=-f(x,y)$ and $f(x,-y)=-xy=-f(x,y)$, we find

$$d_{mn}=\frac{4}{\pi^2}\int_0^{\pi}\int_0^{\pi} xy\sin mx\sin ny\,dx\,dy$$
$$=(-1)^{(m+n)}\frac{4}{mn}$$

Thus, the double Fourier series for f in $-\pi<x<\pi, -\pi<y<\pi$ is

$$f(x,y)=4\sum_{m=1}^{\infty}\sum_{n=1}^{\infty}(-1)^{m+n}\frac{\sin mx\sin ny}{mn}$$

Exercises

1. State whether or not the following functions are piecewise continuous on the given intervals. In each case, determine the right and left hand limits

a. $f(x)=\begin{cases} x & 0\leqslant x<1 \\ 0 & 1<x\leqslant 2\end{cases}$

b. $f(x)=\begin{cases} 0 & -1\leqslant x\leqslant 0 \\ 1/x & 0<x\leqslant 1\end{cases}$

c. $f(x)=\begin{cases} 2 & 0\leqslant x<1 \\ x^2 & 1<x\leqslant 2\end{cases}$

d. $f(x)=\begin{cases} 1-x & 1\leqslant x<2 \\ x/(x-2) & 2<x\leqslant 3\end{cases}$

2. f is said to be *almost differentiable* in $[a,b]$, if at every point in (a,b), $f'(x+)$, $f'(x-)$, and $f'(a+)$ and $f'(b-)$ exist. Prove that if f is piecewise smooth in $[a,b]$, it is almost differentiable there. Give a counterexample to show that the converse is not true.

3. Are the following functions even, odd or neither even nor odd?

a. $x+2x^2+3x^3$
b. x^2+4x^4
c. $x\ln x$
d. $1/x$
e. e^{x^2}
f. $x^2\sin x$
g. $x^2\cos x$
h. $\sinh x$

4. If the subscripts e and o denote even and odd respectively, show that

a. $f_e g_e = h_e$
b. $f_0 g_0 = h_e$
c. $f_e g_0 = f_0 g_e = h_0$
d. $f_e + g_e = h_e$
e. $f_0 + g_0 = h_0$

5. Let $f(x)$ be continuously differentiable on the interval $[0,l)$. Let $f'(0+)$ exist and let $f(0+) = f(0) = 0$. Prove that F_o, the odd extension of f to $(-l,l)$ is continuously differentiable on $(-l,l)$.

6. Determine which of the following functions are periodic and find the least positive period of those which are periodic.

a. $\sin 3x$
b. $\cos x/2$
c. $\cos 2\pi x$
d. $x \cos x$
e. $\cos x + \sin x$
f. $x^2 \cos x$

7. If $f(x)$ is periodic with period p, show that

$$\int_a^{a+p} f(x)\,dx = \int_b^{b+p} f(x)\,dx$$

8. Show that the polynomials $1, x, (3x^2 - 1)/2$ are orthogonal to each other in the interval $[-1,1]$.

9. Prove that every orthogonal set of functions is linearly independent.

10. Prove the *Uniqueness theorem*: if f_1 and f_2 are continuous in $-\pi \leqslant x \leqslant \pi$ and have the same Fourier series, then $f_1 = f_2$.

11. Prove the *Weierstrass Approximation theorem*: if f is a continuous function on the interval $-\pi \leqslant x \leqslant \pi$ and if $f(-\pi) = f(\pi)$, then for any $\varepsilon > 0$, there exists a trigonometric polynomial

$$T(x) = \frac{a_0}{2} + \sum_{k=1}^{n} (a_k \cos kx + b_k \sin kx)$$

such that

$$|f(x) - T(x)| < \varepsilon$$

for all x in $[-\pi, \pi]$.

12. Find the Fourier series of the following functions:

a. $f(x)=\begin{cases} x & -\pi<x<0 \\ h \quad \text{a constant} & 0<x<\pi \end{cases}$

b. $f(x)=\begin{cases} 1 & -\pi<x<0 \\ x^2 & 0<x<\pi \end{cases}$

c. $f(x)=x+\sin x \qquad -\pi<x<\pi$

d. $f(x)=1+x \qquad -\pi<x<\pi$

e. $f(x)=e^x \qquad -\pi<x<\pi$

f. $f(x)=1+x+x^2 \qquad -\pi<x<\pi$

13. Determine the Fourier sine series of the following functions:

a. $f(x)=\pi-x \qquad 0<x<\pi$

b. $f(x)=\begin{cases} 1 & 0<x<\pi/2 \\ 2 & \pi/2<x<\pi \end{cases}$

c. $f(x)=x^2 \qquad 0<x<\pi$

d. $f(x)=\cos x \qquad 0<x<\pi$

e. $f(x)=x^3 \qquad 0<x<\pi$

f. $f(x)=e^x \qquad 0<x<\pi$

14. Obtain the Fourier cosine series representations for the following functions:

a. $f(x)=\pi+x \qquad 0<x<\pi$

b. $f(x)=x \qquad 0<x<\pi$

c. $f(x)=x^2 \qquad 0<x<\pi$

d. $f(x)=\sin 3x \qquad 0<x<\pi$

e. $f(x)=e^x \qquad 0<x<\pi$

f. $f(x)=\cosh x \qquad 0<x<\pi$

15. Expand the following functions in a Fourier series:

a. $f(x)=x^2+x \qquad -1<x<1$

b. $f(x)=\begin{cases} 1 & 0<x<3 \\ 0 & 3<x<6 \end{cases}$

c. $f(x)=\sin \pi x/l \qquad 0<x<l$

d. $f(x)=x^3 \qquad -2<x<2$

e. $f(x)=e^{-x} \qquad 0<x<1$

f. $f(x)=\sinh x \qquad -1<x<1$

16. Find the following functions in a complex Fourier series:

a. $f(x)=e^{2x} \qquad -\pi<x<\pi$

b. $f(x)=\cosh x \qquad -\pi<x<\pi$

c. $f(x)=\begin{cases} 1 & -\pi<x<0 \\ \cos x & 0<x<\pi \end{cases}$

d. $f(x)=x \qquad -1<x<1$

e. $f(x)=x^2 \qquad -\pi<x<\pi$

f. $f(x)=\sinh \pi x/2 \qquad -2<x<2$

17. Determine the Fourier series expansion of each of the following functions by performing the differentiation of the appropriate Fourier series.

a. $\sin^2 x$ $\quad 0<x<\pi$
b. $\cos^2 x$ $\quad 0<x<\pi$
c. $\sin x \cos x$ $\quad 0<x<\pi$
d. $\cos x+\cos 2x$ $\quad 0<x<\pi$
e. $\cos x \cos 2x$ $\quad 0<x<\pi$

18. Find the functions represented by the new series which are obtained by termwise integration of the following series from 0 to x:

a. $\sum_{k=1}^{\infty} \frac{(-1)^{k+1}}{k} \sin kx = x/2 \qquad -\pi<x<\pi$

b. $\frac{3}{2}+\frac{1}{\pi} \sum_{k=1}^{\infty} \frac{1-(-1)^k}{k} \sin kx = \begin{cases} 1 & -\pi<x<0 \\ 2 & 0<x<\pi \end{cases}$

c. $\sum_{k=1}^{\infty} (-1)^{k+1} \frac{\cos kx}{k} = \ln(2\cos \frac{x}{2}) \qquad -\pi<x<\pi$

d. $\sum_{k=1}^{\infty} \frac{\sin(2k+1)x}{(2k+1)^3} = \frac{\pi^2 x-\pi x^2}{8} \qquad 0<x<2\pi$

e. $\frac{4}{\pi} \sum_{k=1}^{\infty} \frac{\sin(2k-1)x}{(2k-1)} = \begin{cases} -1 & -\pi<x<0 \\ 1 & 0<x<\pi \end{cases}$

19. a. Find the Fourier series expansion of the function

$$f(x)=\begin{cases} 0 & -\pi<x<0 \\ x/2 & -0<x<\pi \end{cases}$$

b. With the use of this series, show that

$$\frac{\pi^2}{8}=1+\frac{1}{3^2}+\frac{1}{5^2}+\frac{1}{7^2}+\ldots$$

20. a. Determine the Fourier series of the function

$$f(x)=x^2 \qquad -l<x<l$$

b. With the use of this series, show that

$$\frac{\pi^2}{12}=1-\frac{1}{2^2}+\frac{1}{3^2}-\frac{1}{4^2}+\ldots$$

21. Plot the function f and the partial sums s_1 and s_2 against x in Exercise 12.

22. Let f and g be piecewise continuous in $[-\pi,\pi]$ and periodic with period 2π. If a_k, b_k and c_k, d_k are the Fourier coefficients of f and g respectively, prove that

$$\frac{1}{\pi}\int_{-\pi}^{\pi} f(x)g(x)dx=\frac{a_0 c_0}{2}+\sum_{k=1}^{\infty}(a_k c_k+b_k d_k)$$

23. Improve the convergence of the series

$$f(x)=\sum_{n=2}^{\infty}(-1)^n\frac{n^3}{n^4-1}\sin nx \qquad -\pi<x<\pi$$

[Hint: Since $n^3/(n^4-1)=(1/n)+[1/(n^5-n)]$, show that

$$f(x)=-\frac{x}{2}+\sin x+\sum_{n=2}^{\infty}\frac{(-1)^n\sin nx}{n^5-n}$$

Note that the series with coefficient $1/(n^5-n)$ converges more rapidly than the series with coefficient $n^3/(n^4-1)$.

24. Improve the convergence of the series

$$f(x)=\sum_{n=1}^{\infty}\frac{n^3+n+1}{n(n^3+1)}\sin nx$$

[Hint:

$$\frac{n^3+n+1}{n(n^3+1)}=\frac{1}{n}+\frac{1}{n^3+1}$$

and

$$\frac{\pi}{2}-\frac{x}{2}=\sum_{n=1}^{\infty}\frac{\sin nx}{n} \qquad 0<x<2\pi.$$

25. Considering the function $f(x)$ as the sum of two functions, determine the Fourier series of f

a. $$f(x)=\begin{cases}\dfrac{1}{4} & -\pi<x<0\\ \dfrac{1}{4}(\pi x-1) & 0<x<\pi\end{cases}$$

[Hint: $f(x)=g(x)-\frac{1}{4}h(x)$ where

$$g(x)=\begin{cases}0 & -\pi<x<0\\ \dfrac{1}{4}\pi x & 0<x<\pi\end{cases} \qquad h(x)=\begin{cases}-1 & -\pi<x<0\\ 1 & 0<x<\pi\end{cases}$$

b. $$f(x)=\begin{cases}0 & -\pi<x<0\\ 1 & 0<x<\pi\end{cases}$$

[Hint: $f(x)=\frac{1}{2}+\frac{1}{2}g(x)$ where

$$g(x)=\begin{cases}-1 & -\pi<x<0\\ 1 & 0<x<\pi\end{cases}$$

26. a. Sketch the graph of the function

$$f(x)=\begin{cases}1 & -2<x<-1\\ -1 & -1<x<1\\ 1 & 1<x<2\end{cases}$$

over the intervals $[-4,-2]$ and $[2,4]$.

b. Determine the value of the Fourier series for f when

i. $x=2k$

ii. $x=2k+1$ where k is a positive integer.

27. Prove the *Schwarz inequality*

$$\left(\sum_{k=1}^{n} A_k B_k\right)^2 \leqslant \sum_{k=1}^{n} A_k^2 \sum_{k=1}^{n} B_k^2$$

28. If $f(x)$ is piecewise continuous in $[-\pi,\pi]$ and periodic with period 2π, prove that the same is true for the function

$$f(s+x)\frac{\sin(n+1/2)s}{2\sin(s/2)}$$

29. Prove Theorem 5.12.2.

30. A function f is said to satisfy a *Lipchitz condition* of order a at x_0, if there exist positive constants M and δ, such that

$$|f(x)-f(x_0)|<M|x-x_0|^a \text{ if} |x-x_0|<\delta$$

Prove that, if f is continuous and satisfies the Lipchitz condition at x_0, then the Fourier series of f converges to $f(x_0)$.

31. The conditions that $f(x)$ be periodic, piecewise continuous and has a finite number of relative maxima and minima in $[-\pi,\pi]$ are called *the Dirichlet conditions*. Prove that if $f(x)$ satisfies the Dirichlet conditions, then the Fourier series of f converges to $\frac{1}{2}[f(x+)+f(x-)]$.

32. Determine the double Fourier series of the following functions:

a. $f(x,y)=1$	$0<x<\pi$	$0<y<\pi$
b. $f(x,y)=xy^2$	$0<x<\pi$	$0<y<\pi$
c. $f(x,y)=x^2y^2$	$0<x<\pi$	$0<y<\pi$
d. $f(x,y)=x^2+y$	$-\pi<x<\pi$	$-\pi<y<\pi$
e. $f(x,y)=x\sin y$	$-\pi<x<\pi$	$-\pi<y<\pi$
f. $f(x,y)=e^{x+y}$	$-\pi<x<\pi$	$-\pi<y<\pi$

33. Deduce the general double Fourier series expansion formula for the function $f(x,y)$ in the rectangle $-a<x<a, -b<y<b$.

CHAPTER SIX

Method of Separation of Variables

6.1 Separation of Variables

We have so far been mainly concerned with initial value problems. In this section, we shall introduce one of the most common and elementary methods called the method of separation of variables for solving initial-boundary value problems. The class of problems for which this method is applicable contains a wide range of problems of mathematical physics.

We now describe the method of separation of variables and examine the conditions of applicability of the method to problems which involve second-order partial differential equations in two independent variables.

We consider the second-order homogeneous equation

$$a^* u_{x^*x^*} + b^* u_{x^*y^*} + c^* u_{y^*y^*} + d^* u_{x^*} + e^* u_{y^*} + f^* u = 0 \tag{6.1.1}$$

where a^*, b^*, c^*, d^*, e^*, and f^* are functions of x^* and y^*.

We have stated in Chapter 3 that by the transformation

$$\begin{aligned} x &= x(x^*,y^*) \\ y &= y(x^*,y^*) \qquad \frac{\partial(x,y)}{\partial(x^*,y^*)} \neq 0 \end{aligned} \tag{6.1.2}$$

we can always transform Eq. (6.1.1) into canonical form

$$a(x,y)u_{xx} + c(x,y)u_{yy} + d(x,y)u_x + e(x,y)u_y + f(x,y)u = 0 \tag{6.1.3}$$

which when

i. $a = -c$ is hyperbolic
ii. $a = 0$ or $c = 0$ is parabolic
iii. $a = c$ is elliptic

We assume a solution in the form

$$u(x,y)=X(x)\,Y(y) \tag{6.1.4}$$

where X and Y are, respectively, functions of x and y alone and are twice continuously differentiable. Substituting Eq. (6.1.4) into Eq. (6.1.3), we obtain

$$aX''Y+cXY''+dX'Y+eXY'+fXY=0 \tag{6.1.5}$$

where the primes denote differentiation with the respective variables. Let there exist a function $p(x,y)$, such that, if we divide Eq. (6.1.5) by $p(x,y)$, we obtain

$$a_1(x)X''Y+b_1(y)XY''+a_2(x)X'Y+b_2(y)XY' + \left[a_3(x)+b_3(y)\right]XY \quad =0 \tag{6.1.6}$$

Dividing Eq. (6.1.6) again by XY, we obtain

$$\left[a_1\frac{X''}{X}+a_2\frac{X'}{X}+a_3\right]=-\left[b_1\frac{Y''}{Y}+b_2\frac{Y'}{Y}+b_3\right] \tag{6.1.7}$$

The left side of Eq. (6.1.7) is a function of x only. The right side of Eq. (6.1.7) depends only upon y. Thus, we differentiate Eq. (6.1.7) with respect to x to obtain

$$\frac{d}{dx}\left[a_1\frac{X''}{X}+a_2\frac{X'}{X}+a_3\right]=0 \tag{6.1.8}$$

Integration of Eq. (6.1.8) yields

$$a_1\frac{X''}{X}+a_2\frac{X'}{X}+a_3=\lambda \tag{6.1.9}$$

where λ is a separation constant. From Eqs. (6.1.7) and (6.1.9), we have

$$b_1\frac{Y''}{Y}+b_2\frac{Y'}{Y}+b_3=-\lambda \tag{6.1.10}$$

We may rewrite Eqs. (6.1.9) and (6.1.10) in the form

$$a_1X''+a_2X'+(a_3-\lambda)X=0 \tag{6.1.11}$$

and

$$b_1Y''+b_2Y'+(b_3+\lambda)Y=0 \tag{6.1.12}$$

Thus, $u(x,y)$ is the solution of Eq. (6.1.3) if $X(x)$ and $Y(y)$ are the solutions of the ordinary differential equations (6.1.11) and (6.1.12) respectively.

If the coefficients in Eq. (6.1.1) are constant, then the reduction of Eq. (6.1.1) to canonical form is no longer necessary. To illustrate this, let us consider the second-order equation

$$Au_{xx}+Bu_{xy}+Cu_{yy}+Du_x+Eu_y+Fu=0 \tag{6.1.13}$$

where A, B, C, D, E and F are constants which are not all zero.

As before, we assume a solution in the form

$$u(x,y)=X(x)\,Y(y)$$

Substituting this in Eq. (6.1.13), we obtain

$$AX''Y+BX'Y'+CXY''+DX'Y+EXY'+FXY=0 \tag{6.1.14}$$

Division of this equation by AXY yields

$$\frac{X''}{X}+\frac{B}{A}\frac{X'}{X}\frac{Y'}{Y}+\frac{C}{A}\frac{Y''}{Y}+\frac{D}{A}\frac{X'}{X}+\frac{E}{A}\frac{Y'}{Y}+\frac{F}{A}=0, \quad A\neq 0 \tag{6.1.15}$$

We differentiate this equation with respect to x, and obtain

$$\left(\frac{X''}{X}\right)'+\frac{B}{A}\left(\frac{X'}{X}\right)'\frac{Y'}{Y}+\frac{D}{A}\left(\frac{X'}{X}\right)'=0 \tag{6.1.16}$$

Thus, we have

$$\frac{\left(\dfrac{X''}{X}\right)'}{\dfrac{B}{A}\left(\dfrac{X'}{X}\right)'}+\frac{D}{B}=-\frac{Y'}{Y} \tag{6.1.17}$$

This equation is separated, so that both sides must be equal to a constant λ. Therefore, we obtain

$$Y'+\lambda Y=0 \tag{6.1.18}$$

$$\left(\frac{X''}{X}\right)'+\left(\frac{D}{B}-\lambda\right)\frac{B}{A}\left(\frac{X'}{X}\right)'=0 \tag{6.1.19}$$

Integrating Eq. (6.1.19) with respect to x, we obtain

$$\frac{X''}{X}+\left(\frac{D}{B}-\lambda\right)\frac{B}{A}\left(\frac{X'}{X}\right)=-\beta \tag{6.1.20}$$

where β is a constant to be determined. By substituting Eq. (6.1.18) into the original Eq. (6.1.15), we obtain

$$X''+\left(\frac{D}{B}-\lambda\right)\frac{B}{A}X'+\left(\lambda^2-\frac{E}{C}\lambda+\frac{F}{C}\right)\frac{C}{A}X=0 \tag{6.1.21}$$

Comparing Eqs. (6.1.20) and (6.1.21), we clearly find

$$\beta=\left(\lambda^2-\frac{E}{C}\lambda+\frac{F}{C}\right)\frac{C}{A}$$

Therefore, $u(x,y)$ is a solution of Eq. (6.1.16) if $X(x)$ and $Y(y)$ satisfy the ordinary differential equations (6.1.21) and (6.1.18) respectively.

We have just described the conditions on the separability of a given partial differential equation. Now, we shall take a look at the boundary conditions involved. There are several types of boundary conditions. The ones that appear most frequently in problems of mathematical physics are the *i*. Dirichlet condition: u is prescribed on a boundary

ii. Neumann condition: $\dfrac{\delta u}{\delta n}$ is prescribed on a boundary

iii. Mixed condition: $\dfrac{\delta u}{\delta n}+hu$ is prescribed on a boundary where $\dfrac{\delta u}{\delta n}$ is the directional derivative of u along the outward normal to the boundary, and h is a given continuous function on the boundary. For details, see Chapter 8.

Besides these three boundary conditions, also known as, the first, second and third condition, there are other conditions, such as the Robin condition; one condition is prescribed on one portion of a boundary and the other is given on the remainder of the boundary. We shall consider a variety of conditions as we treat problems later.

To separate boundary conditions, such as the ones listed above, it is best to choose a coordinate system suitable to a boundary. For instance, we choose the Cartesian coordinate system (x,y) for a rectangular region such that the boundary is described by the coordinate lines $x=$constant and $y=$constant, and the polar coordinate system (r,θ) for a circular region so that the boundary is described by the lines $r=$constant and $\theta=$constant.

Another condition that must be imposed on the separability of boundary conditions is that boundary conditions, say at $x=x_0$, must contain the derivatives of u with respect to x only, and their coefficients must depend only on x. For example, the boundary condition

$$[u+u_y]_{x=x_0}=0$$

cannot be separated. It is needless to say that mixed condition, such as u_x+u_y, cannot be prescribed on an axis.

6.2 The Vibrating String Problem

As a first example, we shall consider the problem of the vibrating string stretched along the x-axis from 0 to l, fixed at its end points. We have seen earlier in Chapter 4, that the problem is given by

$$u_{tt}-c^2u_{xx}=0 \qquad 0<x<l \qquad t>0 \tag{6.2.1}$$

$$u(x,0)=f(x) \qquad 0\leqslant x\leqslant l \tag{6.2.2}$$

$$u_t(x,0)=g(x) \qquad 0\leqslant x\leqslant l \tag{6.2.3}$$

$$u(0,t)=0 \qquad t\geqslant 0 \tag{6.2.4}$$

$$u(l,t)=0 \qquad t\geqslant 0 \tag{6.2.5}$$

where f and g are the initial displacement and initial velocity respectively.

By the method of separation of variables, we assume a solution in the form

$$u(x,t)=X(x)T(t) \tag{6.2.6}$$

If we substitute Eq. (6.2.6) into Eq. (6.2.1), we obtain

$$XT'' = c^2 X'' T$$

and hence

$$\frac{X''}{X} = \frac{1}{c^2}\frac{T''}{T} \tag{6.2.7}$$

whenever $XT \neq 0$. Since the left side of Eq. (6.2.7) is independent of t and the right side is independent of x, we must have

$$\frac{X''}{X} = \frac{1}{c^2}\frac{T''}{T} = \lambda$$

where λ is a separation constant. Thus,

$$X'' - \lambda X = 0 \tag{6.2.8}$$

$$T'' - \lambda c^2 T = 0 \tag{6.2.9}$$

We now separate the boundary conditions. From Eqs. (6.2.4) and (6.2.6), we obtain

$$u(0,t) = X(0)T(t) = 0$$

We know that $T(t) \neq 0$ for all values of t, and therefore

$$X(0) = 0 \tag{6.2.10}$$

In a similar manner, boundary condition (6.2.5) implies

$$X(l) = 0 \tag{6.2.11}$$

To determine $X(x)$ we first solve the *eigenvalue problem* (eigenvalue problems are treated in Chapter 7)

$$\begin{aligned} X'' + \lambda X &= 0 \\ X(0) &= 0 \\ X(l) &= 0 \end{aligned} \tag{6.2.12}$$

We look for values of λ which give us nontrivial solutions. We investigate three possible cases

$$\lambda > 0, \lambda = 0, \lambda < 0$$

Case 1. $\lambda > 0$ The general solution in this case is of the form

$$X(x) = Ae^{-\sqrt{\lambda}x} + Be^{\sqrt{\lambda}x}$$

where A and B are arbitrary constants. To satisfy the boundary conditions we must have

$$\begin{aligned} A + B &= 0 \\ Ae^{-\sqrt{\lambda}\,l} + Be^{\sqrt{\lambda}l} &= 0 \end{aligned} \tag{6.2.13}$$

We see that the determinant of the system (6.2.13) is different from zero. Consequently, A and B must both be zero, and hence the general solution $X(x)$ is identically zero. The solution is trivial.

Case 2. $\lambda=0$ Here, the general solution is

$$X(x)=A+Bx$$

Applying the boundary conditions, we have

$$A=0$$
$$A+Bl=0$$

Hence $A=B=0$. The solution is thus identically zero.

Case 3. $\lambda<0$ In this case, the general solution assumes the form

$$X(x)=A\cos\sqrt{\lambda}\,x+B\sin\sqrt{\lambda}\,x$$

From the condition $X(0)=0$, we obtain $A=0$. The condition $X(l)=0$ gives

$$B\sin\sqrt{\lambda}l=0$$

If $B=0$, the solution is trivial. For nontrivial solutions,

$$\sin\sqrt{\lambda}l=0$$

This equation is satisfied when

$$\sqrt{\lambda}l=n\pi \qquad \text{for } n=1,2,3,\ldots$$

or

$$\lambda_n=(n\pi/l)^2 \tag{6.2.14}$$

For this infinite set of discrete values of λ, the problem has a nontrivial solution. These values of λ_n are called the *eigenvalues* of the problem, and the functions

$$\sin(n\pi/l)x, \qquad n=1,2,3,\ldots$$

are the corresponding *eigenfunctions*.

We note that it is not necessary to consider negative values of n since

$$\sin(-n)\pi x/l=-\sin n\pi x/l$$

No new solution is obtained in this way.

The solutions of problem (6.2.12) are, therefore,

$$X_n(x)=B_n\sin n\pi x/l \tag{6.2.15}$$

For $\lambda=\lambda_n$, the general solution of Eq. (6.2.9) may be written in the form

$$T_n(t)=C_n\cos\frac{n\pi c}{l}t+D_n\sin\frac{n\pi c}{l}t \tag{6.2.16}$$

where C_n and D_n are arbitrary constants.

Thus, the functions

$$u_n(x,t)=X_n(x)T_n(t)=\left(a_n\cos\frac{n\pi c}{l}t+b_n\sin\frac{n\pi c}{l}t\right)\sin\frac{n\pi x}{l}$$

satisfy Eq. (6.2.1) and the boundary conditions (6.2.4) and (6.2.5), where $a_n=B_nC_n$ and $b_n=B_nD_n$.

Since Eq. (6.2.1) is linear and homogeneous, by the superposition principle, the infinite series

$$u(x,t)=\sum_{n=1}^{\infty}\left(a_n\cos\frac{n\pi c}{l}t+b_n\sin\frac{n\pi c}{l}t\right)\sin\frac{n\pi x}{l} \tag{6.2.17}$$

is also a solution, provided it converges and twice continuously differentiable with respect to x and t. Since each term of the series satisfies the boundary conditions (6.2.4) and (6.2.5), the series satisfies these conditions. There remain two more initial conditions to be satisfied. From these conditions, we shall determine the constants a_n and b_n.

First we differentiate the series (6.2.17) with respect to t. We have

$$u_t=\sum_{n=1}^{\infty}\frac{n\pi c}{l}\left(-a_n\sin\frac{n\pi c}{l}t+b_n\cos\frac{n\pi c}{l}t\right)\sin\frac{n\pi x}{l} \tag{6.2.18}$$

Then applying the initial conditions (6.2.2) and (6.2.3), we obtain

$$\begin{aligned} u(x,0)=f(x) &= \sum_{n=1}^{\infty} a_n\sin\frac{n\pi x}{l} \\ u_t(x,0)=g(x) &= \sum_{n=1}^{\infty} b_n\left(\frac{n\pi c}{l}\right)\sin\frac{n\pi x}{l} \end{aligned}$$

These equations will be satisfied if $f(x)$ and $g(x)$ can be represented by Fourier sine series. The coefficients are given by

$$\begin{aligned} a_n&=\frac{2}{l}\int_0^l f(x)\sin\frac{n\pi x}{l}\,dx \\ b_n&=\frac{2}{n\pi c}\int_0^l g(x)\sin\frac{n\pi x}{l}\,dx \end{aligned} \tag{6.2.19}$$

The solution of the vibrating string problem is therefore given by the series (6.2.17) where the coefficients a_n and b_n are determined by the formulae (6.2.19).

EXAMPLE 6.2.1. The Plucked String

As a special case of the problem just treated, consider a stretched string fixed at both ends. Suppose the string is raised to a height h at $x=a$ and then released. The string will oscillate freely. The initial conditions, as shown in Fig. 6.2.1, may be written

$$\begin{aligned} u(x,0)=f(x) &= \begin{cases} hx/a, & 0\leqslant x\leqslant a \\ h(l-x)/(l-a), & a\leqslant x\leqslant l \end{cases} \\ u_t(x,0)=g(x) &= 0 \end{aligned}$$

Since $g(x)=0$ the coefficients b_n are identically equal to zero. The

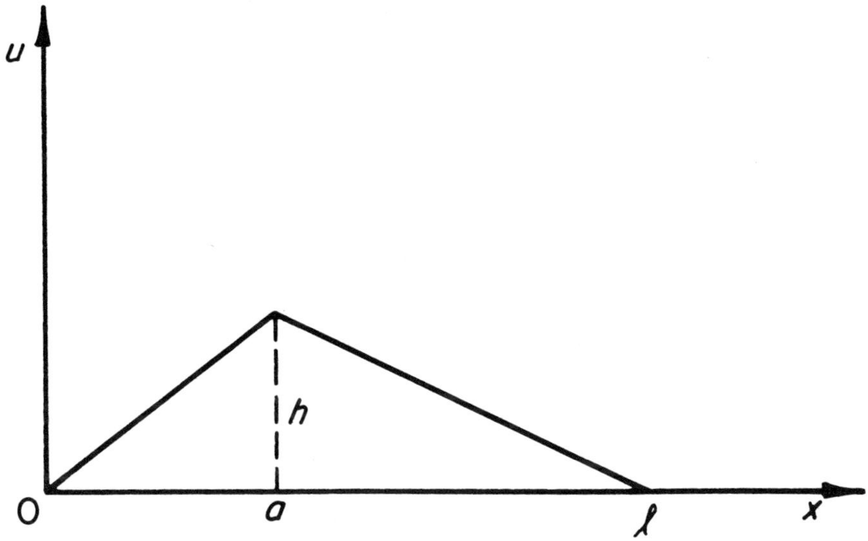

Figure 6.2.1

coefficients a_n, according to Eq. (6.2.19), are given by

$$\begin{aligned} a_n &= \frac{2}{l}\int_0^l f(x)\sin\frac{n\pi x}{l}\,dx \\ &= \frac{2}{l}\int_0^a \frac{hx}{a}\sin\frac{n\pi x}{l}\,dx + \frac{2}{l}\int_a^l \frac{h(l-x)}{(l-a)}\sin\frac{n\pi x}{l}\,dx \end{aligned}$$

Integration and simplification yields

$$a_n = \frac{2hl^2}{\pi^2 a(l-a)}\frac{1}{n^2}\sin\frac{n\pi a}{l}$$

Thus, the displacement of the plucked string is

$$u(x,t) = \frac{2hl^2}{\pi^2 a(l-a)}\sum_{n=1}^{\infty}\frac{1}{n^2}\sin\frac{n\pi a}{l}\sin\frac{n\pi x}{l}\cos\frac{n\pi c}{l}t$$

EXAMPLE 6.2.2. The Struck String

Here, we consider the string with no initial displacement. Let the string be struck at $x=a$ so that the initial velocity is given by

$$u_t(x,0) = \begin{cases} \dfrac{v_0}{a}x, & 0 \leqslant x \leqslant a \\ v_0(l-x)/(l-a), & a \leqslant x \leqslant l \end{cases}$$

Since $u(x,0)=0$, we have $a_n=0$. By applying Eq. (6.2.19), we find that

$$b_n=\frac{2}{n\pi c}\int_0^a \frac{v_0}{a}x\sin\frac{n\pi x}{l}\,dx+\frac{2}{n\pi c}\int_a^l v_0\frac{(l-x)}{(l-a)}\sin\frac{n\pi x}{l}\,dx$$

$$=\frac{2v_0 l^3}{\pi^3 ca(l-a)}\frac{1}{n^3}\sin\frac{n\pi a}{l}$$

Hence the displacement of the struck string is

$$u(x,t)=\frac{2v_0 l^3}{\pi^3 ca(l-a)}\sum_{n=1}^{\infty}\frac{1}{n^3}\sin\frac{n\pi a}{l}\sin\frac{n\pi x}{l}\sin\frac{n\pi c}{l}t$$

6.3 Existence and Uniqueness of Solution of the Vibrating String Problem

In the preceding section, we found that for the initial-boundary value problem

$$\begin{aligned}
&u_{tt}=c^2u_{xx} && 0<x<l && t>0\\
&u(x,0)=f(x) && 0\leqslant x\leqslant l\\
&u_t(x,0)=g(x) && 0\leqslant x\leqslant l\\
&u(0,t)=0 && && t\geqslant 0\\
&u(l,t)=0 && && t\geqslant 0
\end{aligned}\tag{6.3.1}$$

there exists a formal solution

$$u(x,t)=\sum_{n=1}^{\infty}\left(a_n\cos\frac{n\pi c}{l}t+b_n\sin\frac{n\pi c}{l}t\right)\sin\frac{n\pi x}{l}\tag{6.3.2}$$

where a_n and b_n are the coefficients in the series expansions

$$f(x)=\sum_{n=1}^{\infty}a_n\sin\frac{n\pi x}{l}\tag{6.3.3}$$

$$g(x)=\sum_{n=1}^{\infty}b_n\frac{n\pi c}{l}\sin\frac{n\pi x}{l}\tag{6.3.4}$$

on the interval $[0,l]$.

We shall now show that the expression (6.3.2) is the solution of problem (6.3.1) under certain conditions.

First we see that

$$u_1(x,t)=\sum_{n=1}^{\infty}a_n\cos\frac{n\pi c}{l}t\sin\frac{n\pi x}{l}\tag{6.3.5}$$

is the formal solution of the problem

$$\begin{aligned} &u_{tt} = c^2 u_{xx} && 0<x<l && t>0 \\ &u(x,0) = f(x) && 0 \leqslant x \leqslant l \\ &u_t(x,0) = 0 \\ &u(0,t) = 0 \\ &u(l,t) = 0 \end{aligned} \tag{6.3.6}$$

and

$$u_2(x,t) = \sum_{n=1}^{\infty} b_n \sin\frac{n\pi c}{l} t \sin\frac{n\pi x}{l} \tag{6.3.7}$$

is the formal solution of the problem

$$\begin{aligned} &u_{tt} = c^2 u_{xx} && 0<x<l && t>0 \\ &u(x,0) = 0 \\ &u_t(x,0) = g(x) && 0 \leqslant x \leqslant l \\ &u(0,t) = 0 \\ &u(l,t) = 0 \end{aligned} \tag{6.3.8}$$

By the linearity of the problem (6.3.1), the solution (6.3.2) may be considered as the sum of the two formal solutions (6.3.5) and (6.3.7).

We first assume that $f(x)$ and $f'(x)$ are continuous on $[0,l]$, and $f(0) = f(l) = 0$. Then by Theorem 5.12.1, the series for the function $f(x)$ given by Eq. (6.3.3) converges absolutely and uniformly on the interval $[0,l]$.

Using the trigonometric identity

$$\sin\frac{n\pi x}{l}\cos\frac{n\pi c}{l}t = \frac{1}{2}\sin\frac{n\pi}{l}(x-ct) + \frac{1}{2}\sin\frac{n\pi}{l}(x+ct), \tag{6.3.9}$$

$u_1(x,t)$ may be written as

$$u_1(x,t) = \frac{1}{2}\sum_{n=1}^{\infty} a_n \sin\frac{n\pi}{l}(x-ct) + \frac{1}{2}\sum_{n=1}^{\infty} a_n \sin\frac{n\pi}{l}(x+ct)$$

Define

$$F(x) = \sum_{n=1}^{\infty} a_n \sin\frac{n\pi x}{l} \tag{6.3.10}$$

and assume that $F(x)$ is the odd periodic extension of $f(x)$, that is,

$$\begin{aligned} F(x) &= f(x) && 0 \leqslant x \leqslant l \\ F(-x) &= -F(x) && \text{for all } x \\ F(x \pm 2l) &= F(x) \end{aligned}$$

We can now rewrite u_1 in the form

$$u_1(x,t)=\frac{1}{2}\left[F(x-ct)+F(x+ct)\right] \tag{6.3.11}$$

To show that the boundary conditions are satisfied, we note that

$$\begin{aligned} u_1(0,t)&=\frac{1}{2}\left[F(-ct)+F(ct)\right]\\ &=\frac{1}{2}\left[-F(ct)+F(ct)\right]\\ &=0\\ u_1(l,t)&=\frac{1}{2}\left[F(l-ct)+F(l+ct)\right]\\ &=\frac{1}{2}\left[F(-l-ct)+F(l+ct)\right]\\ &=\frac{1}{2}\left[-F(l+ct)+F(l+ct)\right]\\ &=0 \end{aligned}$$

Since

$$\begin{aligned} u_1(x,0)&=\frac{1}{2}\left[F(x)+F(x)\right]\\ &=F(x)\\ &=f(x) \qquad 0\leqslant x\leqslant l \end{aligned}$$

we see that the first initial condition is satisfied. Thus the first three conditions in Eq. (6.3.6) are satisfied. Since f' is continuous on $[0,l]$, F' exists and is continuous for all x. Thus, if we differentiate u_1 with respect to t, we obtain

$$\frac{\partial u_1}{\partial t}=\frac{1}{2}\left[-cF'(x-ct)+cF'(x+ct)\right]$$

and

$$\frac{\partial u_1}{\partial t}(x,0)=\frac{1}{2}\left[-cF'(x)+cF'(x)\right]=0$$

We therefore see that the fourth condition is also satisfied.

In order to show that $u_1(x,t)$ satisfies the differential equation (6.3.6), we impose additional restrictions on f. Let f'' be continuous on $[0,l]$ and let $f''(0)=f''(l)=0$. Then F'' exists and is continuous everywhere, and therefore

$$\frac{\partial^2 u_1}{\partial t^2}=\frac{1}{2}c^2\left[F''(x-ct)+F''(x+ct)\right]$$

$$\frac{\partial^2 u_1}{\partial x^2}=\frac{1}{2}\left[F''(x-ct)+F''(x+ct)\right]$$

We find therefore that

$$\frac{\partial^2 u_1}{\partial t^2} = c^2 \frac{\partial^2 u_1}{\partial x^2}$$

Next, we shall state the assumptions which must be imposed on g to make $u_2(x,t)$ the solution of problem (6.3.8). Let g and g' be continuous on $[0,l]$ and let $g(0)=g(l)=0$. Then the series for the function $g(x)$ given by Eq. (6.3.4) converges absolutely and uniformly in the interval $[0,l]$. Introducing the new coefficient $c_n=(n\pi c/l)b_n$, we have

$$u_2(x,t) = \frac{l}{\pi c}\sum_{n=1}^{\infty} \frac{c_n}{n} \sin\frac{n\pi c}{l} t \sin\frac{n\pi x}{l} \tag{6.3.12}$$

We shall see that term-by-term differentiation with respect to t is permitted, and hence

$$\frac{\partial u_2}{\partial t} = \sum_{n=1}^{\infty} c_n \cos\frac{n\pi c}{l} t \sin\frac{n\pi x}{l} \tag{6.3.13}$$

Using the trigometric identity (6.3.9) we obtain

$$\frac{\partial u_2}{\partial t} = \frac{1}{2}\sum_{n=1}^{\infty} c_n \sin\frac{n\pi}{l}(x-ct) + \frac{1}{2}\sum_{n=1}^{\infty} c_n \sin\frac{n\pi}{l}(x+ct) \tag{6.3.14}$$

These series are absolutely and uniformly convergent because of the assumptions on g, and hence the series (6.3.12) and (6.3.13) converge absolutely and uniformly on $[0,l]$. Thus the term-by-term differentiation is justified.

Let

$$G(x) = \sum_{n=1}^{\infty} c_n \sin\frac{n\pi x}{l}$$

be the odd periodic extension of the function $g(x)$. Then Eq. (6.3.14) can be written in the form

$$\frac{\partial u_2}{\partial t} = \frac{1}{2}\left[G(x-ct)+G(x+ct)\right]$$

Integration yields

$$\begin{aligned} u_2(x,t) &= \frac{1}{2}\int_0^t G(x-ct')\,dt' + \frac{1}{2}\int_0^t G(x+ct')\,dt' \\ &= \frac{1}{2c}\int_{x-ct}^{x+ct} G(\tau)\,d\tau \end{aligned} \tag{6.3.15}$$

It immediately follows that $u_2(x,0)=0$, and

$$\begin{aligned} \frac{\partial u_2}{\partial t}(x,0) &= G(x) \\ &= g(x) \qquad 0 \leqslant x \leqslant l \end{aligned}$$

Moreover,

$$\begin{aligned} u_2(0,t) &= \frac{1}{2}\int_0^t G(-ct')\,dt' + \frac{1}{2}\int_0^t G(ct')\,dt' \\ &= -\frac{1}{2}\int_0^t G(ct')\,dt' + \frac{1}{2}\int_0^t G(ct')\,dt' \\ &= 0 \end{aligned}$$

and

$$\begin{aligned} u_2(l,t) &= \frac{1}{2}\int_0^t G(l-ct')\,dt' + \frac{1}{2}\int_0^t G(l+ct')\,dt' \\ &= \frac{1}{2}\int_0^t G(-l-ct')\,dt' + \frac{1}{2}\int_0^t G(l+ct')\,dt' \\ &= -\frac{1}{2}\int_0^t G(l+ct')\,dt' + \frac{1}{2}\int_0^t G(l+ct')\,dt' \\ &= 0 \end{aligned}$$

Finally, $u_2(x,t)$ must satisfy the differential equation. Since g' is continuous on $[0,l]$, G' exists so that

$$\frac{\partial^2 u_2}{\partial t^2} = \frac{c}{2}\left[-G'(x-ct) + G'(x+ct)\right]$$

Differentiating $u_2(x,t)$ represented by Eq. (6.3.12) with respect to x, we obtain

$$\begin{aligned} \frac{\partial u_2}{\partial x} &= \frac{1}{c}\sum_{n=1}^{\infty} c_n \sin\frac{n\pi c}{l}t \cos\frac{n\pi x}{l} \\ &= \frac{1}{2c}\sum_{n=1}^{\infty} c_n\left[-\sin\frac{n\pi}{l}(x-ct) + \sin\frac{n\pi}{l}(x+ct)\right] \\ &= \frac{1}{2c}\left[-G(x-ct) + G(x+ct)\right] \end{aligned}$$

Differentiating again with respect to x, we obtain

$$\frac{\partial^2 u_2}{\partial x^2} = \frac{1}{2c}\left[-G'(x-ct) + G'(x+ct)\right]$$

It is quite evident that

$$\frac{\partial^2 u_2}{\partial t^2} = c^2\frac{\partial^2 u_2}{\partial x^2}$$

Thus, the solution of the initial-boundary value problem (6.3.1) is established.

Theorem 6.3.1 (Uniqueness) *There exists at most one solution of the wave equation*

$$u_{tt}=c^2u_{xx},\qquad 0<x<l,\qquad t>0$$

satisfying the initial conditions

$$u(x,0)=f(x),\qquad 0\leqslant x\leqslant l$$
$$u_t(x,0)=g(x),\qquad 0\leqslant x\leqslant l$$

and the boundary conditions

$$u(0,t)=0,\qquad t\geqslant 0$$
$$u(l,t)=0,\qquad t\geqslant 0$$

where $u(x,t)$ is a twice continuously differentiable function with respect to both x and t.

PROOF. Suppose that there are two solutions u_1 and u_2 and let $v=u_1-u_2$. It can readily be seen that $v(x,t)$ is the solution of the problem

$$\begin{aligned} &v_{tt}=c^2v_{xx}, && 0<x<l, && t>0\\ &v(0,t)=0 && && t\geqslant 0\\ &v(l,t)=0 && && t\geqslant 0\\ &v(x,0)=0 && 0\leqslant x\leqslant l\\ &v_t(x,0)=0 && 0\leqslant x\leqslant l \end{aligned}$$

We shall prove that the function $v(x,t)$ is identically zero. To do so, consider the function

$$I(t)=\frac{1}{2}\int_0^l\left(c^2v_x^2+v_t^2\right)dx \tag{6.3.16}$$

which physically represents the total energy of the vibrating string at time t.

Since the function $v(x,t)$ is twice continuously differentiable, we differentiate $I(t)$ with respect to t. Thus,

$$\frac{dI}{dt}=\int_0^l\left(c^2v_xv_{xt}+v_tv_{tt}\right)dx \tag{6.3.17}$$

Integrating by parts, we have

$$\int_0^l c^2v_xv_{xt}\,dx=\left[c^2v_xv_t\right]_0^l-\int_0^l c^2v_tv_{xx}\,dx$$

But from the condition $v(0,t)=0$ we have $v_t(0,t)=0$, and similarly $v_t(l,t)=0$ for $x=l$. Hencc the expression in the square bracket vanishes, and Eq. (6.3.17) becomes

$$\frac{dI}{dt}=\int_0^l v_t\left(v_{tt}-c^2v_{xx}\right)dx \tag{6.3.18}$$

Since $v_{tt} - c^2 v_{xx} = 0$, Eq. (6.3.18) reduces to

$$\frac{dI}{dt} = 0$$

which means

$$I(t) = \text{constant} = C$$

Since $v(x,0) = 0$ we have $v_x(x,0) = 0$. Taking into account the condition $v_t(x,0) = 0$, we evaluate C to obtain

$$I(0) = C = \int_0^t \left[c^2 v_x^2 + v_t^2\right]_{t=0} dx = 0$$

This implies that $I(t) = 0$ which can happen only when $v_x = 0$ and $v_t = 0$ for $t > 0$. To satisfy both of these conditions, we must have $v(x,t) = \text{constant}$. Employing the condition $v(x,0) = 0$, we then find $v(x,t) = 0$. Therefore $u_1(x,t) = u_2(x,t)$ and the solution $u(x,t)$ is unique. □

6.4 The Heat Conduction Problem

We consider a homogeneous rod of length l. The rod is sufficiently thin so that the heat is distributed equally over the cross section at time t. The surface of the rod is insulated, and therefore there is no heat loss through the boundary. The temperature distribution of the rod is given by the solution of the initial-boundary value problem

$$\begin{aligned} u_t &= k u_{xx}, \quad 0 < x < l, \quad t > 0 \\ u(0,t) &= 0, \quad t \geqslant 0 \\ u(l,t) &= 0, \quad t \geqslant 0 \\ u(x,0) &= f(x) \quad 0 \leqslant x \leqslant l \end{aligned} \tag{6.4.1}$$

If we assume a solution in the form

$$u(x,t) = X(x)T(t)$$

Eq. (6.4.1) yields

$$XT' = kX''T$$

Thus, we have

$$\frac{X''}{X} = \frac{T'}{kT} = -\alpha^2$$

where α is a positive constant. Hence X and T must satisfy

$$X'' + \alpha^2 X = 0 \tag{6.4.2}$$

$$T' + \alpha^2 k T = 0 \tag{6.4.3}$$

From the boundary conditions, we have

$$u(0,t) = X(0)T(t) = 0$$

$$u(l,t) = X(l)T(t) = 0$$

Thus,

$$X(0)=0, \qquad X(l)=0$$

for arbitrary $T(t)$. Hence we must solve the eigenvalue problem

$$X''+\alpha^2 X=0$$
$$X(0)=0, \qquad X(l)=0$$

The solution of Eq. (6.4.2)

$$X(x)=A\cos\alpha x+B\sin\alpha x$$

Since $X(0)=0$, $A=0$. To satisfy the second condition, we have

$$X(l)=B\sin\alpha l=0$$

Since $B=0$ yields a trivial solution, we must have

$$\sin\alpha l=0$$

Thus,

$$\alpha=\frac{n\pi}{l} \qquad \text{for } n=1,2,3,\ldots$$

Substituting these eigenvalues, we have

$$X_n(x)=B_n\sin\frac{n\pi x}{l}$$

Next, we consider Eq. (6.4.3), namely,

$$T'+\alpha^2 kT=0$$

the solution of which is

$$T(t)=Ce^{-\alpha^2 kt}$$

Substituting $\alpha=n\pi/l$, we have

$$T_n(t)=C_n e^{-\left(\frac{n\pi}{l}\right)^2 kt}$$

Hence the nontrivial solution of the heat equation which satisfies the two boundary conditions is

$$u_n(x,t)=X_n(x)T_n(t)=a_n e^{-\left(\frac{n\pi}{l}\right)^2 kt}\sin\frac{n\pi x}{l}, \qquad n=1,2,3\ldots$$

where $a_n=B_nC_n$ is an arbitrary constant.

We formally form a series

$$u(x,t)=\sum_{n=1}^{\infty} a_n e^{-\left(\frac{n\pi}{l}\right)^2 kt}\sin\frac{n\pi x}{l} \tag{6.4.4}$$

which satisfies the initial condition if

$$u(x,0)=f(x)=\sum_{n=1}^{\infty} a_n\sin\frac{n\pi x}{l}$$

This holds true if $f(x)$ can be represented by a Fourier sine series with

Fourier coefficients

$$a_n = \frac{2}{l}\int_0^l f(x)\sin\frac{n\pi x}{l}\,dx$$

Hence

$$u(x,t) = \sum_{n=1}^{\infty}\left[\frac{2}{l}\int_0^l f(\tau)\sin\frac{n\pi\tau}{l}\,d\tau\right]e^{-\left(\frac{n\pi}{l}\right)^2 kt}\sin\frac{n\pi x}{l} \tag{6.4.5}$$

is the formal solution of the heat conduction problem.

EXAMPLE 6.4.1. (a) Suppose the initial temperature distribution is $f(x) = x(l-x)$. Then from Eq. (6.4.5), we have

$$a_n = \frac{8l^2}{n^3\pi^3}, \qquad n = 1, 3, 5, \ldots$$

Thus, the solution is

$$u(x,t) = \frac{8l^2}{\pi^3}\sum_{n=1,3,5,\ldots}^{\infty}\frac{1}{n^3}e^{-\left(\frac{n\pi}{l}\right)^2 kt}\sin\frac{n\pi x}{l}$$

(b) Suppose the temperature at one end of the rod is held constant, that is,

$$u(l,t) = u_0, \qquad t \geqslant 0$$

The problem here is

$$\begin{aligned} u_t &= ku_{xx}, \quad 0<x<l, \quad t>0 \\ u(0,t) &= 0 \\ u(l,t) &= u_0 \\ u(x,0) &= f(x) \quad 0<x<l \end{aligned} \tag{6.4.6}$$

Let

$$u(x,t) = v(x,t) + \frac{u_0 x}{l}$$

Substitution of $u(x,t)$ in problem (6.4.6) yields

$$\begin{aligned} v_t &= kv_{xx}, & 0<x<l, \quad t>0 \\ v(0,t) &= 0 \\ v(l,t) &= 0 \\ v(x,0) &= f(x) - \frac{u_0 x}{l}, & 0<x<l \end{aligned}$$

Hence with the knowledge of solution (6.4.5), we obtain

$$u(x,t) = \sum_{n=1}^{\infty}\left[\frac{2}{l}\int_0^l\left(f - \frac{u_0 x}{l}\right)\sin\frac{n\pi\tau}{l}\,d\tau\right]e^{-\left(\frac{n\pi}{l}\right)^2 kt}\sin\frac{n\pi x}{l} + \frac{u_0 x}{l}$$

6.5 Existence and Uniqueness of Solution of the Heat Conduction Problem

In the preceding section, we found that

$$u(x,t)=\sum_{n=1}^{\infty} a_n e^{-\left(\frac{n\pi}{l}\right)^2 kt}\sin\frac{n\pi x}{l} \tag{6.5.1}$$

is the formal solution of the heat conduction problem

$$\begin{aligned} u_t &= ku_{xx}, \quad 0<x<l, \quad t>0 \\ u(0,t)&=0 \\ u(l,t)&=0 \\ u(x,0)&=f(x), \quad 0\leqslant x\leqslant l \end{aligned} \tag{6.5.2}$$

where the coefficients a_n are determined from the series expansion

$$f(x)=\sum_{n=1}^{\infty} a_n \sin\frac{n\pi x}{l} \tag{6.5.3}$$

We shall show that this formal solution is the solution if $f(x)$ is continuous in $[0,l]$ and $f(0)=f(l)=0$, and $f'(x)$ is piecewise continuous in $(0,l)$. Since $f(x)$ is bounded, we have

$$|a_n|=\frac{2}{l}\left|\int_0^l f(x)\sin\frac{n\pi x}{l}\,dx\right|\leqslant\frac{2}{l}\int_0^l|f(x)|\,dx\leqslant C$$

where C is a positive constant. Thus, for any $t_0>0$

$$\left|a_n e^{-\left(\frac{n\pi}{l}\right)^2 kt}\sin\frac{n\pi x}{l}\right|\leqslant Ce^{-\left(\frac{n\pi}{l}\right)^2 kt_0} \text{ when } t\geqslant t_0$$

According to the ratio test, the series of the constant terms $\exp[-(n\pi/l)^2kt_0]$ converges. Hence by the Weierstrass M-test, the series (6.5.1) converges uniformly with respect to x and t whenever $t\geqslant t_0$ and $0\leqslant x\leqslant l$.

Differentiating Eq. (6.5.1) termwise with respect to t, we obtain

$$u_t=-\sum_{n=1}^{\infty} a_n\left(\frac{n\pi}{l}\right)^2 ke^{-\left(\frac{n\pi}{l}\right)^2 kt}\sin\frac{n\pi x}{l} \tag{6.5.4}$$

We note that

$$\left|-a_n\left(\frac{n\pi}{l}\right)^2 ke^{-\left(\frac{n\pi}{l}\right)^2 kt}\sin\frac{n\pi x}{l}\right|\leqslant C\left(\frac{n\pi}{l}\right)^2 ke^{-\left(\frac{n\pi}{l}\right)^2 kt_0}$$

when $t\geqslant t_0$, and the series of the constant terms $C(n\pi/l)^2k\exp[-(n\pi/l)^2kt_0]$ converges by the ratio test. Hence Eq. (6.5.4) is uniformly convergent in the region $0\leqslant x\leqslant l$, $t\geqslant t_0$. In a similar manner, the series

(6.5.1) can be differentiated twice with respect to x, and as a result

$$u_{xx}=-\sum_{n=1}^{\infty} a_n\left(\frac{n\pi}{l}\right)^2 e^{-\left(\frac{n\pi}{l}\right)^2 kt}\sin\frac{n\pi x}{l} \tag{6.5.5}$$

Evidently, from Eqs. (6.5.4) and (6.5.5),

$$u_t = ku_{xx}$$

Hence Eq. (6.5.1) is a solution of the one-dimensional heat equation in the region $0\leqslant x\leqslant l$, $t>0$.

Next, we show that the boundary conditions are satisfied. Here, we note that the series (6.5.1) representing the function $u(x,t)$ converges uniformly in the region $0\leqslant x\leqslant l$, $t>0$. Since the function represented by a uniformly convergent series of continuous functions is continuous, $u(x,t)$ is continuous at $x=0$ and $x=l$. As a consequence, when $x=0$ and $x=l$, Eq. (6.5.1) reduces to

$$u(0,t)=0$$
$$u(l,t)=0$$

for all $t>0$.

It remains to show that $u(x,t)$ satisfies the initial condition

$$u(x,0)=f(x), \qquad 0\leqslant x\leqslant l.$$

Under the assumptions stated earlier,

$$f(x)=\sum_{n=1}^{\infty} a_n\sin\frac{n\pi x}{l}$$

is uniformly and absolutely convergent. According to Abel's test [6], the series formed by the product of the terms of a uniformly convergent series

$$\sum_{n=1}^{\infty} a_n\sin\frac{n\pi x}{l}$$

and the members of a uniformly bounded and monotone sequence $\exp[-(n\pi/l)^2kt]$ converges uniformly with respect to t. Hence

$$u(x,t)=\sum_{n=1}^{\infty} a_n e^{-\left(\frac{n\pi}{l}\right)^2 kt}\sin\frac{n\pi x}{l}$$

converges uniformly for $0\leqslant x\leqslant l$, $t\geqslant 0$, and by the same reasoning as before, $u(x,t)$ is continuous for $0\leqslant x\leqslant l$, $t\geqslant 0$. Thus, the initial condition

$$u(x,0)=f(x) \qquad 0\leqslant x\leqslant l$$

is satisfied. The solution is therefore established.

In the above discussion the condition imposed on $f(x)$ is stronger than necessary. The solution can be obtained with a less stringent condition on $f(x)$ (see Ref. 47).

Theorem 6.5.1 (Uniqueness) *Let $u(x,t)$ be a continuously differentiable function. If $u(x,t)$ satisfies the differential equation*

$$u_t = ku_{xx}, \qquad 0<x<l, \qquad t>0$$

the initial condition

$$u(x,0)=f(x), \qquad 0\leqslant x\leqslant l$$

and the boundary conditions

$$u(0,t)=0, \qquad t\geqslant 0$$
$$u(l,t)=0, \qquad t\geqslant 0$$

then it is unique.

PROOF. Suppose there are two solutions $u_1(x,t)$ and $u_2(x,t)$. Let

$$v(x,t)=u_1(x,t)-u_2(x,t).$$

Then

$$\begin{aligned} v_t &= kv_{xx}, & 0<x<l, \quad & t>0 \\ v(0,t)&=0 & & t\geqslant 0 \\ v(l,t)&=0 & & t\geqslant 0 \\ v(x,0)&=0 & 0\leqslant x\leqslant l & \end{aligned} \tag{6.5.6}$$

Consider the function

$$J(t)=\frac{1}{2k}\int_0^l v^2 dx$$

Differentiating with respect to t, we have

$$J'(t)=\frac{1}{k}\int_0^l vv_t dx=\int_0^l vv_{xx}dx$$

by virtue of Eq. (6.5.6). Integrating by parts, we have

$$\int_0^l vv_{xx}dx=[vv_x]_0^l-\int_0^l v_x^2 dx$$

Since

$$v(0,t)=v(l,t)=0,$$
$$J'(t)=-\int_0^l v_x^2 dx\leqslant 0$$

From the condition $v(x,0)=0$, we have $J(0)=0$. This condition and $J'(t)\leqslant 0$ implies that $J(t)$ is a nonincreasing function of t. Thus,

$$J(t)\leqslant 0$$

But by the definition of $J(t)$,

$$J(t)\geqslant 0$$

Hence

$$J(t)=0 \qquad \text{for } t\geqslant 0$$

Since $v(x,t)$ is continuous, $J(t)=0$ implies

$$v(x,t)=0$$

in $0\leqslant x\leqslant l$, $t\geqslant 0$. Therefore, $u_1=u_2$ and the solution is unique. □

6.6 The Laplace and Beam Equations

EXAMPLE 6.6.1. Consider the steady state temperature distribution in a thin rectangular slab. Two sides are insulated, one side is maintained at zero temperature, and the temperature of the remaining side is prescribed to be $f(x)$. Thus, we are required to solve

$$\begin{aligned} &\nabla^2 u=0, && 0<x<a, \quad 0<y<b \\ &u(x,0)=f(x), && 0\leqslant x\leqslant a \\ &u(x,b)=0 \\ &u_x(0,y)=0 \\ &u_x(a,y)=0 \end{aligned}$$

Let $u(x,y)=X(x)Y(y)$. Substitution of this into the Laplace equation yields

$$X''-\lambda X=0$$
$$Y''+\lambda Y=0$$

Since the boundary conditions are homogeneous on $x=0$ and $x=a$, we have $\lambda=-\alpha^2$ with $\alpha\geqslant 0$ for nontrivial solutions of the eigenvalue problem

$$X''+\alpha^2 X=0$$
$$X'(0)=X'(a)=0$$

The solution is

$$X(x)=A\cos\alpha x+B\sin\alpha x$$

Application of the boundary conditions then yields $B=0$ and $\alpha=n\pi/a$ with $n=0,1,2,\ldots$ Hence

$$X_n(x)=A_n\cos\frac{n\pi x}{a}$$

The solution of the Y equation is clearly

$$Y(y)=C\cosh\alpha y+D\sinh\alpha y$$

which can be written in the form

$$Y(y)=E\sinh\alpha(y+F)$$

where $E=(D^2-C^2)^{1/2}$ and $F=[\tanh^{-1}(C/D)]/\alpha$.

Applying the homogeneous boundary condition $Y(b)=0$, we obtain

$$Y(b)=E\sinh\alpha(b+F)=0$$

which implies

$$F=-b, \qquad E\neq 0$$

for nontrivial solutions. Hence we have

$$u(x,y)=\frac{(b-y)}{b}\frac{a_0}{2}+\sum_{n=1}^{\infty}a_n\cos\frac{n\pi x}{a}\sinh\frac{n\pi}{a}(y-b).$$

Now we apply the remaining nonhomogeneous condition. We obtain

$$u(x,0)=f(x)=\frac{a_0}{2}+\sum_{n=1}^{\infty}a_n\cos\frac{n\pi x}{a}\sinh\left(-\frac{n\pi b}{a}\right)$$

Since this is a Fourier cosine series, the coefficients are given by

$$a_0=\frac{2}{a}\int_0^a f(x)\,dx$$

$$a_n=\frac{-2}{a\sinh\dfrac{n\pi b}{a}}\int_0^a f(x)\cos\frac{n\pi x}{a}\,dx \qquad n=1,2,\ldots$$

Thus the formal solution is

$$u(x,y)=\left(\frac{b-y}{b}\right)\frac{a_0}{2}+\sum_{n=1}^{\infty}a_n^*\frac{\sinh\dfrac{n\pi}{a}(b-y)}{\sinh\dfrac{n\pi b}{a}}\cos\frac{n\pi x}{a}$$

where

$$a_n^*=\frac{2}{a}\int_0^a f(x)\cos\frac{n\pi x}{a}\,dx$$

If for example $f(x)=x$ in $0<x<\pi$, $0<y<\pi$, then we find (note that $a=\pi$)

$$a_0=\pi$$

$$a_n^*=\frac{2}{\pi n^2}\left[(-1)^n-1\right], \qquad n=1,2,\ldots$$

and hence

$$u(x,y)=\tfrac{1}{2}(\pi-y)+\sum_{n=1}^{\infty}\frac{2}{\pi n^2}\left[(-1)^n-1\right]\frac{\sinh n(\pi-y)}{\sinh n\pi}\cos nx$$

EXAMPLE 6.6.2. As another example, we consider the transverse vibration of a beam. The equation of motion is governed by

$$u_{tt}+a^2u_{xxxx}=0, \qquad 0<x<l, \qquad t>0$$

where $u(x,t)$ is the displacement and a is the physical constant. Note that the equation is of the fourth order in x. Let the initial and boundary conditions be

$$\begin{aligned} u(x,0)&=f(x) && 0\leqslant x\leqslant l\\ u_t(x,0)&=g(x) && 0\leqslant x\leqslant l\\ u(0,t)&=u(l,t)=0 &&\\ u_{xx}(0,t)&=u_{xx}(l,t)=0 && \end{aligned} \tag{6.6.1}$$

The boundary conditions represent the beam being simple supported, that is, the displacements and the bending moments at the ends are zero.

Assume a solution in the form

$$u(x,t)=X(x)T(t)$$

which transforms the equation of motion into

$$T''+a^2\alpha^4T=0, \qquad \alpha>0$$

$$X''-\alpha^4X=0$$

The equation for $X(x)$ has the general solution

$$X(x)=A\cosh\alpha x+B\sinh\alpha x+C\cos\alpha x+D\sin\alpha x$$

The boundary conditions require that

$$X(0)=X(l)=0$$

$$X''(0)=X''(l)=0$$

Differentiating X twice with respect to x, we obtain

$$X''(x)=A\alpha^2\cosh\alpha x+B\alpha^2\sinh\alpha x-C\alpha^2\cos\alpha x-D\alpha^2\sin\alpha x$$

Now applying the conditions $X(0)=X''(0)=0$, we obtain

$$A+C=0$$

$$\alpha^2(A-C)=0$$

and hence

$$A=C=0$$

The conditions $X(l)=X''(l)=0$ yields

$$B\sinh\alpha l+D\sin\alpha l=0$$

$$B\sinh\alpha l-D\sin\alpha l=0$$

These equations are satisfied if

$$B\sinh\alpha l=0, \qquad D\sin\alpha l=0$$

Since $\sinh\alpha l\neq0$, B must vanish. For nontrivial solution,

$$\sin\alpha l=0, \qquad D\neq0$$

and hence

$$\alpha=\frac{n\pi}{l} \qquad n=1,2,3,\ldots$$

We then obtain

$$X_n(x)=D_n\sin\frac{n\pi x}{l}$$

The general solution for $T(t)$ is

$$T(t)=E\cos a\alpha^2 t+F\sin a\alpha^2 t$$

Inserting the values of α^2, we obtain

$$T_n(t)=E_n\cos a\left(\frac{n\pi}{l}\right)^2 t+F_n\sin a\left(\frac{n\pi}{l}\right)^2 t$$

Thus, the general solution of the equation for transverse vibration of a beam is

$$u(x,t)=\sum_{n=1}^{\infty}\left[a_n\cos a\left(\frac{n\pi}{l}\right)^2 t+b_n\sin a\left(\frac{n\pi}{l}\right)^2 t\right]\sin\frac{n\pi x}{l} \tag{6.6.2}$$

To satisfy the initial condition $u(x,0)=f(x)$, we must have

$$u(x,0)=f(x)=\sum_{n=1}^{\infty}a_n\sin\frac{n\pi x}{l}$$

from which we find

$$a_n=\frac{2}{l}\int_0^l f(x)\sin\frac{n\pi x}{l}\,dx \tag{6.6.3}$$

Now the application of the second initial condition gives

$$u_t(x,0)=g(x)=\sum_{n=1}^{\infty}b_n a\left(\frac{n\pi}{l}\right)^2\sin\frac{n\pi x}{l}$$

and hence

$$b_n=\frac{2}{al}\left(\frac{l}{n\pi}\right)^2\int_0^l g(x)\sin\frac{n\pi x}{l}\,dx \tag{6.6.4}$$

Thus, the solution of the initial-boundary value problem is given by Eqs. (6.6.2)–(6.6.4).

6.7 Nonhomogeneous Problems

The partial differential equations considered so far in this chapter are homogeneous. In practice, there is a very important class of problems involving nonhomogeneous equations. First, we shall illustrate a problem involving time-independent nonhomogeneous equation.

EXAMPLE 6.7.1. Consider the initial-boundary value problem

$$\begin{aligned}
&u_{tt}=c^2u_{xx}+F(x), && 0<x<l, && t>0\\
&u(x,0)=f(x) && 0\leqslant x\leqslant l &&\\
&u_t(x,0)=g(x) && 0\leqslant x\leqslant l &&\\
&u(0,t)=A && && t>0\\
&u(l,t)=B && && t>0
\end{aligned} \tag{6.7.1}$$

We assume a solution in the form

$$u(x,t)=v(x,t)+U(x)$$

Substitution of $u(x,t)$ in Eq. (6.7.1) yields

$$v_{tt}=c^2(v_{xx}+U_{xx})+F(x)$$

so that if $U(x)$ satisfies

$$c^2U_{xx}+F(x)=0$$

then

$$v_{tt}=c^2v_{xx}$$

In a similar manner, if $u(x,t)$ is inserted in the initial and boundary conditions, we obtain

$$\begin{aligned} u(x,0)&=v(x,0)+U(x) &&=f(x)\\ u_t(x,0)&=u_t(x,0) &&=g(x)\\ u(0,t)&=v(0,t)+U(0) &&=A\\ u(l,t)&=v(l,t)+U(l) &&=B \end{aligned}$$

Thus, if $U(x)$ is the solution of the problem

$$\begin{aligned} c^2U_{xx}+F&=0\\ U(0)&=A\\ U(l)&=B \end{aligned}$$

then $v(x,t)$ must satisfy

$$\begin{aligned} v_{tt}&=c^2v_{xx}\\ v(x,0)&=f(x)-U(x)\\ v_t(x,0)&=g(x)\\ v(0,t)&=0\\ v(l,t)&=0 \end{aligned} \tag{6.7.2}$$

Now $v(x,t)$ can be solved easily as $U(x)$ is known. It can be seen that

$$U(x)=A+(B-A)\frac{x}{l}+\frac{x}{l}\int_0^l\left[\frac{1}{c^2}\int_0^\eta F(\xi)\,d\xi\right]d\eta$$
$$-\int_0^x\left[\frac{1}{c^2}\int_0^\eta F(\xi)\,d\xi\right]d\eta$$

As a specific example, consider the problem

$$\begin{aligned} u_{tt}&=c^2u_{xx}+h, \quad h \text{ is a constant}\\ u(x,0)&=0\\ u_t(x,0)&=0\\ u(0,t)&=0\\ u(l,t)&=0 \end{aligned} \tag{6.7.3}$$

Then the solution of

$$c^2U_{xx} + h = 0$$
$$U(0) = 0$$
$$U(l) = 0$$

is

$$U(x) = \frac{h}{2c^2}(lx - x^2)$$

The function $v(x,t)$ must satisfy

$$v_{tt} = c^2 v_{xx}$$
$$v(x,0) = -\frac{h}{2c^2}(lx - x^2)$$
$$v_t(x,0) = 0$$
$$v(0,t) = 0$$
$$v(l,t) = 0$$

The solution is given [Sec. 6.2 with $g(x) = 0$] by

$$v(x,t) = \sum_{n=1}^{\infty} a_n \cos\frac{n\pi c}{l} t \sin\frac{n\pi x}{l}$$

and the coefficient is

$$a_n = \frac{2}{l}\int_0^l \left[-\frac{h}{2c^2}(lx - x^2) \right] \sin\frac{n\pi x}{l}\, dx$$
$$= -\frac{4l^2h}{n^3\pi^3c^2} \qquad \text{for } n \text{ odd}$$
$$= 0 \qquad \text{for } n \text{ even}$$

The solution of the given initial-boundary value problem is, therefore, given by

$$u(x,t) = v(x,t) + U(x)$$
$$+ \frac{hx}{2c^2}(l-x) \sum_{n=1}^{\infty} \left(-\frac{4l^2h}{c^2\pi^3} \right) \frac{\cos(2n-1)\pi ct/l}{(2n-1)^3} \sin(2n-1)\pi x/l \qquad (6.7.4)$$

Let us now consider the problem of a finite string with an external force acting on it. If the ends are fixed, we have

$$\begin{aligned} u_{tt} - c^2u_{xx} &= h(x,t), & 0<x<l, \quad & t>0 \\ u(x,0) &= f(x), & 0 \leqslant x \leqslant l & \\ u_t(x,0) &= g(x), & 0 \leqslant x \leqslant l & \\ u(0,t) &= 0, & & t \geqslant 0 \\ u(l,t) &= 0, & & t \geqslant 0 \end{aligned} \qquad (6.7.5)$$

We assume a solution in the form[6]

$$u(x,t)=\sum_{n=1}^{\infty} u_n(t)\sin\frac{n\pi x}{l} \tag{6.7.6}$$

where the functions $u_n(t)$ are to be determined. It is evident that the boundary conditions are satisfied. Let us also assume that

$$h(x,t)=\sum_{n=1}^{\infty} h_n(t)\sin\frac{n\pi x}{l} \tag{6.7.7}$$

Thus,

$$h_n(t)=\frac{2}{l}\int_0^l h(x,t)\sin\frac{n\pi x}{l}\,dx \tag{6.7.8}$$

We assume that the series (6.7.6) is suitably convergent. We then find u_{tt} and u_{xx} from (6.7.6), and substitution of these values into (6.7.5) yields

$$\sum_{n=1}^{\infty}\left[u_n''(t)+\lambda_n^2 u_n(t)\right]\sin\frac{n\pi x}{l}=\sum_{n=1}^{\infty} h_n(t)\sin\frac{n\pi x}{l}$$

where $\lambda_n = n\pi c/l$. Multiplying both sides of this equation by $\sin m\pi x/l$, $m=1,2,3,\ldots$, and integrating from $x=0$ to $x=l$, we obtain

$$u_n''(t)+\lambda_n^2 u_n(t)=h_n(t)$$

the solution of which is

$$u_n(t)=a_n\cos\lambda_n t+b_n\sin\lambda_n t+\frac{1}{\lambda_n}\int_0^t h_n(\tau)\sin[\lambda_n(t-\tau)]\,d\tau \tag{6.7.9}$$

Hence, the formula solution is

$$u(x,t)=\sum_{n=1}^{\infty}\left\{a_n\cos\lambda_n t+b_n\sin\lambda_n t+\frac{1}{\lambda_n}\int_0^t h_n(\tau)\sin[\lambda_n(t-\tau)]\,d\tau\right\}\cdot\sin\frac{n\pi x}{l} \tag{6.7.10}$$

Applying the initial conditions, we have

$$u(x,0)=f(x)=\sum_{n=1}^{\infty} a_n\sin\frac{n\pi x}{l}$$

Thus,

$$a_n=\frac{2}{l}\int_0^l f(x)\sin\frac{n\pi x}{l}\,dx \tag{6.7.11}$$

---

[6]The functions $\sin n\pi x/l$ are the eigenfunctions of the associated eigenvalue problem. See Chapter 9, Sec. 10.

Similarly,

$$u_t(x,0)=g(x)=\sum_{n=1}^{\infty} b_n\lambda_n \sin\frac{n\pi x}{l}$$

Thus,

$$b_n=\frac{2}{l\lambda_n}\int_0^l g(x)\sin\frac{n\pi x}{l}\,dx \tag{6.7.12}$$

Hence the formal solution of the initial-boundary value problem (6.7.5) is given by (6.7.10) with a_n given by (6.7.11) and b_n given by (6.7.12).

EXAMPLE 6.7.2. Determine the solution of the initial-boundary value problem

$$\begin{aligned}
u_{tt}-u_{xx}&=h, && 0<x<1, \quad t>0, \quad h=\text{constant}\\
u(x,0)&=x(1-x), && 0\leqslant x\leqslant 1\\
u_t(x,0)&=0, && 0\leqslant x\leqslant 1\\
u(0,t)&=0, && t\geqslant 0\\
u(1,t)&=0, && t\geqslant 0
\end{aligned} \tag{6.7.13}$$

In this case, $b_n=0$ and a_n is given by

$$\begin{aligned}
a_n&=2\int_0^1 x(1-x)\sin n\pi x\,dx\\
&=\frac{4}{(n\pi)^3}\left[1-(-1)^n\right]
\end{aligned}$$

We also have

$$\begin{aligned}
h_n&=2\int_0^1 h\sin\frac{n\pi x}{l}\,dx\\
&=\frac{2h}{n\pi}\left[1-(-1)^n\right]
\end{aligned}$$

Hence

$$\phi_n(t)=\frac{1}{\lambda_n}\int_0^t h_n\sin[\lambda_n(t-\tau)]\,d\tau=\frac{2h}{n\pi\lambda_n^2}\left[1-(-1)^n\right](1-\cos\lambda_n t)$$

The solution is thus given by

$$\begin{aligned}
u(x,t)=\sum_{n=1}^{\infty}\Bigg\{&\frac{4}{n^3\pi^3}\left[1-(-1)^n\right]\cos n\pi t\\
&+\frac{2h}{n^3\pi^3}\left[1-(-1)^n\right](1-\cos n\pi t)\Bigg\}\cdot\sin n\pi x
\end{aligned} \tag{6.7.14}$$

We have treated the initial-boundary problem with the fixed end conditions. Problems with other boundary conditions can also be solved in a similar manner.

We will now consider the initial-boundary value problem with the time-dependent boundary conditions, namely,

$$\begin{aligned}
u_{tt}-c^2u_{xx}&=h(x,t), && 0<x<l, && t>0\\
u(x,0)&=f(x), && 0\leqslant x\leqslant l\\
u_t(x,0)&=g(x), && 0\leqslant x\leqslant l\\
u(0,t)&=p(t), && && t\geqslant 0\\
u(l,t)&=q(t), && && t\geqslant 0
\end{aligned}\tag{6.7.15}$$

We assume a solution in the form

$$u(x,t)=v(x,t)+U(x,t)\tag{6.7.16}$$

Substituting this into Eq. (6.7.15), we obtain

$$v_{tt}-c^2v_{xx}=h-U_{tt}+c^2U_{xx}$$

From the initial and boundary conditions, we have

$$\begin{aligned}
v(x,0)&=f(x)-U(x,0)\\
v_t(x,0)&=g(x)-U_t(x,0)\\
v(0,t)&=p(t)-U(0,t)\\
v(l,t)&=q(t)-U(l,t)
\end{aligned}$$

In order to make the boundary conditions homogeneous, we set

$$\begin{aligned}
U(0,t)&=p(t)\\
U(l,t)&=q(t)
\end{aligned}$$

Thus, $U(x,t)$ must take the form

$$U(x,t)=p(t)+\frac{x}{l}\left[q(t)-p(t)\right]\tag{6.7.17}$$

The problem now is to find the function $v(x,y)$ which satisfies

$$\begin{aligned}
v_{tt}-c^2v_{xx}&=h-U_{tt}=H(x,t)\\
v(x,0)&=f(x)-U(x,0)=F(x)\\
v_t(x,0)&=g(x)-U_t(x,0)=G(x)\\
v(0,t)&=0\\
v(l,t)&=0
\end{aligned}\tag{6.7.18}$$

This is the same type of problem with the homogeneous boundary conditions that has previously been treated.

EXAMPLE 6.7.3. Find the solution of the problem

$$\begin{aligned}
u_{tt}-u_{xx}&=h, && 0<x<1, && t>0, && h=\text{constant}\\
u(x,0)&=x(1-x), && 0\leqslant x\leqslant 1\\
u_t(x,0)&=0, && 0\leqslant x\leqslant 1\\
u(0,t)&=t, && && t\geqslant 0\\
u(1,t)&=\sin t, && && t\geqslant 0
\end{aligned}\tag{6.7.19}$$

In this case,

$$U(x,t)=t+x[\sin t-t] \tag{6.7.20}$$

Then v must satisfy

$$\begin{aligned} v_{tt}-v_{xx}&=h+x\sin t\\ v(x,0)&=x(1-x)\\ v_t(x,0)&=1\\ v(0,t)&=0\\ v(1,t)&=0 \end{aligned} \tag{6.7.21}$$

From (6.7.8)

$$\begin{aligned} h_n(t)&=2\int_0^1 (h+x\sin t)\sin n\pi x\,dx\\ &=\frac{2h}{n\pi}\left[1-(-1)^n\right]+\frac{2(-1)^{n+1}}{n\pi}\sin t\\ &=a+b\sin t\,(\text{say}) \end{aligned} \tag{6.7.22}$$

We also find

$$\begin{aligned} a_n&=2\int_0^1 x(1-x)\sin n\pi x\,dx\\ &=\frac{4}{(n\pi)^3}\left[1-(-1)^n\right] \end{aligned}$$

and

$$\begin{aligned} b_n&=\frac{2}{n\pi}\int_0^1 \sin n\pi x\,dx\\ &=\frac{2}{(n\pi)^2}\left[1-(-1)^n\right] \end{aligned}$$

Then we determine

$$\begin{aligned} \phi_n(t)&=\frac{1}{n\pi}\int_0^t (a+b\sin\tau)\sin[n\pi(t-\tau)]\,d\tau\\ &=\frac{1}{n\pi}\left\{\frac{a}{n\pi}(1-\cos n\pi t)+\frac{b}{4}\left[(\sin 2t-2t)\cos n\pi t\right.\right.\\ &\quad\left.\left.-(\cos 2t-1)\sin n\pi t\right]\right\} \end{aligned} \tag{6.7.23}$$

Hence the solution of the problem (6.7.21) is

$$v(x,t)=\sum_{n=1}^{\infty} a_n\cos n\pi t+b_n\sin n\pi t+\phi_n(t) \tag{6.7.24}$$

Thus, the solution of problem (6.7.19) is given by

$$u(x,t)=v(x,t)+U(x,t)$$

where $v(x,t)$ is given by (6.7.24) and $U(x,t)$ is given by (6.7.20).

6.8 Finite Fourier Transforms

The finite Fourier transforms are often used in determining solutions of nonhomogeneous problems. These finite transforms, namely, the sine and cosine transforms, follow immediately from the theory of Fourier series.

Let $f(x)$ be a piecewise continuous function in a finite interval, say, $(0,\pi)$. This interval is introduced for convenience, and the change of interval can be made without difficulty.

The finite Fourier sine transform denoted by $F_s(n)$ of the function $f(x)$ may be defined by

$$\mathcal{F}[f(x)]=\frac{2}{\pi}\int_0^\pi f(x)\sin nx\,dx \qquad n=1,2,3,\ldots \tag{6.8.1}$$

and the inverse of the transform follows at once from the Fourier sine series; that is

$$f(x)=\sum_{n=1}^{\infty} F_s(n)\sin nx \tag{6.8.2}$$

The finite Fourier cosine transform $F_c(n)$ of $f(x)$ may be defined by

$$\mathcal{F}[f(x)]=\frac{2}{\pi}\int_0^\pi f(x)\cos nx\,dx \qquad n=0,1,2,\ldots \tag{6.8.3}$$

The inverse of the transform is given by

$$f(x)=\frac{F_c(0)}{2}+\sum_{n=1}^{\infty} F_c(n)\cos nx \tag{6.8.4}$$

Theorem 6.8.1 *Let $f'(x)$ be continuous and $f''(x)$ be piecewise continuous in $[0,\pi]$. If $F_s(n)$ is the finite Fourier transform, then*

$$\mathcal{F}[f''(x)]=\frac{2n}{\pi}[f(0)-(-1)^n f(\pi)]-n^2F_s(n) \tag{6.8.5}$$

PROOF.

$$\begin{aligned}\mathcal{F}[f''(x)]&=\frac{2}{\pi}\int_0^\pi f''(x)\sin nx\,dx\\&=\frac{2}{\pi}[f'(x)\sin nx]_0^\pi-\frac{2n}{\pi}\int_0^\pi f'(x)\cos nx\,dx\\&=-\frac{2n}{\pi}[f(x)\cos nx]_0^\pi-\frac{2n^2}{\pi}\int_0^\pi f(x)\sin nx\,dx\\&=-\frac{2n}{\pi}[f(\pi)(-1)^n-f(0)]-n^2F_s(n)\end{aligned}$$

□

The transforms of higher-ordered derivatives can be derived in a similar manner.

Theorem 6.8.2 *Let $f'(x)$ be continuous and $f''(x)$ be piecewise continuous in $[0,\pi]$. If $F_c(n)$ is the finite Fourier transform, then*

$$\mathfrak{F}[f''(x)]=\frac{2}{\pi}[(-1)^n f'(\pi)-f'(0)]-n^2F_c(n) \tag{6.8.6}$$

The proof is left to the reader.

EXAMPLE 6.8.1. Consider the motion of a string of length π due to a force acting on it. Let the string be fixed at both ends. The motion is thus governed by

$$\begin{aligned} u_{tt}&=c^2u_{xx}+f(x,t), \qquad 0<x<\pi, \qquad t>0\\ u(x,0)&=0\\ u_t(x,0)&=0\\ u(0,t)&=0\\ u(\pi,t)&=0 \end{aligned} \tag{6.8.7}$$

Transforming the equation of motion with respect to x, we obtain

$$\mathfrak{F}[u_{tt}-c^2u_{xx}-f(x,t)]=0$$

Due to its linearity (see Exercise 23), this can be written in the form

$$\mathfrak{F}[u_{tt}]-c^2\mathfrak{F}[u_{xx}]=\mathfrak{F}[f(x,t)] \tag{6.8.8}$$

Let $U(n,t)$ be the finite sine transform of $u(x,t)$. Then we have

$$\begin{aligned} \mathfrak{F}[u_{tt}]&=\frac{2}{\pi}\int_0^\pi u_{tt}\sin nx\,dx\\ &=\frac{d^2}{dt^2}\left[\frac{2}{\pi}\int_0^\pi u(x,t)\sin nx\,dx\right]\\ &=\frac{d^2U}{dt^2} \end{aligned}$$

We also have from *Theorem* 6.8.1

$$\mathfrak{F}[u_{xx}]=\frac{2n}{\pi}[u(0,t)-(-1)^n u(\pi,t)]-n^2U(n,t)$$

Because of the boundary conditions

$$u(0,t)=u(\pi,t)=0$$

$\mathfrak{F}[u_{xx}]$ becomes

$$\mathfrak{F}[u_{xx}]=-n^2U(n,t)$$

If we denote the transform of $f(x,t)$ by

$$F(n,t)=\frac{2}{\pi}\int_0^{\pi} f(x,t)\sin nx\,dx$$

then Eq. (6.8.8) takes the form

$$\frac{d^2U}{dt^2}+n^2c^2U=F(n,t)$$

This is the second order ordinary differential equation, the solution of which is given by

$$U(n,t)=A\cos nct+B\sin nct+\frac{1}{nc}\int_0^t F(n,\tau)\sin nc(t-\tau)\,d\tau$$

Applying the initial conditions

$$\begin{aligned}\mathcal{F}[u(x,0)]&=\frac{2}{\pi}\int_0^{\pi} u(x,0)\sin nx\,dx\\&=U(n,0)\end{aligned}$$

and

$$\mathcal{F}[u_t(x,0)]=\frac{dU}{dt}(n,0)$$

we have

$$U(n,t)=\frac{1}{nc}\int_0^t F(n,\tau)\sin nc(t-\tau)\,d\tau$$

Thus the inverse transform of $U(n,t)$ is

$$\begin{aligned}u(x,t)&=\sum_{n=1}^{\infty} U(n,t)\sin nx\\&=\sum_{n=1}^{\infty}\left[\frac{1}{nc}\int_0^t F(n,\tau)\sin nc(t-\tau)\,d\tau\right]\sin nx\end{aligned}$$

In the case when $f(x,t)$ is a constant, say h, then

$$\begin{aligned}\mathcal{F}[h]&=\frac{2}{\pi}\int_0^{\pi} h\sin nx\,dx\\&=\frac{2h}{n\pi}[1-(-1)^n]\end{aligned}$$

Now, we evaluate

$$\begin{aligned}U(n,t)&=\frac{1}{nc}\int_0^t \frac{2h}{n\pi}[1-(-1)^n]\sin nc(t-\tau)\,d\tau\\&=\frac{2h}{n^3\pi c^2}[1-(-1)^n](1-\cos nct)\end{aligned}$$

Hence the solution is given by

$$u(x,t)=\frac{2h}{\pi c^2}\sum_{n=1}^{\infty}\frac{[1-(-1)^n]}{n^3}(1-\cos nct)\sin nx$$

EXAMPLE 6.8.2. Find the temperature distribution in a rod of length π. The heat is generated in the rod at the rate $g(x,t)$ per unit time. The ends are insulated. The initial temperature distribution is given by $f(x)$.

The problem is to find $u(x,t)$ of

$$\begin{aligned} &u_t=u_{xx}+g(x,t), && 0<x<\pi, \quad t>0\\ &u(x,0)=f(x) && 0\leqslant x\leqslant\pi\\ &u_x(0,t)=0 && t\geqslant 0\\ &u_x(\pi,t)=0 && t\geqslant 0 \end{aligned}$$

Let $U(n,t)$ be the finite cosine transform of $u(x,t)$. As before, transformation of the heat equation with respect to x, using the boundary conditions, yields

$$\frac{dU}{dt}=-n^2U+G(n,t)$$

where

$$G(n,t)=\frac{2}{\pi}\int_0^{\pi}g(x,t)\cos nx\,dx$$

Rewriting this equation, we obtain

$$\frac{d}{dt}(e^{n^2t}U)=Ge^{n^2t}$$

Thus, the solution is

$$U(n,t)=\int_0^t e^{-n^2(t-\tau)}G(n,\tau)d\tau+Ae^{-n^2t}$$

Transformation of the initial condition gives

$$\begin{aligned} U(n,0)&=\frac{2}{\pi}\int_0^{\pi}u(x,0)\cos nx\,dx\\ &=\frac{2}{\pi}\int_0^{\pi}f(x)\cos nx\,dx \end{aligned}$$

Hence $U(n,t)$ takes the form

$$U(n,t)=\int_0^t e^{-n^2(t-\tau)}G(n,\tau)d\tau+U(n,0)e^{-n^2t}$$

The solution $u(x,t)$, therefore, is

$$u(x,t)=\frac{U(0,0)}{2}+\sum_{n=1}^{\infty}U(n,t)\cos nx$$

EXAMPLE 6.8.3. A rod with diffusion constant κ contains a fuel which produces neutrons by fissions. The ends of the rod are perfectly reflecting. If the initial neutron distribution is $f(x)$, find the distribution at any subsequent time t.

The problem is governed by

$$u_t = \kappa u_{xx} + bu, \qquad 0<x<l, \qquad t>0$$
$$u(x,0) = f(x)$$
$$u_x(0,t) = u_x(l,t) = 0$$

If $U(n,t)$ is the cosine transform of $u(x,t)$, then by transforming the equation and using the boundary conditions, we obtain

$$U_t + (\kappa n^2 - b)U = 0$$

The solution of this equation is

$$U(n,t) = Ce^{-(\kappa n^2 - b)t}$$

where C is a constant. Then applying the initial condition, we obtain

$$U(n,t) = U(n,0)e^{-(\kappa n^2 - b)t}$$

where

$$U(n,0) = \frac{2}{l}\int_0^l f(x)\cos nx\,dx$$

Thus, the solution takes the form

$$u(x,t) = \frac{U(0,0)}{2} + \sum_{n=1}^{\infty} U(n,t)\cos nx$$

If for instance $f(x) = x$ in $0<x<\pi$ then

$$U(0,0) = \pi$$

and

$$U(n,0) = \frac{2}{n^2\pi}\left[(-1)^n - 1\right] \qquad n = 1,2,3,\ldots$$

Hence the solution is given by

$$u(x,t) = \frac{\pi}{2} + \sum_{n=1}^{\infty} \frac{2}{n^2\pi}\left[(-1)^n - 1\right]e^{-(\kappa n^2 - b)t}\cos nx$$

Exercises

1. Solve the following initial-boundary value problems:

a. $u_{tt} = c^2 u_{xx}$ $\qquad 0<x<1 \quad t>0$
$u(x,0) = x(1-x) \qquad 0 \leqslant x \leqslant 1$
$u_t(x,0) = 0$
$u(0,t) = u(1,t) = 0$

b.
$$\begin{aligned} u_{tt} &= c^2 u_{xx} && 0<x<\pi \quad t>0 \\ u(x,0) &= 3\sin x && 0\leqslant x\leqslant \pi \\ u_t(x,0) &= 0 \\ u(0,t) &= u(\pi,t)=0 \end{aligned}$$

2. Determine the solutions of the following initial-boundary value problems:

a.
$$\begin{aligned} u_{tt} - c^2 u_{xx} &= 0 && 0<x<\pi \quad t>0 \\ u(x,0) &= 0 \\ u_t(x,0) &= 8\sin^2 x && 0\leqslant x\leqslant \pi \\ u(0,t) &= u(\pi,t)=0 \end{aligned}$$

b.
$$\begin{aligned} u_{tt} - c^2 u_{xx} &= 0 && 0<x<1 \quad t>0 \\ u(x,0) &= 0 \\ u_t(x,0) &= x\sin x && 0\leqslant x\leqslant 1 \\ u(0,t) &= u(1,t)=0 \end{aligned}$$

3. Find the solution of each of the following problems:

a.
$$\begin{aligned} u_{tt} - c^2 u_{xx} &= 0 && 0<x<1 \quad t>0 \\ u(x,0) &= x(1-x) && 0\leqslant x\leqslant 1 \\ u_t(x,0) &= x-\tan\frac{\pi x}{4} && 0\leqslant x\leqslant 1 \\ u(0,t) &= u(1,t)=0 \end{aligned}$$

b.
$$\begin{aligned} u_{tt} - c^2 u_{xx} &= 0 && 0<x<\pi \quad t>0 \\ u(x,0) &= \sin x && 0\leqslant x\leqslant \pi \\ u_t(x,0) &= x^2-\pi x && 0\leqslant x\leqslant \pi \\ u(0,t) &= u(\pi,t)=0 \end{aligned}$$

4. Solve the following problems:

a.
$$\begin{aligned} u_{tt} - c^2 u_{xx} &= 0 && 0<x<\pi \quad t>0 \\ u(x,0) &= x+\sin x, && 0\leqslant x\leqslant \pi \\ u_t(x,0) &= 0 \\ u(0,t) &= 0 \\ u_x(\pi,t) &= 0 \end{aligned}$$

b.
$$\begin{aligned} u_{tt} - c^2 u_{xx} &= 0 && 0<x<\pi \quad t>0 \\ u(x,0) &= \cos x && 0\leqslant x\leqslant \pi \\ u_t(x,0) &= 0 \\ u_x(0,t) &= 0 \\ u_x(\pi,t) &= 0 \end{aligned}$$

5. By the method of separation of variables, solve the telegraph problem:

$$\begin{aligned} u_{tt} + au_t + bu &= c^2 u_{xx} && 0<x<l \qquad t>0 \\ u(x,0) &= f(x) \\ u_t(x,0) &= 0 \\ u(0,t) &= u(l,t)=0 \end{aligned}$$

6. Obtain the solution of the damped wave motion problem:

$$\begin{aligned} &u_{tt}+au_t=c^2u_{xx} \qquad 0<x<l \qquad t>0 \\ &u(x,0)=0 \\ &u_t(x,0)=g(x) \\ &u(0,t)=u(l,t)=0 \end{aligned}$$

7. The torsional oscillations of a shaft of circular cross section is governed by the partial differential equation

$$\theta_{tt}=a^2\theta_{xx}$$

where $\theta(x,t)$ is the angular displacement of the cross section and a is a physical constant. The ends of the shaft are fixed elastically, that is,

$$\theta_x(0,t)-h\theta(0,t)=0 \qquad \theta_x(l,t)+h\theta(l,t)=0$$

Determine the angular displacement if the initial angular displacement is $f(x)$.

8. Solve the initial-boundary value problem of the longitudinal vibration of a truncated cone of length l and base radius a. The equation of motion is given by

$$\left(1-\frac{x}{h}\right)^2\frac{\partial^2 u}{\partial t^2}=c^2\frac{\partial}{\partial x}\left[\left(1-\frac{x}{h}\right)^2\frac{\partial u}{\partial x}\right], \quad 0<x<l, \quad t>0$$

where $c^2=E/\rho$, E is the elastic modulus, ρ is the density of the material and $h=la/(a-r)$. The two ends are rigidly fixed. If the initial displacement is $f(x)$, find $u(x,t)$.

9. Establish the validity of the formal solution of the initial-boundary value problem:

$$\begin{aligned} u_{tt}&=c^2u_{xx} \quad 0<x<\pi \quad t>0 \\ u(x,0)&=f(x) \quad 0\leqslant x\leqslant\pi \\ u_t(x,0)&=g(x) \quad 0\leqslant x\leqslant\pi \\ u_x(0,t)&=0 \\ u_x(\pi,t)&=0 \end{aligned}$$

10. Prove the uniqueness of the solution of the initial-boundary value problem:

$$\begin{aligned} u_{tt}&=c^2u_{xx} \quad 0<x<\pi \quad t>0 \\ u(x,0)&=f(x) \quad 0\leqslant x\leqslant\pi \\ u_t(x,0)&=g(x) \quad 0\leqslant x\leqslant\pi \\ u_x(0,t)&=0 \\ u_x(\pi,t)&=0 \end{aligned}$$

11. Determine the solution of

$$
\begin{aligned}
u_{tt} &= c^2 u_{xx} + A\sinh x \qquad 0<x<l \qquad t>0\\
u(x,0) &= 0\\
u_t(x,0) &= 0\\
u(0,t) &= h, \qquad t>0\\
u(l,t) &= k, \qquad t>0
\end{aligned}
$$

where h, k, and A are constants.

12. Solve the problem:

$$
\begin{aligned}
u_{tt} &= c^2 u_{xx} + A_x, \qquad 0<x<1, \qquad t>0, \qquad A=\text{constant}\\
u(x,0) &= 0\\
u_t(x,0) &= 0\\
u(0,t) &= 0\\
u(1,t) &= 0
\end{aligned}
$$

13. Solve the problem:

$$
\begin{aligned}
u_{tt} &= c^2 u_{xx} + x^2 \qquad 0<x<1 \qquad t>0\\
u(x,0) &= x \qquad 0\leqslant x\leqslant 1\\
u_t(x,0) &= 0\\
u(0,t) &= 0\\
u(1,t) &= 1 \qquad t\geqslant 0
\end{aligned}
$$

14. Find the solution of the problem

$$
\begin{aligned}
u_t &= ku_{xx} + h, \qquad 0<x<1, \quad t>0, \quad h=\text{constant}\\
u(x,0) &= u_0(1-\cos\pi x), \qquad 0\leqslant x\leqslant 1, \qquad u_0=\text{constant}\\
u(0,t) &= 0\\
u(1,t) &= 2u_0, \qquad t\geqslant 0
\end{aligned}
$$

15. Obtain the solution of each of the following initial-boundary value problems:

a.
$$
\begin{aligned}
u_t &= 4u_{xx} \qquad 0<x<1 \quad t>0\\
u(x,0) &= x^2(1-x) \qquad 0\leqslant x\leqslant 1\\
u(0,t) &= 0\\
u(1,t) &= 0
\end{aligned}
$$

b.
$$
\begin{aligned}
u_t &= ku_{xx} \qquad 0<x<\pi \quad t>0\\
u(x,0) &= \sin^2 x \qquad 0\leqslant x\leqslant \pi\\
u(0,t) &= 0\\
u(\pi,t) &= 0
\end{aligned}
$$

c.
$$\begin{aligned} u_t &= u_{xx} && 0<x<2 \quad t>0 \\ u(x,0) &= x, && 0 \leqslant x \leqslant 2 \\ u(0,t) &= 0 && \\ u_x(2,t) &= 1 && \quad t \geqslant 0 \end{aligned}$$

d.
$$\begin{aligned} u_t &= ku_{xx} && 0<x<l \quad t>0 \\ u(x,0) &= \sin \pi x/2l && 0 \leqslant x \leqslant l \\ u(0,t) &= 0 && \\ u(l,t) &= 1 && \quad t \geqslant 0 \end{aligned}$$

16. Find the temperature distribution in a rod of length l. The faces are insulated, and the initial temperature distribution is given by $x(l-x)$.

17. Find the temperature distribution in a rod of length π, one end of which is kept at zero temperature and the other end of which loses heat at a rate proportional to the temperature at that end $x=\pi$. The initial temperature distribution is given by $f(x)=x$.

18. The voltage distribution in a transmission line is given by

$$v_t = kv_{xx} \qquad 0<x<l \qquad t>0$$

The voltage equal to zero is maintained at $x=l$, while at the end $x=0$, the voltage varies according to the law

$$v(0,t) = Ct \qquad t>0$$

where C is a constant. Find $v(x,t)$ if the initial voltage distribution is zero.

19. Establish the validity of the formal solution of the initial boundary value problem:

$$\begin{aligned} u_t &= ku_{xx} && 0<x<l \quad t>0 \\ u(x,0) &= f(x) && 0 \leqslant x \leqslant l \\ u(l,t) &= 0 && \\ u_x(l,t) &= 0 && \end{aligned}$$

20. Prove the uniqueness of the solution of the problem:

$$\begin{aligned} u_t &= ku_{xx} && 0<x<l \quad t>0 \\ u(x,0) &= f(x) && 0 \leqslant x \leqslant l \\ u_x(0,t) &= 0 && \\ u_x(l,t) &= 0 && \end{aligned}$$

21. Solve the radioactive decay problem:

$$\begin{aligned} u_t - ku_{xx} &= Ae^{-ax} && 0<x<\pi \quad t>0 \\ u(x,0) &= \sin x && 0 \leqslant x \leqslant \pi \\ u(0,t) &= 0 && \\ u(\pi,t) &= 0 && \end{aligned}$$

22. Determine the solution of the problem

$$
\begin{aligned}
u_t &= ku_{xx} - hu, && -\pi < x < \pi, && t > 0, \quad h = \text{constant} \\
u(x,0) &= f(x) && -\pi \leqslant x \leqslant \pi && \\
u(-\pi,t) &= u(\pi,t) && && t \geqslant 0 \\
u_x(-\pi,t) &= u_x(\pi,t) && && t \geqslant 0
\end{aligned}
$$

[Use Substitution $u(x,t) = e^{-ht}v(x,t)$]

23. Prove that the Fourier sine and cosine transforms are linear.

24. If $F_s(n)$ is the Fourier sine transform of $f(x)$ on $0 \leqslant x \leqslant l$, show that

$$\mathcal{F}[f''] = \frac{2n\pi}{l^2}[f(0) - (-1)^n f(l)] - \left(\frac{n\pi}{l}\right)^2 F_s(n)$$

25. If $F_c(n)$ is the Fourier cosine transform of $f(x)$ on $0 \leqslant x \leqslant l$, show that

$$\mathcal{F}[f''] = \frac{2}{l}[(-1)^n f'(l) - f'(0)] - \left(\frac{n\pi}{l}\right)^2 F_c(n)$$

When $l = \pi$, show that

$$\mathcal{F}[f''] = \frac{2}{\pi}[(-1)^n f'(\pi) - f'(0)] - n^2 F_c(n)$$

26. By the transform method, solve

$$
\begin{aligned}
u_t &= u_{xx} + g(x,t) && 0 < x < \pi \quad t > 0 \\
u(x,0) &= f(x) && 0 \leqslant x \leqslant \pi \\
u(0,t) &= 0 \\
u(\pi,t) &= 0
\end{aligned}
$$

27. By the transform method, solve

$$
\begin{aligned}
u_t &= u_{xx} + g(x,t) && 0 < x < \pi \quad t > 0 \\
u(x,0) &= 0 \\
u(0,t) &= 0 \\
u_x(\pi,t) + hu(\pi,t) &= 0
\end{aligned}
$$

28. By the transform method, solve

$$
\begin{aligned}
u_t &= u_{xx} + g(x,t) && 0 < x < \pi \quad t > 0 \\
u(x,0) &= 0 \\
u(0,t) &= 0 \\
u_x(\pi,t) &= 0
\end{aligned}
$$

29. By the transform method, solve

$$
\begin{aligned}
u_t &= u_{xx} - hu && 0 < x < \pi \quad t > 0 \\
u(x,0) &= \sin x && 0 \leqslant x \leqslant \pi \\
u(0,t) &= 0 \\
u(\pi,t) &= 0
\end{aligned}
$$

30. By the transform method, solve

$$u_{tt}=u_{xx}+h, \qquad 0<x<\pi,\ t>0, \qquad h=\text{constant}$$
$$u(x,0)=0$$
$$u_t(x,0)=0$$
$$u_x(0,t)=0$$
$$u_x(\pi,t)=0$$

31. By the transform method, solve

$$u_{tt}=u_{xx}+g(x) \qquad 0<x<\pi, \qquad t>0$$
$$u(x,0)=0$$
$$u_t(x,0)=0$$
$$u(0,t)=0$$
$$u(\pi,t)=0$$

32. By the transform method, solve

$$u_{tt}+c^2u_{xxxx}=0 \qquad 0<x<\pi \qquad t>0$$
$$u(x,0)=0$$
$$u_t(x,0)=0$$
$$u(0,t)=0$$
$$u(\pi,t)=0$$
$$u_{xx}(0,t)=0$$
$$u_{xx}(\pi,t)=\sin t, \qquad t\geqslant 0$$

33. Determine the solution of the initial-boundary value problem

$$u_t-ku_{xx}=h(x,t), \quad 0<x<l, \quad t>0, \quad k=\text{constant}$$
$$u(x,0)=f(x), \quad 0\leqslant x\leqslant l$$
$$u(0,t)=p(t), \quad t\geqslant 0$$
$$u(l,t)=q(t), \quad t\geqslant 0$$

34. Determine the solution of the initial-boundary value problem

$$u_t-ku_{xx}=h(x,t), \quad 0<x<l, \quad t>0$$
$$u(x,0)=f(x), \quad 0\leqslant x\leqslant l$$
$$u(0,t)=p(t), \quad t\geqslant 0$$
$$u_x(l,t)=q(t), \quad t\geqslant 0$$

35. Solve the problem

$$u_t-ku_{xx}=0, \quad 0<x<1, \quad t>0$$
$$u(x,0)=x(1-x), \quad 0\leqslant x\leqslant 1$$
$$u(0,t)=t, \quad t\geqslant 0$$
$$u(1,t)=\sin t, \quad t\geqslant 0$$

36. Solve the problem

$$\begin{aligned} u_t - 4u_{xx} &= xt, & 0<x<1, \quad & t>0 \\ u(x,0) &= \sin \pi x, & 0\leqslant x\leqslant 1 \quad & \\ u(0,t) &= t, & & t\geqslant 0 \\ u(1,t) &= t^2, & & t\geqslant 0 \end{aligned}$$

37. Solve the problem

$$\begin{aligned} u_t - ku_{xx} &= x\cos t, & 0<x<\pi, \quad & t>0 \\ u(x,0) &= \sin x, & 0\leqslant x\leqslant \pi \quad & \\ u(0,t) &= t^2, & & t\geqslant 0 \\ u(\pi,t) &= 2t, & & t\geqslant 0 \end{aligned}$$

38. Solve the problem

$$\begin{aligned} u_t - u_{xx} &= 2x^2t, & 0<x<1, \quad & t>0 \\ u(x,0) &= \cos 3\pi x/2, & 0\leqslant x\leqslant 1 \quad & \\ u(0,t) &= 1, & & t\geqslant 0 \\ u_x(1,t) &= \frac{3\pi}{2}, & & t\geqslant 0 \end{aligned}$$

39. Solve the problem

$$\begin{aligned} u_t - 2u_{xx} &= h, & 0<x<1, \quad & t>0, \quad h=\text{constant} \\ u(x,0) &= x, & 0\leqslant x\leqslant 1 \quad & \\ u(0,t) &= \sin t, & & t\geqslant 0 \\ u_x(1,t)+u(1,t) &= 2, & & t\geqslant 0 \end{aligned}$$

40. Determine the solution of the initial-boundary value problem

$$\begin{aligned} u_{tt} - c^2u_{xx} &= h(x,t), & 0<x<l, \quad & t>0 \\ u(x,0) &= f(x), & 0\leqslant x\leqslant l \quad & \\ u_t(x,0) &= g(x), & 0\leqslant x\leqslant l \quad & \\ u(0,t) &= p(t), & & t\geqslant 0 \\ u_x(l,t) &= q(t), & & t\geqslant 0 \end{aligned}$$

41. Determine the solution of the initial-boundary value problem

$$\begin{aligned} u_{tt} - c^2u_{xx} &= h(x,t), & 0<x<l, \quad & t>0 \\ u(x,0) &= f(x), & 0\leqslant x\leqslant l \quad & \\ u_t(x,0) &= g(x), & 0\leqslant x\leqslant l \quad & \\ u_x(0,t) &= p(t), & & t\geqslant 0 \\ u_x(l,t) &= q(t), & & t\geqslant 0 \end{aligned}$$

42. Solve

$$\begin{aligned} u_{tt}-u_{xx}&=0, && 0<x<1, && t>0\\ u(x,0)&=x, && 0\leqslant x\leqslant 1\\ u_t(x,0)&=0, && 0\leqslant x\leqslant 1\\ u(0,t)&=t^2, && && t\geqslant 0\\ u(1,t)&=\cos t, && && t\geqslant 0 \end{aligned}$$

43. Solve

$$\begin{aligned} u_{tt}-4u_{xx}&=xt, && 0<x<1, && t>0\\ u(x,0)&=x, && 0\leqslant x\leqslant 1\\ u_t(x,0)&=0, && 0\leqslant x\leqslant 1\\ u(0,t)&=0, && && t\geqslant 0\\ u_x(l,t)&=1+t, && && t\geqslant 0 \end{aligned}$$

44. Solve

$$\begin{aligned} u_{tt}-9u_{xx}&=0, && 0<x<1, && t>0\\ u(x,0)&=\sin\frac{\pi x}{2}, && 0\leqslant x\leqslant 1\\ u_t(x,0)&=1+x, && 0\leqslant x\leqslant 1\\ u_x(0,t)&=\pi/2, && && t\geqslant 0\\ u_x(1,t)&=0, && && t\geqslant 0 \end{aligned}$$

45. Find the solution of the problem

$$\begin{aligned} u_{tt}+2ku_t-c^2u_{xx}&=0, && 0<x<l, && t>0\\ u(x,0)&=0, && 0\leqslant x\leqslant l\\ u_t(x,0)&=0, && 0\leqslant x\leqslant 1\\ u_x(0,t)&=0, && && t\geqslant 0\\ u(l,t)&=h, && && t\geqslant 0, \quad h=\text{constant} \end{aligned}$$

46. Solve

$$\begin{aligned} u_t-c^2u_{xx}+hu&=hu_0, && -\pi<x<\pi, && t>0\\ u(x,0)&=f(x), && -\pi\leqslant x\leqslant\pi\\ u(-\pi,t)&=u(\pi,t), && && t\geqslant 0\\ u_x(-\pi,t)&=u_x(\pi,t), && && t\geqslant 0 \end{aligned}$$

where u_0 is a constant.

CHAPTER SEVEN

Eigenvalue Problems

7.1 Sturm-Liouville Systems

In the preceding chapter, we have determined the solutions of partial differential equations by the method of separation of variables. Under separable conditions we transformed the second order homogeneous partial differential equation into two ordinary differential equations (6.1.11) and (6.1.12) which are of the form

$$a_1(x)\frac{d^2y}{dx^2}+a_2(x)\frac{dy}{dx}+\left[a_3(x)+\lambda\right]y=0 \tag{7.1.1}$$

If we introduce

$$p(x)=\exp\left[\int^x\frac{a_2(t)}{a_1(t)}dt\right],\quad q(x)=\frac{a_3(x)}{a_1(x)}p(x),\quad s(x)=\frac{p(x)}{a_1(x)} \tag{7.1.2}$$

into Eq. (7.1.1), we obtain

$$\frac{d}{dx}\left(p\frac{dy}{dx}\right)+(q+\lambda s)y=0 \tag{7.1.3}$$

which is known as the *Sturm-Liouville equation*. In terms of the operator

$$L=\frac{d}{dx}\left(p\frac{d}{dx}\right)+q$$

Eq. (7.1.3) can be written as

$$L[y]+\lambda s(x)y=0 \tag{7.1.4}$$

where λ is a parameter independent of x, and p, q, and s are real-valued functions of x. To ensure the existence of solutions, we let q and s be continuous and p be continuously differentiable in a closed finite interval $[a,b]$.

The Sturm-Liouville equation is called *regular* in the interval $[a,b]$ if the functions $p(x)$ and $s(x)$ are positive in the interval $[a,b]$. Thus, for a given λ, there exist two linearly independent solutions of a regular Sturm-Liouville equation in the interval $[a,b]$.

The Sturm-Liouville equation

$$L[y]+\lambda s(x)y=0, \quad a\leqslant x\leqslant b$$

together with the separated end conditions

$$\begin{aligned} a_1 y(a)+a_2 y'(a)&=0\\ b_1 y(b)+b_2 y'(b)&=0 \end{aligned} \tag{7.1.5}$$

where the constants a_1 and a_2, and likewise b_1 and b_2, are not both zero and are given real numbers, is called a *regular Sturm-Liouville system.*

The values of λ for which the Sturm-Liouville system has a nontrivial solution are called the *eigenvalues*, and the corresponding solutions are called the *eigenfunctions*.

EXAMPLE 7.1.1. Consider the Sturm-Liouville system

$$\begin{aligned} &y''+\lambda y=0, \qquad 0\leqslant x\leqslant \pi\\ &y(0)=0\\ &y'(\pi)=0 \end{aligned}$$

When $\lambda\leqslant 0$, it can be readily shown that λ is not an eigenvalue. However, when $\lambda>0$, the solution of the Sturm-Liouville equation is

$$y(x)=A\cos\sqrt{\lambda}\,x+B\sin\sqrt{\lambda}\,x$$

Applying the condition $y(0)=0$, we obtain $A=0$. The condition $y'(\pi)=0$ yields

$$B\sqrt{\lambda}\cos\sqrt{\lambda}\,\pi=0$$

Since $\lambda\neq 0$ and $B=0$ yields a trivial solution, we must have

$$\cos\sqrt{\lambda}\,\pi=0, \quad B\neq 0$$

This equation is satisfied, if

$$\sqrt{\lambda}=\frac{2n-1}{2}, \quad n=1,2,3,\ldots$$

and hence the eigenvalues are $\lambda_n=(2n-1)^2/4$, and the corresponding eigenfunctions are

$$\sin\left(\frac{2n-1}{2}\right)x, \quad n=1,2,3,\ldots$$

EXAMPLE 7.1.2. Consider the Euler equation

$$x^2y''+xy'+\lambda y=0, \quad 1\leqslant x\leqslant e$$

with the end conditions

$$y(1)=0, \quad y(e)=0$$

By using the transformation (7.1.2), the Euler equation can be put into Sturm-Liouville form:

$$\frac{d}{dx}\left(x\frac{dy}{dx}\right)+\frac{1}{x}\lambda y=0$$

The solution of the Euler equation is

$$y(x)=c_1x^{i\sqrt{\lambda}}+c_2x^{-i\sqrt{\lambda}}$$

Noting that $x^{ia}=e^{ia\ln x}=\cos(a\ln x)+i\sin(a\ln x)$, $y(x)$ becomes

$$y(x)=A\cos(\sqrt{\lambda}\,\ln x)+B\sin(\sqrt{\lambda}\,\ln x)$$

where A and B are constants related to c_1 and c_2. The end condition $y(1)=0$ gives $A=0$, and the end condition $y(e)=0$ gives

$$\sin\sqrt{\lambda}=0, \quad B\neq 0$$

which in turn yields the eigenvalues

$$\lambda_n=n^2\pi^2, \quad n=1,2,3,\ldots$$

and the corresponding eigenfunctions

$$\sin(n\pi\ln x), \quad n=1,2,3,\ldots$$

Another type of problem that often occurs in practice is the periodic Sturm-Liouville system.

The Sturm-Liouville equation

$$\frac{d}{dx}\left(p(x)\frac{dy}{dx}\right)+\left[q(x)+\lambda s(x)\right]y=0, \quad a\leqslant x\leqslant b$$

in which $p(a)=p(b)$, together with the periodic end conditions

$$y(a)=y(b)$$
$$y'(a)=y'(b)$$

is called a *periodic Sturm-Liouville system.*

EXAMPLE 7.1.3. Consider the periodic Sturm-Liouville system

$$y''+\lambda y=0 \quad -\pi\leqslant x\leqslant\pi$$
$$y(-\pi)=y(\pi)$$
$$y'(-\pi)=y'(\pi)$$

Here we note that $p(x)=1$ and hence $p(-\pi)=p(\pi)$. When $\lambda>0$, we see that the solution of the Sturm-Liouville equation is

$$y(x)=A\cos\sqrt{\lambda}\,x+B\sin\sqrt{\lambda}\,x$$

Application of the periodic end conditions yields

$$(2\sin\sqrt{\lambda}\,\pi)B=0$$

$$(2\sqrt{\lambda}\,\sin\sqrt{\lambda}\,\pi)A=0$$

Thus to obtain a nontrivial solution, we must have

$$\sin\sqrt{\lambda}\,\pi=0,\quad A\neq 0,\quad B\neq 0$$

Consequently,

$$\lambda_n=n^2,\quad n=1,2,3,\ldots.$$

Since $\sin\sqrt{\lambda}\,\pi=0$ is satisfied for arbitrary A and B, we obtain two linearly independent eigenfunctions $\cos nx$, $\sin nx$ corresponding to the same eigenvalue n^2.

It can be readily shown that if $\lambda<0$, the solution of the Sturm-Liouville equation does not satisfy the periodic end conditions. However, when $\lambda=0$ the corresponding eigenfunction is 1. Thus, the eigenvalues of the periodic Sturm-Liouville system are 0, $\{n^2\}$, and the corresponding eigenfunctions are 1, $\{\cos nx\}$, $\{\sin nx\}$, where n is a positive integer.

7.2 Eigenvalues and Eigenfunctions

In Examples 7.1.1 and 7.1.2 of the regular Sturm-Liouville systems in the preceding section, we see that there exists only one linearly independent eigenfunction corresponding to the eigenvalue λ, which is called an eigenvalue of multiplicity one (or a simple eigenvalue). An eigenvalue is said to be of multiplicity k if there exist k linearly independent eigenfunctions corresponding to the same eigenvalue. In Example 7.1.3 of the periodic Sturm-Liouville system, the eigenfunctions $\cos nx$, $\sin nx$ correspond to the same eigenvalue n^2. Thus, this eigenvalue is of multiplicity two.

In the preceding examples, we see that the eigenfunctions are $\cos nx$ and $\sin nx$ for $n=1,2,3,\ldots$. It can be easily shown by using trigonometric identities that

$$\int_{-\pi}^{\pi}\cos mx\cos nx\,dx=0,\quad m\neq n$$

$$\int_{-\pi}^{\pi}\cos mx\sin nx\,dx=0,\quad \text{for all integers } m,n,$$

$$\int_{-\pi}^{\pi}\sin mx\sin nx\,dx=0,\quad m\neq n.$$

We say that these functions are orthogonal to each other in the interval

$[-\pi,\pi]$. The orthogonality relation holds in general for the eigenfunctions of Sturm-Liouville systems.

Let $\phi(x)$ and $\psi(x)$ be any real-valued integrable functions on an interval I. Then ϕ and ψ are said to be *orthogonal* on I with respect to a weight function $\rho(x)>0$, if and only if,

$$(\phi,\psi)=\int_I \phi(x)\psi(x)\rho(x)\,dx=0 \tag{7.2.1}$$

The interval I may be of infinite extent, or it may be either open or closed at one or both ends of the finite interval.

When $\phi=\psi$ in (7.2.1) we have the *norm* of ϕ

$$\|\phi\|=\left[\int_I \phi^2(x)\rho(x)\,dx\right]^{1/2} \tag{7.2.2}$$

Theorem 7.2.1 *Let the coefficients p, q, and s in the Sturm-Liouville system be continuous in $[a,b]$. Let the eigenfunctions ϕ_j and ϕ_k, corresponding to λ_j and λ_k, be continuously differentiable. Then ϕ_j and ϕ_k are orthogonal with respect to the weight function s in $[a,b]$.*

PROOF. Since ϕ_j corresponding to λ_j satisfies the Sturm-Liouville equation, we have

$$\frac{d}{dx}(p\phi_j')+(q+\lambda_j s)\phi_j=0 \tag{7.2.3}$$

and for the same reason

$$\frac{d}{dx}(p\phi_k')+(q+\lambda_k s)\phi_k=0 \tag{7.2.4}$$

Multiplying Eq. (7.2.3) by ϕ_k and Eq. (7.2.4) by ϕ_j, and subtracting, we obtain

$$\begin{aligned}(\lambda_j-\lambda_k)s\phi_j\phi_k&=\phi_j\frac{d}{dx}(r\phi_k')-\phi_k\frac{d}{dx}(p\phi_j')\\&=\frac{d}{dx}\left[(p\phi_k')\phi_j-(p\phi_j')\phi_k\right]\end{aligned}$$

and integration yields

$$\begin{aligned}(\lambda_j-\lambda_k)\int_a^b s\phi_j\phi_k\,dx&=\left[p(\phi_j\phi_k'-\phi_j'\phi_k)\right]_a^b\\&=p(b)\left[\phi_j(b)\phi_k'(b)-\phi_j'(b)\phi_k(b)\right]\\&\quad-p(a)\left[\phi_j(a)\phi_k'(a)-\phi_j'(a)\phi_k(a)\right]\end{aligned} \tag{7.2.5}$$

the right side of which is called the boundary term of the Sturm-Liouville system. The end conditions for the eigenfunctions ϕ_j and ϕ_k are

$$b_1\phi_j(b)+b_2\phi_j'(b)=0$$
$$b_1\phi_k(b)+b_2\phi_k'(b)=0$$

If $b_2\neq 0$, we multiply the first condition by $\phi_k(b)$ and the second condition by $\phi_j(b)$, and subtract to obtain

$$\left[\phi_j(b)\phi_k'(b)-\phi_j'(b)\phi_k(b)\right]=0 \tag{7.2.6}$$

In a similar manner, if $a_2\neq 0$, we obtain

$$\left[\phi_j(a)\phi_k'(a)-\phi_j'(a)\phi_k(a)\right]=0 \tag{7.2.7}$$

We see by virtue of (7.2.6) and (7.2.7) that

$$(\lambda_j-\lambda_k)\int_a^b s\phi_j\phi_k\,dx=0 \tag{7.2.8}$$

If λ_j and λ_k are distinct eigenvalues, then

$$\int_a^b s\phi_j\phi_k\,dx=0 \tag{7.2.9}$$

□

Theorem 7.2.2 *The eigenfunctions of a periodic Sturm-Liouville system in* $[a,b]$ *are orthogonal with respect to the weight function s in* $[a,b]$.

PROOF. The periodic conditions for the eigenfunctions ϕ_j and ϕ_k are

$$\phi_j(a)=\phi_j(b)\quad \phi_j'(a)=\phi_j'(b)$$
$$\phi_k(a)=\phi_k(b)\quad \phi_k'(a)=\phi_k'(b)$$

Substitution of these into Eq. (7.2.5) yields

$$(\lambda_j-\lambda_k)\int_a^b s\phi_j\phi_k\,dx=\left[p(b)-p(a)\right]\left[\phi_j(a)\phi_k'(a)-\phi_j'(a)\phi_k(a)\right]$$

Since $p(a)=p(b)$, we have

$$(\lambda_j-\lambda_k)\int_a^b s\phi_j\phi_k\,dx=0 \tag{7.2.10}$$

For distinct eigenvalues $\lambda_j\neq\lambda_k$, and thus

$$\int_a^b s\phi_j\phi_k\,dx=0 \tag{7.2.11}$$

□

Theorem 7.2.3 *All the eigenvalues of a regular Sturm-Liouville system with* $s(x)>0$ *are real.*

PROOF. Suppose that there is a complex eigenvalue $\lambda_j=\alpha+i\beta$ with eigenfunction $\phi_j=u+iv$. Then, because the coefficients of the equation are real, the complex conjugate of the eigenvalue is also an eigenvalue. Thus, there exists an eigenfunction $\phi_k = u - iv$ corresponding to the eigenvalue $\lambda_k = \alpha - i\beta$.

By using the relation (7.2.8), we have

$$2i\beta \int_a^b s(u^2 + v^2)\, dx = 0$$

This implies that β must vanish for $s > 0$, and hence the eigenvalues are real. □

Theorem 7.2.4 *If $\phi_1(x)$ and $\phi_2(x)$ are any two solutions of $L[y] + \lambda sy = 0$ on $[a,b]$, then $p(x)W(x;\phi_1,\phi_2) = \text{constant}$, where W is the Wronskian.*

PROOF. Since ϕ_1 and ϕ_2 are solutions of $L[y] + \lambda sy = 0$, we have

$$\frac{d}{dx}\left(p\frac{d\phi_1}{dx}\right) + (q + \lambda s)\phi_1 = 0$$

$$\frac{d}{dx}\left(p\frac{d\phi_2}{dx}\right) + (q + \lambda s)\phi_2 = 0$$

Multiplying the first equation by ϕ_2 and the second by ϕ_1, and subtracting, we obtain

$$\phi_1\frac{d}{dx}\left(p\frac{d\phi_2}{dx}\right) - \phi_2\frac{d}{dx}\left(p\frac{d\phi_1}{dx}\right) = 0$$

Integrating this equation from a to x, we obtain

$$p(x)[\phi_1(x)\phi_2'(x) - \phi_1'(x)\phi_2(x)] = p(a)[\phi_1(a)\phi_2'(a) - \phi_1'(a)\phi_2(a)]$$
$$= \text{constant}, \qquad (7.2.12)$$

which is called *Abel's formula*. □

Theorem 7.2.5 *An eigenfunction of a regular Sturm-Liouville system is unique except for a constant factor.*

PROOF. Let $\phi_1(x)$ and $\phi_2(x)$ be eigenfunctions corresponding to an eigenvalue λ. Then according to Abel's formula (7.2.12), we have

$$p(x)W(x;\phi_1,\phi_2) = \text{constant}, \quad p(x) > 0$$

where W is the Wronskian. Thus, if W vanishes at a point in $[a,b]$, it must vanish for all $x \in [a,b]$.

Since ϕ_1 and ϕ_2 satisfy the end condition at $x = a$, we have

$$a_1\phi_1(a) + a_2\phi_1'(a) = 0,$$
$$a_1\phi_2(a) + a_2\phi_2'(a) = 0$$

Since a_1 and a_2 are not both zero, we have

$$\begin{vmatrix} \phi_1(a) & \phi_1'(a) \\ \phi_2(a) & \phi_2'(a) \end{vmatrix} = +W(a;\phi_1,\phi_2)=0$$

Therefore $W(x;\phi_1,\phi_2)=0$ for all $x\in[a,b]$, which is a sufficient condition for the linear dependence of two functions ϕ_1 and ϕ_2. Hence $\phi_1(x)$ differs from $\phi_2(x)$ only by a constant factor. □

Theorem 7.2.3 states that all eigenvalues of a regular Sturm-Liouville system are real, but it does not guarantee that any eigenvalue exists. However, it can be proved that a self-adjoint regular Sturm-Liouville system has a denumerably infinite number of eigenvalues. To illustrate this, let us consider the following example.

EXAMPLE 7.2.1. Consider the Sturm-Liouville system

$$y''+\lambda y=0, \quad 0\leqslant x\leqslant 1,$$
$$y(0)=0,$$
$$y(1)+hy'(1)=0, \quad h>0 \text{ a constant.}$$

Here $p=1$, $q=0$, $s=1$. The solution of the Sturm-Liouville equation is

$$y(x)=A\cos\sqrt{\lambda}\,x+B\sin\sqrt{\lambda}\,x$$

Since $y(0)=0$ gives $A=0$, we have

$$y(x)=B\sin\sqrt{\lambda}\,x$$

Applying the second end condition, we have

$$\sin\sqrt{\lambda}+h\sqrt{\lambda}\cos\sqrt{\lambda}=0 \quad \text{for} \quad B\neq 0$$

which can be rewritten as

$$\tan\sqrt{\lambda}=-h\sqrt{\lambda}$$

If $\alpha=\sqrt{\lambda}$ is introduced in this equation, we have

$$\tan\alpha=-h\alpha$$

This equation does not possess an explicit solution. Thus, we determine the solution graphically by plotting the functions $\xi=\tan\alpha$ and $\xi=-h\alpha$ against α, as shown in Fig. 7.2.1. The roots are given by the intersection of two curves, and as is evident from the graph, there are infinitely many roots α_n for $n=1,2,3,\ldots$. To each root α_n, there corresponds an eigenvalue

$$\lambda_n=\alpha_n^2, \quad n=1,2,3,\ldots.$$

Thus, there exists a sequence of eigenvalues

$$\lambda_1<\lambda_2<\lambda_3<\ldots$$

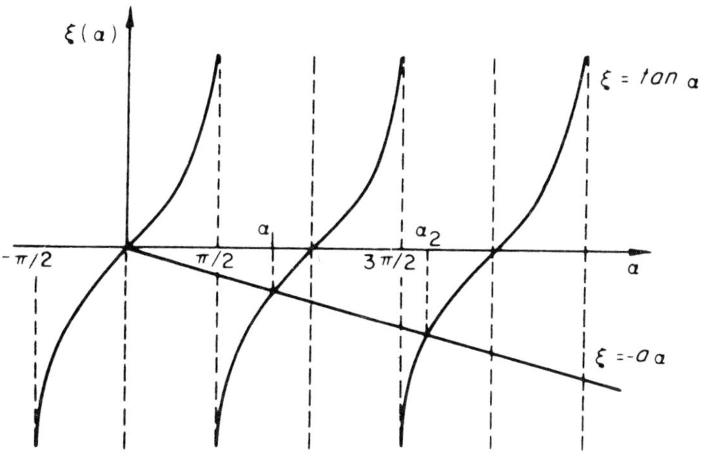

Figure 7.2.1

with

$$\lim_{n\to\infty} \lambda_n = \infty.$$

The corresponding eigenfunctions are $\sin\sqrt{\lambda_n}\,x$.

Theorem 7.2.6 *A self-adjoint regular Sturm-Liouville system has an infinite sequence of real eigenvalues*

$$\lambda_0 < \lambda_1 < \lambda_2 < \lambda_3 < \ldots$$

with

$$\lim_{n\to\infty} \lambda_n = \infty$$

For each n the corresponding eigenfunction $\phi_n(x)$, uniquely determined up to a constant factor, has exactly n zeros in the interval (a,b).

Proof of this theorem can be found in [67].

7.3 Eigenfunction Expansions

A real-valued function $\phi(x)$ is said to be *square-integrable* with respect to a weight function $\rho(x) > 0$, if on an interval I,

$$\int_I \phi^2(x)\rho(x)\,dx < +\infty \qquad (7.3.1)$$

An immediate consequence of this definition is the *Schwarz inequality*

$$\left|\int_I \phi(x)\psi(x)\rho(x)\,dx\right|^2 \leqslant \int_I \phi^2(x)\rho(x)\,dx \int_I \psi^2(x)\rho(x)\,dx \tag{7.3.2}$$

for square-integrable functions $\phi(x)$ and $\psi(x)$ (see Exercises in 12).

Let $\{\phi_n(x)\}$, for a positive integer n, be an orthogonal set of square-integrable functions with a positive weight function $\rho(x)$ on an interval I. Let $f(x)$ be a given function that can be represented by a uniformly convergent series of the form

$$f(x) = \sum_{n=1}^{\infty} c_n \phi_n(x) \tag{7.3.3}$$

where the coefficients c_n are constants. Now multiplying both sides of (7.3.3) by $\phi_m(x)\rho(x)$ and integrating term by term over the interval I (uniform convergence of the series is a sufficient condition for this), we obtain

$$\int_I f(x)\phi_m(x)\rho(x)\,dx = \sum_{n=1}^{\infty} \int_I c_n \phi_n(x)\phi_m(x)\rho(x)\,dx$$

and hence

$$\int_I f(x)\phi_n(x)\rho(x)\,dx = c_n \int_I \phi_n^2(x)\rho(x)\,dx$$

Thus

$$c_n = \frac{\int_I f\phi_n\rho\,dx}{\int_I \phi_n^2\rho\,dx} \tag{7.3.4}$$

Hence we have the following theorem:

Theorem 7.3.1 *If f is represented by a uniformly convergent series*

$$f(x) = \sum_{n=1}^{\infty} c_n \phi_n(x)$$

on an interval I, where ϕ_n are square-integrable functions orthogonal with respect to a positive weight function $\rho(x)$, then c_n are determined by

$$c_n = \frac{\int_I f\phi_n\rho\,dx}{\int_I \phi_n^2\rho\,dx}$$

EXAMPLE 7.3.1. The Legendre polynomials $P_n(x)$ are orthogonal with respect to the weight function $\rho(x)=1$ on $(-1,1)$. If we assume that $f(x)$ can be represented by the *Fourier-Legendre series*

$$f(x)=\sum_{n=1}^{\infty} c_n P_n(x)$$

then c_n are given by

$$c_n=\frac{\int_{-1}^{1} f(x)P_n(x)\,dx}{\int_{-1}^{1} P_n^2(x)\,dx}$$

$$=\frac{2n+1}{2}\int_{-1}^{1} f(x)P_n(x)\,dx$$

In the above discussion, we assumed that the given function $f(x)$ is represented by a uniformly convergent series. This is rather restrictive, and we will show in the following section that $f(x)$ can be represented by a mean-square convergent series.

7.4 Convergence In the Mean

Let $\{\phi_n\}$ be the set of square-integrable functions orthogonal with respect to the weight function $\rho(x)$ on $[a,b]$. Let

$$s_n(x)=\sum_{k=1}^{n} c_k\phi_k(x)$$

be the nth partial sum of the series $\sum_{k=1}^{\infty} c_k\phi_k(x)$

Let f be a square-integrable function. The sequence $\{s_n\}$ is said to *converge in the mean* to $f(x)$ on the interval I with respect to the weight function $\rho(x)$ if

$$\lim_{n\to+\infty}\int_I [f(x)-s_n(x)]^2\rho(x)\,dx=0 \tag{7.4.1}$$

We shall now seek the coefficients c_k such that $s_n(x)$ represents the best approximation to $f(x)$ in the sense of least squares, that is, we seek to minimize the integral

$$E(c_k)=\int_I [f(x)-s_n(x)]^2\rho(x)\,dx$$

$$=\int_I f^2\rho\,dx-2\sum_{k=1}^{n} c_k\int_I f\phi_k\rho\,dx+\sum_{k=1}^{n} c_k^2\int_I \phi_k^2\rho\,dx \tag{7.4.2}$$

This is an extremal problem. A necessary condition on the c_k for E to be a minimum is that the first partial derivatives of E with respect to these coefficients vanish. Thus, differentiating (7.4.2) with respect to c_k, we obtain

$$\frac{\partial E}{\partial c_k} = -2\int_I f\phi_k\rho\,dx + 2c_k\int_I \phi_k^2\rho\,dx = 0 \tag{7.4.3}$$

and hence

$$c_k = \frac{\int_I f\phi_k\rho\,dx}{\int_I \phi_k^2\rho\,dx} \tag{7.4.4}$$

Now if we complete the square, the right side of (7.4.2) becomes

$$E = \int_I f^2\rho\,dx + \sum_{k=1}^{n} \int_I \phi_k^2\rho\,dx \left[c_k - \frac{\int_I f\phi_k\rho\,dx}{\int_I \phi_k^2\rho\,dx}\right]^2 - \sum_{k=1}^{n} \frac{\left(\int_I f\phi_k\rho\,dx\right)^2}{\int_I \phi_k^2\rho\,dx}$$

The right side shows that E is a minimum if, and only if, c_k is given by (7.4.4). Therefore, this choice of c_k yields the best approximation to $f(x)$ in the sense of least squares.

For series convergent in the mean to $f(x)$, we conventionally write

$$f(x) \sim \sum_{k=1}^{\infty} c_k\phi_k(x)$$

where the coefficients c_k are the *Fourier coefficients* and the series is the *Fourier series*. This series may or may not be pointwise or uniformly convergent.

7.5 Completeness and Parseval's Equality

Substituting the Fourier coefficients (7.4.4) into (7.4.2), we obtain

$$\int_I \left(f(x) - \sum_{k=1}^{n} c_k\phi_k(x)\right)^2 \rho(x)\,dx = \int_I f^2\rho\,dx - \sum_{k=1}^{n} c_k^2\int_I \phi_k^2\rho\,dx$$

Since the left side is nonnegative, we have

$$\sum_{k=1}^{n} c_k^2\int_I \phi_k^2\rho\,dx \leqslant \int_I f^2\rho\,dx \tag{7.5.1}$$

The integral on the right side is finite, and hence the series on the left side is bounded above for any n. Thus, as $n\to\infty$, the inequality (7.5.1) may be written as

$$\sum_{k=1}^{\infty} c_k^2 \int_I \phi_k^2 \rho \, dx \leqslant \int_I f^2 \rho \, dx \tag{7.5.2}$$

This is called *Bessel's inequality*.

If the series converges in the mean to $f(x)$, that is,

$$\lim_{n\to\infty} \int_I \left(f(x) - \sum_{k=1}^{n} c_k \phi_k(x) \right)^2 \rho(x) \, dx = 0$$

then it follows from the above derivation that

$$\sum_{k=1}^{\infty} c_k^2 \int_I \phi_k^2 \rho \, dx = \int_I f^2 \rho \, dx$$

which is called *Parseval's equality*. Sometimes it is known as the completeness relation. Thus, when every continuous square-integrable function $f(x)$ can be expanded into an infinite series

$$f(x) = \sum_{k=1}^{\infty} c_k \phi_k(x)$$

the sequence of continuous square-integrable functions $\{\phi_k\}$ orthogonal with respect to the weight function ρ is said to be *complete*.

Next we state the following theorem:

Theorem 7.5.1 *The eigenfunctions of any regular Sturm-Liouville system are complete in the space of functions that are piecewise continuous on the interval* $[a,b]$ *with respect to the weight function* $s(x)$. *Moreover, any piecewise smooth function on* $[a,b]$ *that satisfies the end conditions of the regular Sturm-Liouville system can be expanded in an absolutely and uniformly convergent series*

$$f(x) = \sum_{k=1}^{\infty} c_k \phi_k(x)$$

where c_k *are given by*

$$c_k = \int_a^b f \phi_k s \, dx \Big/ \int_a^b \phi_k^2 s \, dx$$

Proof of a more general theorem can be found in Coddington and Levinson [8].

EXAMPLE 7.5.1. Consider a cylindrical wire of length l whose surface is perfectly insulated against the flow of heat. The end $l=0$ is maintained at the zero degree temperature while the other end radiates freely into the surrounding medium of zero degree temperature. Let the initial temperature distribution in the wire be $f(x)$. Find the temperature distribution $u(x,t)$.

The initial-boundary value problem is

$$\begin{aligned} u_t &= ku_{xx}, \quad 0<x<l, \quad t>0 \\ u(x,0) &= f(x), \quad 0<x\leqslant l \\ u(0,t) &= 0, \qquad t>0 \\ hu(l,t)+u'(l,t) &= 0, \qquad t>0, \quad h>0 \end{aligned} \tag{7.5.3}$$

By the method of separation of variables, we assume a solution in the form

$$u(x,t)=X(x)T(t)$$

and substituting it in the heat equation, we obtain

$$XT'=kX''T$$

or

$$\frac{X''}{X}=\frac{T'}{kT}=-\lambda, \quad \lambda>0$$

Thus, we have

$$\begin{aligned} X''+\lambda X&=0 \\ T'+k\lambda T&=0 \end{aligned}$$

The solution of the latter equation is

$$T(t)=Ce^{-k\lambda t}$$

where C is an arbitrary constant. From the first boundary condition $u(0,t)=0$, we have

$$X(0)=0 \quad \text{for} \quad T(t)\neq 0$$

In a similar manner, the second boundary condition reduces to

$$hX(l)+X'(l)=0 \quad \text{for} \quad T(t)\neq 0$$

The problem now is to solve the eigenvalue problem

$$\begin{aligned} X''+\lambda X&=0 \\ X(0)&=0 \\ hX(l)+X'(l)&=0 \end{aligned}$$

which is a Sturm-Liouville system. The solution of the Sturm-Liouville equation is

$$X(x)=A\cos\sqrt{\lambda}\, x+B\sin\sqrt{\lambda}\, x$$

Since $X(0)=0$ gives $A=0$, we have

$$X(x)=B\sin\sqrt{\lambda}\, x$$

Application of the second end condition yields

$$h\sin\sqrt{\lambda}\ l+\sqrt{\lambda}\cos\sqrt{\lambda}\ l=0 \quad \text{for} \quad B\neq 0$$

which can be rewritten as

$$\tan\sqrt{\lambda}\ l=-\sqrt{\lambda}\ /h$$

If $\alpha=\sqrt{\lambda}\ l$ is introduced in the preceding equation, we have

$$\tan\alpha=-a\alpha$$

where $a=1/hl$. As in Example 7.2.1, there exists a sequence of eigenvalues

$$\lambda_1<\lambda_2<\lambda_3<\ldots$$

with $\lim_{n\to\infty}\lambda_n=\infty$. The corresponding eigenfunctions are $\sin\sqrt{\lambda_n}\ x$, and hence

$$X_n(x)=B_n\sin\sqrt{\lambda_n}\ x$$

Therefore, the solution takes the form

$$u_n(x,t)=a_n e^{-k\lambda_n t}\sin\sqrt{\lambda_n}\ x, \quad a_n=B_nC_n$$

which satisfies the heat equation and the boundary conditions. Since the heat equation is linear and homogeneous, we form a series of solutions

$$u(x,t)=\sum_{n=1}^{\infty} a_n e^{-k\lambda_n t}\sin\sqrt{\lambda_n}\ x \tag{7.5.4}$$

which is also a solution, provided it converges and is twice differentiable with respect to x and once differentiable with respect to t. According to Theorem 7.2.1, the eigenfunctions $\sin\sqrt{\lambda_n}\ x$ form an orthogonal system over the interval $(0,l)$. Application of the initial condition yields

$$u(x,0)=f(x)\sim\sum_{n=1}^{\infty} a_n\sin\sqrt{\lambda_n}\ x$$

If we assume that f is a piecewise smooth function on $[a,b]$, then by Theorem 7.5.1 we can expand $f(x)$ in terms of the eigenfunctions, and formally write

$$f(x)=\sum_{n=1}^{\infty} a_n\sin\sqrt{\lambda_n}\ x$$

where the coefficient a_n is given by

$$a_n=\int_0^l f(x)\sin\sqrt{\lambda_n}\ x\,dx\Big/\int_0^l \sin^2\sqrt{\lambda_n}\ x\,dx$$

With this value of a_n, the temperature distribution is given by (7.5.4).

7.6 Bessel's Equation

Bessel's equation frequently occurs in problems of mathematical physics involving cylindrical symmetry.

The standard form of Bessel's equation is given by

$$x^2y'' + xy' + (x^2 - \nu^2)y = 0 \tag{7.6.1}$$

where ν is a nonnegative real number. We shall first restrict our attention to $x>0$. Since $x=0$ is the regular singular point, a solution is taken in accordance with the Frobenius method to be

$$y(x) = \sum_{n=0}^{\infty} a_n x^{s+n} \tag{7.6.2}$$

where the index s is to be determined. Substitution of this series into Eq. (7.6.1) then yields

$$\begin{aligned}&[s^2 - \nu^2]a_0x^s + [(s+1)^2 - \nu^2]a_1x^{s+1}\\ &+ \sum_{n=2}^{\infty} \{[(s+n)^2 - \nu^2]a_n + a_{n-2}\}x^{s+n} = 0\end{aligned} \tag{7.6.3}$$

The requirement that the coefficient of x^s vanish leads to the *indicial equation*

$$(s^2 - \nu^2)a_0 = 0 \tag{7.6.4}$$

from which it follows that $s = \pm\nu$ for arbitrary $a_0 \neq 0$. Since the leading term in the series (7.6.2) is a_0x^s, it is clear that for $\nu>0$ the solution of Bessel's equation corresponding to the choice $s=\nu$ vanishes at the origin, whereas the solution corresponding to $s=-\nu$ is infinite at that point.

We consider first the regular solution of Bessel's equation, that is, the solution corresponding to the choice $s=\nu$. The vanishing of the coefficient of x^{s+1} in Eq. (7.6.3) requires that

$$(2\nu+1)a_1 = 0 \tag{7.6.5}$$

which in turn implies that $a_1 = 0$ (since $\nu \geqslant 0$). From the requirement that the coefficient of x^{s+n} in Eq. (7.6.3) be zero, we obtain the two-term recurrence relation

$$a_n = -\frac{a_{n-2}}{n(2\nu+n)} \tag{7.6.6}$$

Since $a_1 = 0$, it is obvious that $a_n = 0$ for $n = 3, 5, 7, \ldots$. The remaining coefficients are given by

$$a_{2k} = \frac{(-1)^k a_0}{2^{2k}k!(\nu+k)(\nu+k-1)\cdots(\nu+1)} \tag{7.6.7}$$

for $k=1,2,3,\ldots$. This relation may also be written as

$$a_{2k}=\frac{(-1)^k 2^\nu \Gamma(\nu+1)a_0}{2^{2k+\nu}k!\Gamma(\nu+k+1)},\ k=1,2,\ldots \tag{7.6.8}$$

where $\Gamma(\alpha)$ is the gamma function, whose properties are described in Appendix I.

Hence, the regular solution of Bessel's equation takes the form

$$y(x)=a_0\sum_{k=0}^{\infty}\frac{(-1)^k 2^\nu \Gamma(\nu+1)}{2^{2k+\nu}k!\Gamma(\nu+k+1)}x^{2k+\nu} \tag{7.6.9}$$

It is customary to choose

$$a_0=\frac{1}{2^\nu\Gamma(\nu+1)} \tag{7.6.10}$$

and to denote the corresponding solution by $J_\nu(x)$. This solution, called the *Bessel function of the first kind of order* ν, is therefore given by

$$J_\nu(x)=\sum_{k=0}^{\infty}\frac{(-1)^k x^{2k+\nu}}{2^{2k+\nu}k!\Gamma(\nu+k+1)} \tag{7.6.11}$$

To determine the irregular solution of the Bessel equation for $s=-\nu$, we proceed as above. In this way, we obtain as the analogue of Eq. (7.6.5) the relation

$$(-2\nu+1)a_1=0$$

from which it follows, without loss of generality, that $a_1=0$. Using the recurrence relation

$$a_n=-\frac{a_{n-2}}{n(n-2\nu)},\quad n\geqslant 2 \tag{7.6.12}$$

we obtain as the irregular solution the Bessel function of the first kind of order $-\nu$:

$$J_{-\nu}(x)=\sum_{k=0}^{\infty}\frac{(-1)^k x^{2k-\nu}}{2^{2k-\nu}k!\Gamma(-\nu+k+1)} \tag{7.6.13}$$

It can be easily proved that, if ν is not an integer, J_ν and $J_{-\nu}$ converge for all values of x, and are linearly independent. Thus, the general solution of the Bessel equation for nonintegral ν is

$$y(x)=c_1J_\nu(x)+c_2J_{-\nu}(x) \tag{7.6.14}$$

If ν is an integer, say $\nu=n$, then from Eq. (7.6.13), noting that, when gamma functions in the coefficients of the first n terms become infinite, the coefficients become zero, we have

$$J_{-n}(x)=\sum_{k=n}^{\infty}\frac{(-1)^k x^{2k-n}}{2^{2k-\nu}k!\Gamma(-n+k+1)}$$

$$=(-1)^n\sum_{k=0}^{\infty}\frac{(-1)^k x^{2k+n}}{2^{2k+n}k!\Gamma(n+k+1)}$$

$$=(-1)^n J_n(x) \tag{7.6.15}$$

It shows that J_{-n} is not independent of J_n, and therefore a second linearly independent solution is required.

A number of distinct irregular solutions are discussed in the literature, but the one most commonly used, as defined by Weber, is

$$Y_\nu(x)=\frac{(\cos\nu\pi)J_\nu(x)-J_{-\nu}(x)}{\sin\nu\pi} \tag{7.6.16}$$

For nonintegral ν, it is obvious that $Y_\nu(x)$, being a linear combination of $J_\nu(x)$ and $J_{-\nu}(x)$ is linearly independent of $J_\nu(x)$. When ν is a nonnegative integer n, $Y_\nu(x)$ is indeterminate. But

$$Y_n(x)=\lim_{\nu\to\infty}Y_\nu(x)$$

exists and is a solution of the Bessel equation. Moreover, it is linearly independent of $J_n(x)$ (For an extended treatment, see Watson [39]). The function $Y_\nu(x)$ is called *the Bessel function of the second kind of order* ν. Thus, the general solution of the Bessel equation is

$$y(x)=c_1J_\nu(x)+c_2Y_\nu(x),\quad \text{for}\quad \nu\geqslant 0 \tag{7.6.17}$$

Like elementary functions, Bessel's functions are tabulated (see Jahnke and Emde [85]). For illustration, the functions J_0, J_1, Y_0, and Y_1 are plotted for small values of x, as shown in Fig. 7.6.1.

It should be noted that $J_\nu(x)$ for $\nu\geqslant 0$ and $J_{-\nu}(x)$ for a positive integer ν, are finite at the origin, but $J_{-\nu}(x)$ for nonintegral ν and $Y_\nu(x)$ for $\nu\geqslant 0$ approach infinity as x tends to zero.

Some of the useful recurrence relations are

$$J_{\nu-1}(x)+J_{\nu+1}(x)=\frac{2\nu}{x}J_\nu(x) \tag{7.6.18}$$

$$\nu J_\nu(x)+xJ_\nu'(x)=xJ_{\nu-1}(x) \tag{7.6.19}$$

$$J_{\nu-1}(x)-J_{\nu+1}(x)=2J_\nu'(x) \tag{7.6.20}$$

$$\nu J_\nu(x)-xJ_\nu'(x)=xJ_{\nu+1}(x) \tag{7.6.21}$$

$$\frac{d}{dx}\left[x^\nu J_\nu(x)\right]=x^\nu J_{\nu-1}(x) \tag{7.6.22}$$

$$\frac{d}{dx}\left[x^{-\nu}J_\nu(x)\right]=-x^{-\nu}J_{\nu+1}(x) \tag{7.6.23}$$

All of these relations also hold true for $Y_\nu(x)$.

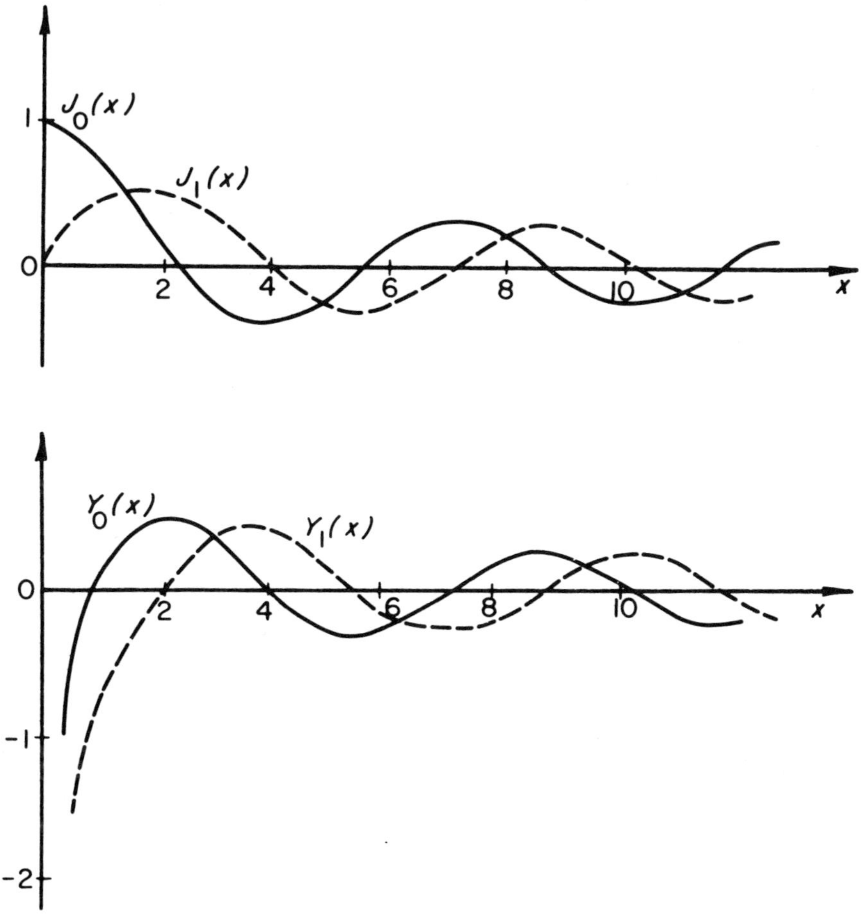

Figure 7.6.1

For $|x| \gg 1$ and $|x| \gg \nu$, the asymptotic expansion of $J_\nu(x)$ is

$$J_\nu(x) \sim \sqrt{\frac{2}{\pi x}} \left[\left\{ 1 - \frac{(4\nu^2-1)(4\nu^2-3^2)}{2!(8x)^2} + \frac{(4\nu^2-1)(4\nu^2-3^2)(4\nu^2-5^2)(4\nu^2-7^2)}{4!(8x)^4} - \dots \right\} \cos\phi - \left\{ \frac{(4\nu^2-1)}{8x} - \frac{(4\nu^2-1)(4\nu^2-3^2)(4\nu^2-5^2)}{3!(8x)^3} + \dots \right\} \sin\phi \right] \tag{7.6.24}$$

where $\phi = x - (\nu + \frac{1}{2})\frac{\pi}{2}$.

For $|x| \gg 1$ and $|x| \gg \nu$, the asymptotic expansion of $Y_\nu(x)$ is

$$Y_\nu(x) \sim \sqrt{\frac{2}{\pi x}}\left[\left\{1 - \frac{(4\nu^2-1^2)(4\nu^2-3^2)}{2!(8x)^2} + \frac{(4\nu^2-1^2)(4\nu^2-3^2)(4\nu^2-5^2)(4\nu^2-7^2)}{4!(8x)^4} - \dots\right\}\sin\phi + \left\{\frac{(4\nu^2-1^2)}{8x} - \frac{(4\nu^2-1^2)(4\nu^2-3^2)(4\nu^2-5^2)}{3!(8)^3} + \dots\right\}\cos\phi\right] \tag{7.6.25}$$

When $\nu = \pm 1/2$, Bessel's functions may be expressed in the form

$$J_{1/2}(x) = \sqrt{\frac{2}{\pi x}}\ \sin x \tag{7.6.26}$$

$$J_{-1/2}(x) = \sqrt{\frac{2}{\pi x}}\ \cos x \tag{7.6.27}$$

Bessel's functions which satisfy the condition

$$J_n(\lambda_{nj}a) + hJ_n'(\lambda_{nj}a) = 0, \quad h, a = \text{constant} \tag{7.6.28}$$

are orthogonal to each other with respect to the weight function x, that is, for a nonnegative integer n,

$$\int_0^a J_n(\lambda_{ni}x)J_n(\lambda_{nj}x)x\,dx = 0 \quad \text{for} \quad i \neq j \tag{7.6.29}$$

When $i = j$, we have the norm

$$\|J_n(\lambda_{nj}x)\|^2 = \frac{1}{2\lambda_{nj}^2}\left[\lambda_{nj}^2x^2(J_n'(\lambda_{nj}x))^2 + (\lambda_{nj}^2x^2 - n^2)(J_n(\lambda_{nj}x))^2\right]_0^a \tag{7.6.30}$$

where λ_{nj} are the roots of (7.6.28).

Closely related to Bessel's functions are *Hankel's functions of the first and second kind*, defined by

$$H_\nu^{(1)}(x) = J_\nu(x) + iY_\nu(x)$$
$$H_\nu^{(2)}(x) = J_\nu(x) - iY_\nu(x) \tag{7.6.31}$$

respectively, where $i = \sqrt{-1}$

Other closely related functions are the modified Bessel functions. Consider Bessel's equation containing a parameter λ, namely,

$$x^2y'' + xy' + (\lambda^2x^2 - \nu^2)y = 0 \tag{7.6.32}$$

The general solution of this equation is

$$y(x) = c_1J_\nu(\lambda x) + c_2Y_\nu(\lambda x)$$

If $\lambda = i$, then

$$y(x) = c_1 J_\nu(ix) + c_2 Y_\nu(ix)$$

We write

$$\begin{aligned} J_\nu(ix) &= \sum_{k=0}^{\infty} \frac{(-1)^k (ix)^{2k+\nu}}{2^{2k+\nu} k! \Gamma(\nu+k+1)} \\ &= i^\nu I_\nu(x) \end{aligned}$$

where

$$I_\nu(x) = \sum_{k=0}^{\infty} \frac{x^{2k+\nu}}{2^{2k+\nu} k! \Gamma(\nu+k+1)} \tag{7.6.33}$$

$I_\nu(x)$ is called *the modified Bessel function of the first kind of order* ν. As in the case of J_ν and $J_{-\nu}$, I_ν, and $I_{-\nu}$ (which is defined in a similar manner) are linearly independent solutions except when ν is an integer. Consequently, we define *the modified Bessel function of the second kind of order* ν as

$$K_\nu(x) = \frac{\pi}{2}\left(\frac{I_{-\nu}(x) - I_\nu(x)}{\sin \nu\pi}\right) \tag{7.6.34}$$

Thus, we obtain the general solution of the modified Bessel equation,

$$x^2 y'' + xy' - (x^2 + \nu^2) y = 0$$

as

$$y(x) = c_1 I_\nu(x) + c_2 K_\nu(x) \tag{7.6.35}$$

We should note that

$$I_\nu(0) = \begin{cases} 1 & \nu = 0 \\ 0 & \nu > 0 \end{cases} \tag{7.6.36}$$

and that K_ν approaches infinity as $x \to 0$.

For a detailed treatment of Bessel and related functions, refer to Watson's Theory of Bessel Functions [82].

The eigenvalue problems which involve Bessel's functions will be described in Sec. 7.8 on singular Sturm-Liouville systems.

7.7 Adjoint Forms: Lagrange Identity

Self-adjoint equations play a very important role in many areas of mathematical physics. Here we will give a brief account of self-adjoint operators and the Lagrange identity.

Let us consider the equation

$$L[y] = a_0(x) y'' + a_1(x) y' + a_2(x) y = 0$$

defined on an interval I. Integrating $zL[y]$ by parts from a to x, we have

$$\int_a^x zL[y]\,dx = [(za_0)y' - (za_0)'y + (za_1)y]_a^x$$
$$+ \int_a^x [(za_0)'' - (za_1)' + (za_2)]\,y\,dx \qquad (7.7.1)$$

Now, if we define the second-order operator L^* by

$$L^*[z] = (za_0)'' - (za_1)' + (za_2) = a_0 z'' + (2a_0' - a_1)z' + (a_0'' - a_1' + a_2)z,$$

the relation (7.7.1) takes the form

$$\int_a^x (zL[y] - yL^*[z])\,dx = [a_0(y'z - yz') + (a_1 - a_0')yz]_a^x \qquad (7.7.2)$$

The operator L^* is called the *adjoint operator* corresponding to the operator L. It can be readily verified that the adjoint of L^* is L itself. If L and L^* are the same, L is said to be *self-adjoint*. The necessary and sufficient condition for this is that

$$a_1 = 2a_0' - a_1$$
$$a_2 = a_0'' - a_1' + a_2$$

which is satisfied if

$$a_1 = a_0'$$

Thus, if L is self-adjoint, we have

$$L[y] = a_0 y'' + a_0' y' + a_2 y$$
$$= (a_0 y')' + a_2 y \qquad (7.7.3)$$

In general $L[y]$ is not self-adjoint. But if we let

$$h(x) = \frac{1}{a_0} \exp\left\{ \int^x \frac{a_1(t)}{a_0(t)}\,dt \right\} \qquad (7.7.4)$$

then $h(x)L[y]$ is self-adjoint. Thus any second-order linear differential equation

$$a_0(x)y'' + a_1(x)y' + a_2(x)y = 0 \qquad (7.7.5)$$

can be made self-adjoint. Multiplying by $h(x)$, given by Eq. (7.7.4), Eq. (7.7.5) is transformed into the self-adjoint form

$$\frac{d}{dx}\left[p(x)\frac{dy}{dx} \right] + q(x)y = 0 \qquad (7.7.6)$$

where

$$p(x) = \exp\left\{ \int^x \frac{a_1(t)}{a_0(t)}\,dt \right\}$$
$$q(x) = \frac{a_2}{a_0} \exp\left\{ \int^x \frac{a_1(t)}{a_0(t)}\,dt \right\} \qquad (7.7.7)$$

For example, the self-adjoint form of the Legendre equation

$$(1-x^2)y''-2xy'+n(n+1)y=0$$

is

$$\frac{d}{dx}\left[(1-x^2)\frac{dy}{dx}\right]+n(n+1)y=0 \tag{7.7.8}$$

and the self-adjoint form of the Bessel equation

$$x^2y''+xy'+(x^2-\nu^2)y=0$$

is

$$\frac{d}{dx}\left(x\frac{dy}{dx}\right)+\left(x-\frac{\nu^2}{x}\right)y=0 \tag{7.7.9}$$

Now, if we differentiate both sides of Eq. (7.7.2), we obtain

$$zL[y]-yL^*[z]=\frac{d}{dx}\left[a_0(y'z-yz')+(a_1-a_0')yz\right] \tag{7.7.10}$$

which is known as the *Lagrange identity* for the operator L.

If we consider the integral from a to b of Eq. (7.7.2), we obtain *Green's identity*,

$$\int_a^b(zL[y]-yL^*[z])\,dx=\left[a_0(y'z-yz')+(a_1-a_0')yz\right]_a^b \tag{7.7.11}$$

When L is self-adjoint, this relation becomes

$$\int_a^b(zL[y]-yL[z])\,dx=\left[a_0(y'z-yz')\right]_a^b \tag{7.7.12}$$

7.8 Singular Sturm-Liouville Systems

A Sturm-Liouville equation is called singular when it is given on a semiinfinite or infinite interval, or when the coefficient $p(x)$ or $s(x)$ vanishes, or when one of the coefficients becomes infinite at one end or both ends of a finite interval. A singular Sturm-Liouville equation together with appropriate linear homogeneous end conditions is called a *singular Sturm-Liouville system*. The conditions imposed in this case are not like the separated end conditions in the regular Sturm-Liouville system. The condition that it is often necessary to prescribe is the boundedness of the function $y(x)$ at the singular end point. To exhibit this, let us consider a problem with a singularity at the end point $x=a$. By the relation (7.7.12), for any twice continuously differentiable functions $y(x)$ and $z(x)$, we have on (a,b)

$$\int_{a+\varepsilon}^b\{zL[y]-yL[z]\}\,dx=p(b)[y'(b)z(b)-y(b)z'(b)]$$
$$-p(a+\varepsilon)[y'(a+\varepsilon)z(a+\varepsilon)-y(a+\varepsilon)z'(a+\varepsilon)]$$

where ε is a small positive number. If the conditions

$$\lim_{x\to a+} p(x)[y'(x)z(x)-y(x)z'(x)]=0 \tag{7.8.1}$$

$$p(b)[y'(b)z(b)-y(b)z'(b)]=0 \tag{7.8.2}$$

are imposed on y and z, it follows that

$$\int_a^b \{zL[y]-yL[z]\}\,dx=0 \tag{7.8.3}$$

For example, when $p(a)=0$, the relations (7.8.1) and (7.8.2) are replaced by the conditions

1. $y(x)$ and $y'(x)$ are finite as $x\to a$
2. $b_1 y(b)+b_2 y'(b)=0$

Thus, we say that the singular Sturm-Liouville system is *self-adjoint*, if any functions $y(x)$ and $z(x)$ that satisfy the end conditions satisfy

$$\int_a^b \{zL[y]-yL[z]\}\,dx=0$$

EXAMPLE 7.8.1. Consider the singular Sturm-Liouville system involving Legendre's equation

$$\frac{d}{dx}\left[(1-x^2)\frac{dy}{dx}\right]+\lambda y=0, \quad -1<x<1$$

with the conditions that y and y' are finite as $x\to\pm 1$.

In this case $p(x)=1-x^2$ and $s(x)=1$, and $p(x)$ vanishes at $x=\pm 1$. The Legendre functions of the first kind, $P_n(x), n=0,1,2,\ldots$, are the eigenfunctions which are finite as $x\to\pm 1$. The corresponding eigenvalues are $\lambda_n=n(n+1)$ for $n=0,1,2,\ldots$. We observe here that the singular Sturm-Liouville system has infinitely many real eigenvalues, and the eigenfunctions $P_n(x)$ are orthogonal to each other.

EXAMPLE 7.8.2. Another example of a singular Sturm-Liouville system is the Bessel equation for fixed ν,

$$\frac{d}{dx}\left(x\frac{dy}{dx}\right)+\left(\lambda x-\frac{\nu^2}{x}\right)y=0, \quad 0<x<a$$

with the end conditions that $y(a)=0$ and y, y' are finite as $x\to 0+$.

Here $p(x)=x$, $q(x)=-\nu^2/x$, $s(x)=x$. Now $p(0)=0$, $q(x)$ becomes infinite as $x\to 0+$, and $s(0)=0$; therefore the system is singular. If $\lambda=k^2$, the eigenfunctions of the system are Bessel functions of the first kind of order ν, namely $J_\nu(k_n x), n=1,2,3,\ldots$, where $k_n a$ is the nth zero of J_ν. The Bessel function J_ν and its derivative are both finite as $x\to 0+$. The eigenvalues are $\lambda_n=k_n^2$. Thus, the system has infinitely many eigenvalues, and the eigenfunctions are orthogonal with respect to the weight function x.

In the preceding examples, we see that the eigenfunctions are orthogonal with respect to the weight function $s(x)$. In general the eigenfunctions of a singular Sturm-Liouville system are orthogonal if they are square-integrable with respect to the weight function $s(x)$.

Theorem 7.8.1 *The square-integrable eigenfunctions corresponding to distinct eigenvalues of a singular Sturm-Liouville system are orthogonal with respect to a weight function $s(x)$.*

PROOF. Proceeding as in Theorem 7.2.1, we arrive at

$$(\lambda_j-\lambda_k)\int_a^b s\phi_j\phi_k\,dx = p(b)\left[\phi_j(b)\phi_k'(b)-\phi_j'(b)\phi_k(b)\right]$$
$$-p(a)\left[\phi_j(a)\phi_k'(a)-\phi_j'(a)\phi_k(a)\right]$$

Suppose the boundary term vanishes, as in the case mentioned earlier, where $p(a)=0$, y and y' are finite as $x\to a$, and at the other end $b_1 y(b)+b_2 y'(b)=0$. Then we have

$$(\lambda_j-\lambda_k)\int_a^b s\phi_j\phi_k\,dx=0$$

This integral exists by virtue of (7.3.2). Thus, for distinct eigenvalues $\lambda_j\neq\lambda_k$, the square-integrable functions ϕ_j and ϕ_k are orthogonal with respect to the weight function $s(x)$. □

EXAMPLE 7.8.3. Consider the singular Sturm-Liouville system involving the *Hermite equation*

$$u''-2xu'+\lambda u=0, \quad -\infty<x<\infty \tag{7.8.4}$$

which is not self-adjoint.

If we let $y(x)=e^{-x^2/2}u(x)$, the Hermite equation takes the self-adjoint form

$$y''+\left[(1-x^2)+\lambda\right]y=0, \quad -\infty<x<\infty$$

Here $p=1, q(x)=1-x^2, s=1$. The eigenvalues are $\lambda_n=2n$ for nonnegative integer n, and the corresponding eigenfunctions are $\phi_n(x)=e^{-x^2/2}H_n(x)$, where $H_n(x)$ are the Hermite polynomials of Eq. (7.8.4). (See Magnus and Oberhettinger [86].)

Now, we impose the end condition that y tends to zero as $x\to\pm\infty$. This is satisfied because $H_n(x)$ are polynomials in x and in fact $x^n e^{-x^2/2}\to 0$ as $x\to\pm\infty$. Since $\phi_n(x)$ are square-integrable, we have

$$\int_{-\infty}^{\infty} H_m(x)H_n(x)e^{-x^2}\,dx=0, \quad m\neq n$$

EXAMPLE 7.8.4. Consider the problem of the transverse vibration of a thin elastic circular membrane

$$\begin{aligned} u_{tt} &= c^2\left(u_{rr} + \frac{1}{r}u_r\right), && r<1, && t>0 \\ u(r,0) &= f(r), && 0\leqslant r\leqslant 1 \\ u_t(r,0) &= 0, && 0\leqslant r\leqslant 1 \\ u(1,t) &= 0, && && t\geqslant 0 \\ \lim_{r\to 0} u(r,t) &< \infty \end{aligned} \tag{7.8.5}$$

By the method of separation of variables, we seek a solution in the form

$$u(r,t) = R(r)T(t)$$

Substitution of this in the wave equation yields

$$\frac{R'' + (1/r)R'}{R} = \frac{1}{c^2}\frac{T''}{T} = -\alpha^2$$

where α is a positive constant. The negative sign in front of α^2 is chosen to obtain the solution periodic in time. Thus, we have

$$rR'' + R' + \alpha^2 rR = 0$$

$$T'' + \alpha^2 c^2 T = 0$$

The solution $T(t)$ is therefore given by

$$T(t) = A\cos\alpha ct + B\sin\alpha ct$$

Next, it is required to determine the solution $R(r)$ of the following singular Sturm-Liouville system

$$\frac{d}{dr}\left[r\frac{dR}{dr}\right] + \alpha^2 rR = 0 \tag{7.8.6}$$

$$R(1) = 0 \tag{7.8.7}$$

$$\lim_{r\to 0} R(r) < \infty \tag{7.8.8}$$

We note that in this case, $p=r$ which vanishes at $r=0$. The condition on the boundedness of the function $R(r)$ is obtained from the fact that

$$\lim_{r\to 0} u(r,t) = \lim_{r\to 0} R(r)T(t) < \infty$$

which implies that

$$\lim_{r\to 0} R(r) < \infty \tag{7.8.9}$$

for arbitrary $T(t)$. Equation (7.8.6) is Bessel's equation of order zero, the solution of which is given by

$$R(r) = CJ_0(\alpha r) + DY_0(\alpha r) \tag{7.8.10}$$

where J_0 and Y_0 are Bessel functions of the first and second kinds respectively of order zero. The condition (7.8.8) requires that $D=0$ since

$Y_0(\alpha r) \to -\infty$ as $r \to 0$. Hence

$$R(r) = CJ_0(\alpha r)$$

The remaining condition $R(1)=0$ yields

$$J_0(\alpha)=0$$

This transcendental equation has infinitely many positive zeros

$$\alpha_1 < \alpha_2 < \alpha_3 < \ldots$$

Thus, the solution of problem (7.8.5) is given by

$$u_n(r,t) = J_0(\alpha_n r)(A_n \cos\alpha_n ct + B_n \sin\alpha_n ct), \quad n=1,2,3,\ldots$$

Since the Bessel equation is linear and homogeneous,

$$u(r,t) = \sum_{n=1}^{\infty} J_0(\alpha_n r)(A_n \cos\alpha_n ct + B_n \sin\alpha_n ct) \tag{7.8.11}$$

is also a solution, provided the series converges and is sufficiently differentiable with respect to r and t. Differentiating (7.8.11) formally with respect to t, we obtain

$$u_t(r,t) = \sum_{n=1}^{\infty} J_0(\alpha_n r)(-A_n \alpha_n c \sin\alpha_n ct + B_n \alpha_n c \cos\alpha_n ct)$$

Application of the initial condition $u_t(r,0)=0$ yields $B_n=0$. Consequently, we have

$$u(r,t) = \sum_{n=1}^{\infty} A_n J_0(\alpha_n r)\cos\alpha_n ct \tag{7.8.12}$$

It now remains to show that $u(r,t)$ satisfies the initial condition $u(r,0)=f(r)$. For this, we have

$$u(r,0) = f(r) \sim \sum_{n=1}^{\infty} A_n J_0(\alpha_n r)$$

If $f(r)$ is piecewise smooth on $[0,1]$, then the eigenfunctions $J_0(\alpha_n r)$ form a complete orthogonal system with respect to the weight function r over the interval $(0,1)$. Hence we can formally expand $f(r)$ in terms of the eigenfunctions. Thus

$$f(r) = \sum_{n=1}^{\infty} A_n J_0(\alpha_n r)$$

where the coefficient A_n is represented by

$$A_n = \int_0^1 rf(r)J_0(\alpha_n r)\,dr \Big/ \int_0^1 r[J_0(\alpha_n r)]^2\,dr \tag{7.8.13}$$

The solution of the problem (7.8.5) is therefore given by (7.8.12) with the coefficient A_n given by (7.8.13).

7.9 Legendre's Equation

Legendre's equation is

$$(1-x^2)y''-2xy'+\nu(\nu+1)y=0 \tag{7.9.1}$$

where ν is a real number. This equation arises in problems with spherical symmetry in mathematical physics. Its coefficients are analytic at $x=0$. Thus, if we expand near the point $x=0$, the coefficients are

$$p(x)=-\frac{2x}{1-x^2}=-2x\sum_{m=0}^{\infty}x^{2m}=\sum_{m=0}^{\infty}(-2)x^{2m+1}$$

and

$$q(x)=\frac{\nu(\nu+1)}{1-x^2}=\nu(\nu+1)\sum_{m=0}^{\infty}x^{2m}=\sum_{m=0}^{\infty}\nu(\nu+1)x^{2m}$$

We see that these series converge for $|x|<1$. Thus the Legendre equation on $|x|<1$ has convergent power series solution at $x=0$.

Now to find the solution near the ordinary point $x=0$, we assume

$$y(x)=\sum_{m=0}^{\infty}a_m x^m$$

Then differentiation yields

$$y'(x)=\sum_{m=0}^{\infty}ma_m x^{m-1}$$

$$y''(x)=\sum_{m=0}^{\infty}m(m-1)a_m x^{m-2}$$

Substituting y, y', and y'' in the Legendre equation, we obtain

$$(1-x^2)\sum_{m=0}^{\infty}m(m-1)a_m x^{m-2}-2x\sum_{m=0}^{\infty}ma_m x^{m-1}+\nu(\nu+1)\sum_{m=0}^{\infty}a_m x^m=0$$

Simplification gives

$$\sum_{m=0}^{\infty}\left[(m+1)(m+2)a_{m+2}+(\nu-m)(\nu+m+1)a_m\right]x^m=0$$

The coefficients in the power series must therefore satisfy the recurrence relation

$$a_{m+2}=-\frac{(\nu-m)(\nu+m+1)}{(m+1)(m+2)}a_m,\quad m\geqslant 0 \tag{7.9.2}$$

This relation determines $a_2, a_4, a_6, \ldots$ in terms of a_0, and $a_3, a_5, a_7, \ldots$ in terms of a_ν. Thus we write

$$a_2 = -\frac{\nu(\nu+1)}{1\times 2}a_0$$

$$a_3 = -\frac{(\nu-1)(\nu+2)}{2\times 3}a_1$$

$$a_4 = -\frac{(\nu-2)(\nu+3)}{3\times 4}a_2$$

$$a_5 = -\frac{(\nu-3)(\nu+4)}{4\times 5}a_3$$

$$a_{2k} = -\frac{(\nu-2k+2)(\nu+2k-1)}{(2k-1)(2k)}a_{2k-2}$$

$$a_{2k+1} = -\frac{(\nu-2k+1)(\nu+2k)}{(2k)(2k+1)}a_{2k-1}$$

Taking the product of the left and right sides of the above equations and cancelling the like terms, we obtain

$$a_{2k} = \frac{(-1)^k\nu(\nu-2)\ldots(\nu-2k+2)(\nu+1)(\nu+3)\ldots(\nu+2k-1)}{(2k)!}a_0$$

and

$$a_{2k+1} = \frac{(-1)^k(\nu-1)(\nu-3)\ldots(\nu-2k+1)(\nu+2)(\nu+4)\ldots(\nu+2k)}{(2k+1)!}a_1$$

Hence the solution of the Legendre equation is

$$y(x) = a_0\left[1 + \sum_{k=1}^{\infty}\frac{(-1)^k\nu(\nu-2)\ldots(\nu-2k+2)(\nu+1)(\nu+3)\ldots(\nu+3k-1)x^{2k}}{(2k)!}\right]$$

$$+a_1\left[x + \sum_{k=1}^{\infty}\frac{(-1)^k(\nu-1)(\nu-3)\ldots(\nu-2k+1)(\nu+2)(\nu+4)\ldots(\nu+2k)x^{2k+1}}{(2k+1)!}\right]$$

$$= a_0\phi_\nu(x) + a_1\psi_\nu(x) \qquad (7.9.3)$$

It can easily be proved that the functions $\phi_\nu(x)$ and $\psi_\nu(x)$ converge for $|x|<1$ and are linearly independent.

Now consider the case in which $\nu = n$, with n a nonnegative integer. It is then evident from the recurrence relation (7.9.2) that when $m = n$,

$$a_{n+2} = a_{n+4} = \ldots = 0$$

Consequently, when n is even, the series $\phi_n(x)$ terminates with x^n, whereas the series for $\psi_n(x)$ does not terminate. When n is odd, it is the series for $\psi_n(x)$ which terminates with x^n, while that for $\phi_n(x)$ does not terminate. In the first case (n even), $\phi_n(x)$ is a polynomial of degree n; the same is true for $\psi_n(x)$ in the second case (n odd).

Thus, for any nonnegative integer n, either $\phi_n(x)$ or $\psi_n(x)$, but not both, is a polynomial of degree n. Consequently, the general solution of the Legendre equation contains a polynomial solution $P_n(x)$ and an infinite-series solution $Q_n(x)$ for a nonnegative integer n. To find the polynomial solution $P_n(x)$ it is convenient to choose a_n so that $P_n(1)=1$. Let this a_n be

$$a_n=\frac{(2n)!}{2^n(n!)^2} \tag{7.9.4}$$

Rewriting the recurrence relation (7.9.2), we have

$$a_{n-2}=-\frac{(n-1)n}{2(2n-1)}a_n$$

Substituting a_n from (7.9.4) into this relation, we obtain

$$a_{n-2}=-\frac{(2n-2)!}{2^n(n-1)!(n-2)!}$$

and

$$a_{n-4}=\frac{(2n-4)!}{2^n 2!(n-2)!(n-4)!}$$

It follows by induction that

$$a_{n-2k}=\frac{(-1)^k(2n-2k)!}{2^n k!(n-k)!(n-2k)!}$$

Hence we may write $P_n(x)$ in the form

$$P_n(x)=\sum_{k=0}^{N}\frac{(-1)^k(2n-2k)!}{2^n k!(n-k)!(n-2k)!}x^{n-2k} \tag{7.9.5}$$

where $N=n/2$ when n is even and $N=(n-1)/2$ when n is odd. This polynomial $P_n(x)$ is called the *Legendre function of the first kind of order n*. It is also known as the *Legendre polynomial of degree n*.

The first few Legendre polynomials are

$$P_0(x)=1$$

$$P_1(x)=x$$

$$P_2(x)=\frac{1}{2}(3x^2-1)$$

$$P_3(x) = \frac{1}{2}(5x^3 - 3x)$$

$$P_4(x) = \frac{1}{8}(35x^4 - 30x^2 + 3)$$

These polynomials are plotted in Fig. 7.9.1 for small values of x.

Recall that for a given nonnegative integer n, only one of the two solutions $\phi_n(x)$ and $\psi_n(x)$ of Legendre's equation is a polynomial, while the other is an infinite series. This infinite series, when appropriately normalized, is called the *Legendre function of the second kind*. It is defined for $|x| < 1$ by

$$Q_n(x) = \begin{cases} \phi_n(1)\psi_n(x) & \text{for } n \text{ even} \\ -\psi_n(1)\phi_n(x) & \text{for } n \text{ odd} \end{cases} \tag{7.9.6}$$

Thus, when n is a nonnegative integer, the general solution of the Legendre equation is given by

$$y(x) = c_1 P_n(x) + c_2 Q_n(x) \tag{7.9.7}$$

Figure 7.9.1

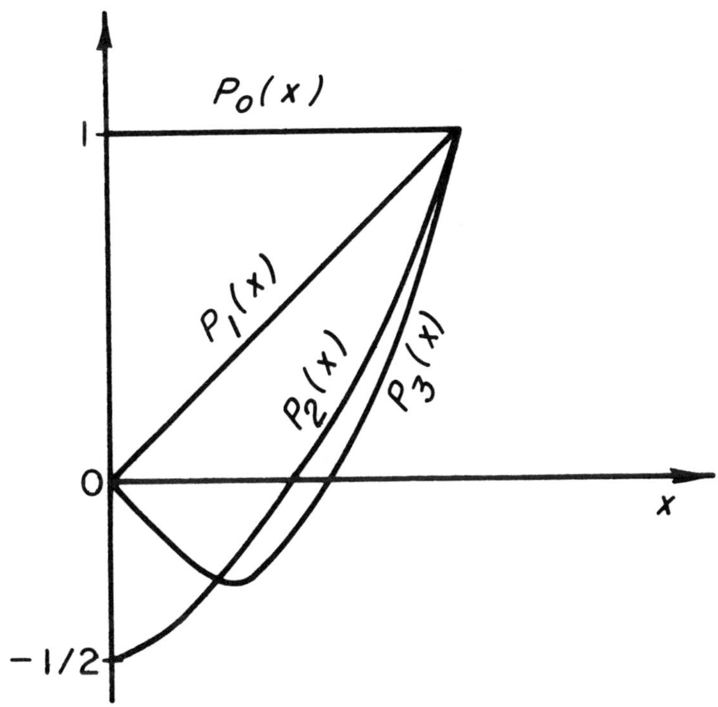

The Legendre polynomial may also be expressed in the form

$$P_n(x)=\frac{1}{2^n n!}\frac{d^n}{dx^n}(x^2-1)^n \tag{7.9.8}$$

This expression is known as the *Rodriguez formula.*

Like Bessel's functions, Legendre polynomials satisfy certain recurrence relations. Some of the important relations are

$$(n+1)P_{n+1}(x)-(2n+1)xP_n(x)+nP_{n-1}(x)=0, \quad n\geqslant 1 \tag{7.9.9}$$

$$(x^2-1)P_n'(x)=nxP_n(x)-nP_{n-1}(x), \quad n\geqslant 1 \tag{7.9.10}$$

$$nP_n(x)+P_{n-1}'(x)-xP_n'(x)=0, \quad n\geqslant 1 \tag{7.9.11}$$

$$P_{n+1}'(x)=xP_n'(x)+(n+1)P_n(x), \quad n\geqslant 0 \tag{7.9.12}$$

In addition,

$$P_{2n}(-x)=P_{2n}(x) \tag{7.9.13}$$

$$P_{2n+1}(-x)=-P_{2n+1}(x) \tag{7.9.14}$$

These indicate that $P_n(x)$ is an even function for n even, and an odd function for n odd.

It can easily be shown that the Legendre polynomials form a sequence of orthogonal functions on the interval $[1,-1]$. Thus, we have

$$\int_{-1}^{1} P_n(x)P_m(x)\,dx=0 \quad \text{for} \quad n\neq m \tag{7.9.15}$$

The norm of the function $P_n(x)$ is given by

$$\|P_n(x)\|^2=\int_{-1}^{1} P_n^2(x)\,dx=\frac{2}{2n+1} \tag{7.9.16}$$

Another important equation in mathematical physics, one which is closely related to the Legendre equation (7.9.1), is *Legendre's associated equation.*

$$(1-x^2)y''-2xy'+\left[n(n+1)-\frac{m^2}{1-x^2}\right]y=0 \tag{7.9.17}$$

where m is an integer. Although this equation is independent of the algebraic sign of the integer m, it is often convenient to have the solutions for negative m differ somewhat from those for positive m.

We consider first the case for a nonnegative integer m. Introducing the change of variable

$$y=(1-x^2)^{m/2}u, \quad |x|<1$$

Legendre's assocated equation becomes

$$(1-x^2)u''-2(m+1)xu'+(n-m)(n+m+1)u=0$$

But this is the same as the equation obtained by differentiating the

Legendre equation (7.9.1) m times. Thus, the general solution of (7.9.17) is given by

$$y(x)=(1-x^2)^{m/2}\frac{d^mY(x)}{dx^m} \tag{7.9.18}$$

where

$$Y(x)=c_1P_n(x)+c_2Q_n(x) \tag{7.9.19}$$

is the general solution of (7.9.1). Hence we have the linearly independent solutions of (7.9.17), known as the *associated Legendre functions of the first and second kind*, respectively given by

$$P_n^m(x)=(1-x^2)^{m/2}\frac{d^mP_n(x)}{dx^m} \tag{7.9.20}$$

and

$$Q_n^m(x)=(1-x^2)^{m/2}\frac{d^mQ_n(x)}{dx^m} \tag{7.9.21}$$

We observe that

$$P_n^0(x)=P_n(x)$$
$$Q_n^0(x)=Q_n(x)$$

and that $P_n^m(x)$ vanishes for $m>n$.

$P_n^{-m}(x)$ and $Q_n^{-m}(x)$ are defined by

$$P_n^{-m}(x)=(-1)^m\frac{(n-m)!}{(n+m)!}P_n^m(x), \quad m=0,1,2,\dots,n \tag{7.9.22}$$

$$Q_n^{-m}(x)=(-1)^m\frac{(n-m)!}{(n+m)!}Q_n^m(x), \quad m=0,1,2,\dots,n \tag{7.9.23}$$

The first few associated Legendre functions are

$$P_1^1(x)=(1-x^2)^{1/2}$$
$$P_2^1(x)=3x(1-x^2)^{1/2}$$
$$P_2^2(x)=3(1-x^2)$$

The associated Legendre functions of the first kind also form a sequence of orthogonal functions in the interval $[-1,1]$. Their orthogonality, as well as their norm, is expressed by the equation

$$\int_{-1}^{1}P_n^m(x)P_k^m(x)\,dx=\frac{2(n+m)!}{(2n+1)(n-m)!}\delta_{nk} \tag{7.9.24}$$

Note that (7.9.15) and (7.9.16) are special cases of (7.9.24), corresponding to the choice $m=0$.

We finally observe that whereas $P_n^m(x)$ is bounded everywhere in the interval $[-1,1]$, $Q_n^m(x)$ is unbounded at the end points $x=\pm 1$.

Problems in which Legendre's polynomials arise will be treated in Chapter 9.

7.10 Boundary Value Problems Involving Ordinary Differential Equations

A *boundary-value problem* consists in finding an unknown solution which satisfies an ordinary differential equation and appropriate boundary conditions at two or more points. This is in contrast to an initial-value problem for which a unique solution exists for an equation satisfying prescribed initial conditions at one point.

The linear two-point boundary-value problem, in general, may be written in the form

$$L[y]=f(x), \quad a<x<b$$

$$U_i[y]=\alpha_i, \quad 1\leqslant i\leqslant n \tag{7.10.1}$$

where L is the linear operator of order n and U_i is the boundary operator defined by

$$U_i[y]=\sum_{j=1}^{n} a_{ij}y^{(j-1)}(a)+\sum_{j=1}^{n} b_{ij}y^{(j-1)}(b) \tag{7.10.2}$$

Here a_{ij}, b_{ij}, and α_i are constants. The treatment of this problem can be found in Coddington and Levinson [59]. More complicated boundary conditions occur in practice. Treating a general differential system is rather complex and difficult.

A large class of boundary-value problems that occur often in physical science consists of the second-order equations of the type

$$y''=f(x,y,y'), \quad a<x<b$$

with the boundary conditions

$$U_1[y]=a_1y(a)+a_2y'(a)=\alpha$$

$$U_2[y]=b_1y(b)+b_2y'(b)=\beta$$

where a_1, a_2, b_1, b_2, α, and β are constants. The existence and uniqueness of solutions to this problem are treated by Keller [64]. Here we are interested in considering a special case where the linear boundary-value problem consists of the differential equation

$$L[y]=y''+p(x)y'+q(x)y=f(x) \tag{7.10.3}$$

and the boundary conditions

$$U_1[y]=a_1y(a)+a_2y'(a)=\alpha$$

$$U_2[y]=b_1y(b)+b_2y'(b)=\beta \tag{7.10.4}$$

where the constants a_1 and a_2, and likewise b_1 and b_2, are not both zero, and α and β are constants.

In general a boundary-value problem may not possess a solution, and if it does, the solution may not be unique. We will illustrate this with a simple problem.

EXAMPLE 7.10.1. We first consider the boundary-value problem

$$y''+y=1$$
$$y(0)=0 \quad y\left(\frac{\pi}{2}\right)=0$$

By the method of variation of parameters, we find a unique solution

$$\phi(x)=1-\cos x-\sin x$$

We observe that the solution of the associated homogeneous boundary-value problem

$$y''+y=0$$
$$y(0)=0 \quad y\left(\frac{\pi}{2}\right)=0$$

is trivial.

Next we consider the boundary-value problem

$$y''+y=1$$
$$y(0)=0 \quad y(\pi)=0$$

The general solution is

$$y(x)=c_1\cos x+c_2\sin x+1$$

Applying the boundary conditions, we see that

$$y(0)=c_1+1=0$$
$$y(\pi)=-c_1+1=0$$

This is not possible, and hence the boundary-value problem has no solution. However, if we consider its associated homogeneous boundary-value problem

$$y''+y=0$$
$$y(0)=0 \quad y(\pi)=0$$

we can easily determine that solutions exist and are given by

$$y(x)=c_2\sin x$$

where c_2 is an arbitrary constant.

We are thus led to the following alternative theorem.

Theorem 7.10.1 *Let $p(x)$, $q(x)$, and $f(x)$ be continuous on $[a,b]$. Then either the boundary-value problem*

$$L[y]=f$$

$$U_1[y]=\alpha \quad U_2[y]=\beta \tag{7.10.5}$$

has a unique solution for any given constants α and β, or else the associated homogeneous boundary-value problem

$$L[y]=0$$

$$U_1[y]=0 \quad U_2[y]=0 \tag{7.10.6}$$

has a nontrivial solution.

Proof of this theorem can be found in [67].

7.11 Green's Functions

In the present section we will introduce Green's functions. Let us consider the linear nonhomogeneous ordinary differential equation of second order:[7]

$$L[y]=-f(x) \tag{7.11.1}$$

in $[a,b]$, where $L=(d/dx)[p(x)(d/dx)]+q(x)$, with the homogeneous boundary conditions

$$a_1 y(a)+a_2 y'(a)=0 \tag{7.11.2}$$

$$b_1 y(b)+b_2 y'(b)=0 \tag{7.11.3}$$

where the constants a_1 and a_2, and likewise b_1 and b_2, are not both zero. We shall assume that f and q are continuous and that p is continuously differentiable and does not vanish in the interval $[a,b]$. We first present a heuristic argument for Green's function.

We consider a flexible cable which is stretched horizontally. Since the flexible cable has no resistance to bending, the only internal force is the tension T acting in the direction of the tangent of the deflection curve.

Let $y(x)$ be the vertical displacement of the cable. Let $\rho(x)$ be the continuously distributed load per unit length. Thus, the load $\rho(x)\Delta x$ supported by the cable in the interval $(x,x+\Delta x)$ is in equilibrium with the vertical component of the tension T. If we assume that the vertical displacement is small compared with the length of the cable, we have the equilibrium equation for an element between x and $x+\Delta x$,

$$T[y'(x+\Delta x)-y'(x)]-\rho(x)\Delta x=0$$

[7]The minus sign is introduced for consistency with the usual notation.

Dividing this equation by Δx, we obtain

$$\frac{y'(x+\Delta x)-y'(x)}{\Delta x}=\frac{\rho(x)}{T}$$

Thus, in the limit as $\Delta x\to 0$, we have

$$y''(x)=\rho(x)/T$$

which may be written as

$$L[y]=y''(x)=-g(x)$$

where $g(x)=-\rho(x)/T$.

If we assume that the ends of the cable are connected to the end supports such that the vertical reactions are proportional to the displacements, the boundary conditions take the form

$$Ty'(a)=-a_1 y(a)$$

$$Ty'(b)=-b_1 y(b)$$

which may be written as

$$a_1 y(a)+a_2 y'(a)=0$$

$$b_1 y(b)+b_2 y'(b)=0$$

where $a_2=b_2=T$. There are other boundary conditions one of which, for example, is a fixed end. If the end at $x=a$ is fixed, then $y(a)=0$.

We will now consider the boundary-value problem (7.2.1–7.2.3). If we denote by $G(x,\xi)$ the deflection of a cable at x due to a unit concentrated force applied at the point ξ, then the deflection at x due to a uniformly distributed force $f(\xi)$ per unit length over an elemental interval $(\xi,\xi+d\xi)$ is given by $f(\xi)G(x,\xi)d\xi$. Thus, because of the linearity of the problem, the deflection of the cable at x due to the distribution of force $f(\xi)$ over the entire interval $[a,b]$ is given by the integral

$$y(x)=\int_a^b G(x,\xi)f(\xi)\,d\xi \tag{7.11.4}$$

The function $G(x,\xi)$ is called the *influence function* or *Green's function*. It is evident that $G(x,\xi)$ must be defined and continuous in $[a,b]$ and must satisfy the prescribed boundary conditions. By its very definition, $G(x,\xi)$ is a solution of the equation

$$L[y]=-f_\varepsilon(x) \tag{7.11.5}$$

where $f_\varepsilon(x)$ is the function which vanishes outside the interval $|x-\xi|<\varepsilon$, and in the interval $|x-\xi|<\varepsilon$, it satisfies the relation

$$\int_{\xi-\varepsilon}^{\xi+\varepsilon} f_\varepsilon(x)\,dx=1 \tag{7.11.6}$$

Thus, for all $x\neq\xi$, $G(x,\xi)$ satisfies the homogeneous equation

$$L[y]=0$$

When $x=\xi$, we have from Eq. (7.11.5) the following relation after integration:

$$\int_{\xi-\varepsilon}^{\xi+\varepsilon}\frac{d}{dx}[p(x)G'(x,\xi)]\,dx+\int_{\xi-\varepsilon}^{\xi+\varepsilon}q(x)G(x,\xi)\,dx=-\int_{\xi-\varepsilon}^{\xi+\varepsilon}f_\varepsilon(x)\,dx$$

or

$$p(x)G'(x,\xi)\Big|_{\xi-\varepsilon}^{\xi+\varepsilon}+\int_{\xi-\varepsilon}^{\xi+\varepsilon}q(x)G(x,\xi)\,dx=-1$$

If we now assume that $G(x,\xi)$ is continuously differentiable except at $x=\xi$, then in the limit as ε approaches zero, we obtain

$$\left.\frac{dG(x,\xi)}{dx}\right|_{x=\xi-}^{x=\xi+}=-\frac{1}{p(\xi)} \tag{7.11.7}$$

which states that the derivative of $G(x,\xi)$ has a jump discontinuity at $x=\xi$.

This heuristic discussion leads us to the following definition of the Green's function.

The *Green's function* for the differential expression $L[y]$ under given homogeneous boundary conditions is the function $G(x,\xi)$ satisfying the following conditions:

i. $G(x,\xi)$ is continuous for all values of x, and its first and second derivatives are continuous for all $x\neq\xi$ in $a\leqslant x,\xi\leqslant b$.
ii. At the point $x=\xi$, the first derivative of $G(x,\xi)$ has a jump discontinuity given by

$$\left.\frac{dG(x,\xi)}{dx}\right|_{x=\xi-}^{x=\xi+}=-\frac{1}{p(\xi)}$$

iii. For fixed ξ, $G(x,\xi)$ satisfies the prescribed boundary conditions. Moreover, $G(x,\xi)$ is the solution of the associated homogeneous equation

$$L[y]=0$$

except at the point $x=\xi$.

With this definition in mind, we state the fundamental theorem for the Green's function.

Theorem 7.11.1 *If $f(x)$ is continuous in the interval $[a,b]$, then the function*

$$y(x)=\int_a^b G(x,\xi)f(\xi)\,d\xi$$

is a solution of the boundary-value problem

$$L[y] = -f(x)$$
$$a_1 y(a) + a_2 y'(a) = 0$$
$$b_1 y(b) + b_2 y'(b) = 0$$

PROOF. First, we differentiate $y(x)$ with respect to x by Leibnitz's Rule,[8] and obtain

$$y'(x) = \int_a^x G'(x,\xi) f(\xi)\, d\xi + G(x, x-) f(x)$$
$$+ \int_x^b G'(x,\xi) f(\xi)\, d\xi - G(x, x+) f(x)$$

Since $G(x,\xi)$ is continuous in ξ, we have

$$G(x, x-) = G(x, x+)$$

Thus, $y'(x)$ may be written

$$y'(x) = \int_a^b G'(x,\xi) f(\xi)\, d\xi$$

Since G satisfies the boundary conditions, we have

$$a_1 y(a) + a_2 y'(a) = \int_a^b [a_1 G(a,\xi) + a_2 G'(a,\xi)] f(\xi)\, d\xi = 0$$

Similarly,

$$b_1 y(b) + b_2 y'(b) = 0$$

Differentiating y' again with respect to x, we have

$$y''(x) = \int_a^x G''(x,\xi) f(\xi)\, d\xi + G'(x, x-) f(x)$$
$$+ \int_x^b G''(x,\xi) f(\xi)\, d\xi - G'(x, x+) f(x)$$
$$= \int_a^b G''(x,\xi) f(\xi)\, d\xi + f(x)[G'(x, x-) - G'(x, x+)]$$

But we may write condition (7.11.7) in an equivalent form (see Exercises in Chapter 16)

$$\frac{dG}{dx}(x,\xi)\Big|_{\xi = x-}^{\xi = x+} = \frac{1}{p(x)} \qquad (7.11.8)$$

---

[8]See W. Kaplan, *Advanced Calculus*, Addison-Wesley, p. 220, 1953.

Hence y'' becomes

$$y''(x)=\int_a^b G''(x,\xi)f(\xi)\,d\xi - f(x)\left[G'(x+,x)-G'(x-,x)\right]$$

$$=\int_a^b G''(x,\xi)f(\xi)\,d\xi - \frac{f(x)}{p(x)}$$

Therefore,

$$L[y]=p(x)y''(x)+p'(x)y'(x)+q(x)y(x)$$

$$=-f(x)+\int_a^b\left[p(x)G''(x,\xi)+p'(x)G'(x,\xi)\right.$$

$$\left.+q(x)G(x,\xi)\right]f(\xi)\,d\xi$$

$$=-f(x)+\int_a^b L[G]f(\xi)\,d\xi$$

Since by the definition of G, $L[G]=0$ for $x\neq\xi$, we have

$$L[y]=-f(x)$$

To express the Green's function in two intervals separated by $x=\xi$, we let

$$G(x,\xi)=\begin{cases} G_1(x,\xi) & \text{for } a<x\leqslant\xi \\ G_2(x,\xi) & \text{for } \xi\leqslant x<b \end{cases} \tag{7.11.9}$$

Then from the continuity condition, we must have

$$G_1(\xi,\xi)=G_2(\xi,\xi)$$

and from the condition (7.11.7)

$$\left.\frac{dG}{dx}(x,\xi)\right|_{x=\xi-}^{x=\xi+}=\left.\frac{dG_2}{dx}(x,\xi)\right|_{x=\xi+}-\left.\frac{dG_1}{dx}(x,\xi)\right|_{x=\xi-}=-\frac{1}{p(\xi)} \tag{7.11.10}$$

Similarly, if we take ξ as the variable, we may define

$$G(x,\xi)=\begin{cases} G_1(x,\xi) & \text{for } a\leqslant\xi<x \\ G_2(x,\xi) & \text{for } x<\xi\leqslant b \end{cases} \tag{7.11.11}$$

where G_1 and G_2 are continuous. Thus,

$$G_1(x,x)=G_2(x,x)$$

It follows from the condition (7.11.8) that

$$\left.\frac{dG}{dx}(x,\xi)\right|_{\xi=x-}^{\xi=x+}=\left.\frac{dG_2}{dx}(x,\xi)\right|_{\xi=x+}-\left.\frac{dG_1}{dx}(x,\xi)\right|_{\xi=x-}=\frac{1}{p(x)} \tag{7.11.12}$$

□

EXAMPLE 7.11.1. Consider the problem

$$y'' = -x$$
$$y(0) = 0$$
$$y(1) = 0 \tag{7.11.13}$$

For a fixed value of ξ, the Green's function $G(x,\xi)$ satisfies the associated homogeneous equation

$$G'' = 0$$

in $0 < x < \xi, \xi < x < 1$, and the boundary conditions

$$G(0,\xi) = 0$$
$$G(1,\xi) = 0$$

In addition, it satisfies

$$\frac{dG}{dx}(x,\xi)\bigg|_{x=\xi-}^{x=\xi+} = -\frac{1}{p(\xi)}$$

Now if we choose $G(x,\xi)$ such that

$$G(x,\xi) = \begin{cases} G_1(x,\xi) = (1-\xi)x & \text{for } 0 \leqslant x \leqslant \xi \\ G_2(x,\xi) = (1-x)\xi & \text{for } \xi \leqslant x \leqslant 1 \end{cases}$$

It can be seen that $G'' = 0$ over the intervals $0 < x < \xi$, $\xi < x < 1$. Also

$$G_1(0,\xi) = 0$$
$$G_2(1,\xi) = 0$$

Moreover

$$G_2'(x,\xi) - G_1'(x,\xi) = -\xi - (1-\xi) = -1$$

which is the value of the jump $-1/p(\xi)$, because in this case $p = 1$. Hence from Theorem 7.11.1, keeping in mind that ξ is the variable in $G(x,\xi)$, the solution of (7.11.13) is

$$\begin{aligned} y(x) &= \int_0^x G(x,\xi)f(\xi)\,d\xi + \int_x^1 G(x,\xi)f(\xi)\,d\xi \\ &= \int_0^x (1-x)\xi^2 d\xi + \int_x^1 x(1-\xi)\,\xi d\xi \\ &= \frac{x}{6}(1-x^2) \end{aligned}$$

7.12 Construction of Green's Function

In the above example, we see that the solution was obtained immediately as soon as the Green's function was selected properly. Thus, the real problem is not that of finding the solution but that of determining the Green's function for the problem. We will now show by construction that there exists a Green's function for $L[y]$ satisfying the prescribed boundary conditions.

We first assume that the associated homogeneous equation satisfying the conditions (7.11.2) and (7.11.3) has the trivial solution only, as in Example 7.11.1. We construct the solution $y_1(x)$ of

$$L[y]=0$$

satisfying $a_1 y(a)+a_2 y'(a)=0$. We see that $c_1 y_1(x)$ is the most general such solution, where c_1 is an arbitrary constant.

In a similar manner, we let $c_2 y_2(x)$, with c_2 as an arbitrary constant, be the most general solution of

$$L[y]=0$$

satisfying $b_1 y(b)+b_2 y'(b)=0$. Thus, y_1 and y_2 exist in the interval (a,b) and are linearly independent. For, if they were linearly dependent, then $y_1=cy_2$, which shows that y_1 would satisfy both the boundary conditions at $x=a$ and $x=b$. This contradicts our assumption about the trivial solution. Consequently, the Green's function can take the form

$$G(x,\xi)=\begin{cases} c_1(\xi)y_1(x) & \text{for } x<\xi \\ c_2(\xi)y_2(x) & \text{for } x>\xi \end{cases} \tag{7.12.1}$$

Since $G(x,\xi)$ is continuous at $x=\xi$, we have

$$c_2(\xi)y_2(\xi)-c_1(\xi)y_1(\xi)=0 \tag{7.12.2}$$

The discontinuity in the derivative of G at that point requires that

$$\frac{dG}{dx}(x,\xi)\bigg|_{x=\xi-}^{x=\xi+}=c_2(\xi)y_2'(\xi)-c_1(\xi)y_1'(\xi)=-\frac{1}{p(\xi)} \tag{7.12.3}$$

Solving Eqs. (7.12.2) and (7.12.3) for c_1 and c_2, we find

$$c_1(\xi)=\frac{-y_2(\xi)}{p(\xi)\,W(y_1,y_2;\xi)}$$

$$c_2(\xi)=\frac{-y_1(\xi)}{p(\xi)\,W(y_1,y_2;\xi)} \tag{7.12.4}$$

where $W(y_1,y_2;\xi)$ is the Wronskian given by $W(y_1,y_2;\xi)=y_1(\xi)y_2'(\xi)-y_2(\xi)y_1'(\xi)$. Since the two solutions are linearly independent, the Wronskian differs from zero.

From Theorem 7.2.4. with $\lambda=0$, we have

$$pW=\text{constant}=C \tag{7.12.5}$$

Hence the Green's function is given by

$$G(x,\xi)=\begin{cases} -y_1(x)y_2(\xi)/C & \text{for } x\leqslant\xi \\ -y_2(x)y_1(\xi)/C & \text{for } x\geqslant\xi \end{cases} \tag{7.12.6}$$

Thus, we state the following theorem:

Theorem 7.12.1 *If the associated homogeneous boundary-value problem of* (7.11.1–7.11.3) *has the trivial solution only, then the Green's function exists and is unique.*

The proof for uniqueness of Green's function is left as an exercise for the reader.

EXAMPLE 7.12.1. Consider the problem

$$\begin{aligned} y''+y&=-1 \\ y(0)&=0 \\ y\left(\frac{\pi}{2}\right)&=0 \end{aligned} \tag{7.12.7}$$

The solution of $L[y]=dy'/dx+y=0$ satisfying $y(0)=0$ is

$$y_1(x)=\sin x,\quad 0\leqslant x<\xi$$

and the solution of $L[y]=0$ satisfying $y(\pi/2)=0$ is

$$y_2(x)=\cos x,\quad \xi<x\leqslant\frac{\pi}{2}$$

The Wronskian of y_1 and y_2 is then given by

$$W(\xi)=y_1(\xi)y_2'(\xi)-y_2(\xi)y_1'(\xi)=-1$$

Since in this case $p=1$, (7.12.6) becomes

$$G(x,\xi)=\begin{cases} \sin x\cos\xi & \text{for } x\leqslant\xi \\ \cos x\sin\xi & \text{for } x\geqslant\xi \end{cases}$$

Therefore the solution of (7.12.7) is

$$\begin{aligned} y(x)&=\int_0^x G(x,\xi)f(\xi)\,d\xi+\int_x^{\pi/2}G(x,\xi)f(\xi)\,d\xi \\ &=\int_0^x \cos x\sin\xi\,d\xi+\int_x^{\pi/2}\sin x\cos\xi\,d\xi \\ &=-1+\sin x+\cos x \end{aligned}$$

It can be seen in the formula (7.12.6) that the Green's function is symmetric in x and ξ. Thus, we state:

Theorem 7.12.2 *The Green's function for a boundary-value problem* (7.11.1–7.11.3) *is symmetric, that is,* $G(x,\xi)=G(\xi,x)$.

Proof of this can be found in [67].

7.13 Nonhomogeneous Boundary Conditions

The solution of the two-point boundary-value problem

$$L[y]=-f(x) \tag{7.13.1}$$

$$U_1[y]=a_1y(a)+a_2y'(a)=\alpha \tag{7.13.2}$$

$$U_2[y]=b_1y(b)+b_2y'(b)=\beta \tag{7.13.3}$$

may be obtained in a very simple manner.

Let $H_1(x)$ and $H_2(x)$ be the respective solutions of

$$\left.\begin{aligned} L[H_1]&=0\\ U_1[H_1]&=1\\ U_2[H_1]&=0 \end{aligned}\right\} \tag{7.13.4}$$

$$\left.\begin{aligned} L[H_2]&=0\\ U_1[H_2]&=0\\ U_2[H_2]&=1 \end{aligned}\right\} \tag{7.13.5}$$

It can be easily verified that the solution of the problem (7.13.1)–(7.13.3) is

$$y(x)=\int_a^b G(x,\xi)f(\xi)\,d\xi+\alpha H_1(x)+\beta H_2(x) \tag{7.13.6}$$

EXAMPLE 7.13.1. Consider the problem

$$\begin{aligned} y''&=-f(x)\\ y(0)&=0\\ y(1)+y'(1)&=2 \end{aligned} \tag{7.13.7}$$

The solution of $y''=0$ that satisfies $y(0)=0$ in $0\leqslant x<\xi$ is

$$y_1(x)=x$$

and the solution of $y''=0$ that satisfies $y(1)+y'(1)=0$ in $\xi<x\leqslant 1$ is

$$y_2(x)=2-x$$

The Wronskian of y_1 and y_2 is given by

$$W=y_1(\xi)y_2'(\xi)-y_2(\xi)y_1'(\xi)=-2$$

Since in this case $p=1$, (7.12.6) becomes

$$G(x,\xi)=\begin{cases}(2-\xi)x/2, & 0\leqslant x<\xi\\ \xi(2-x)/2, & \xi\leqslant x\leqslant 1\end{cases}$$

In this case we need to determine $H_2(x)$ only, since α in (7.13.6) is zero. The function $H_2(x)$ that satisfies the problem

$$y''=0$$
$$y(0)=0$$
$$y(1)+y'(1)=1$$

is readily found to be $H_2(x)=x/2$. Hence the solution of the problem (7.13.7) is given by

$$y(x)=\frac{2-x}{2}\int_0^x \xi f(\xi)\,d\xi+\frac{x}{2}\int_x^1 (2-\xi)f(\xi)\,d\xi+x$$

In the preceding discussions, we have assumed that the associated homogeneous problem has a trivial solution only. However, this is not always the case. For example, if we consider the problem

$$y''+y=-1$$
$$y(0)=0 \quad y(\pi)=0$$

the solution of the associated homogeneous equation that satisfies $y(0)=0$ is obviously $y=\sin x$, which also satisfies $y(\pi)=0$. Hence the Green's function does not exist in the usual sense.

In the succeeding theorem, we introduce a generalized Green's function that is appropriate to nonhomogeneous problems associated with homogeneous problems having nontrivial solutions.

Theorem 7.13.1 *The nonhomogeneous boundary-value problem*

$$L[y]=\frac{d}{dx}[p(x)y']+q(x)y=-f(x) \tag{7.13.8}$$

$$a_1y(a)+a_2y'(a)=0$$
$$b_1y(b)+b_2y'(b)=0 \tag{7.13.9}$$

has a solution

$$y(x)=\int_a^b G(x,\xi)f(\xi)\,d\xi \tag{7.13.10}$$

if and only if

$$\int_a^b f(\xi)y_0(\xi)\,d\xi=0$$

where y_0 is a normalized solution, that is,

$$\int_a^b y_0^2(x)\,dx=1$$

of the associated homogeneous problem. The function $G(x,\xi)$, called the generalized Green's function, is a solution of

$$L[G]=y_0(x)y_0(\xi)$$

except at $x=\xi$, satisfying the boundary conditions (7.13.9). $G(x,\xi)$ is continuous at $x=\xi$, and $G'(x,\xi)$ is continuous everywhere except at $x=\xi$. At $x=\xi$, $G'(x,\xi)$ possesses a jump discontinuity of magnitude $-1/p(\xi)$. In addition, $G(x,\xi)$ satisfies the condition

$$\int_a^b G(x,\xi)y_0(x)\,dx=0$$

Proof of this theorem can be found in Sagan [68].

7.14 Eigenvalue Problems and Green's Function

By means of Green's function, the equation associated with the eigenvalue problem can be replaced by an integral equation. If the solution of that integral equation is determined, then the eigenvalue problem is solved.

To see this, we consider the differential equation

$$L[y]+\lambda s(x)y(x)=g(x) \tag{7.14.1}$$

where λ is a parameter. We assume here that g is piecewise continuous and s is continuous and positive. We suppose that Green's function for $L[y]$ which satisfies the homogeneous boundary conditions (7.11.2) and (7.11.3) exists. Then if we identify $f(x)$ with $\lambda sy-g$, we immediately obtained from formula (7.11.4) the integral equation

$$y(x)=\lambda\int_a^b G(x,\xi)s(\xi)y(\xi)\,d\xi-\int_a^b G(x,\xi)g(\xi)\,d\xi \tag{7.14.2}$$

which is the equivalence of the differential equation (7.14.1). Thus, the determination of the solution y of Eq. (7.14.1) is equivalent to the determination of the solution y of the integral equation (7.14.2).

In the integral equation, we observe that

$$G(x,\xi)s(\xi)\neq G(\xi,x)s(x)$$

provided $s(x)$ is a constant. Thus, the kernel $G(x,\xi)s(\xi)$ is not symmetric. To obtain the symmetric kernel, we introduce the new unknown function

$$v(x)=\sqrt{s(x)}\;y(x) \tag{7.14.3}$$

Substitution of $y(x)$ given by Eq. (7.14.3) into Eq. (7.14.2) yields

$$v(x)=\lambda\int_a^b K(x,\xi)v(\xi)\,d\xi-\int_a^b K(x,\xi)\frac{g(\xi)}{\sqrt{s(\xi)}}\,d\xi$$

where

$$K(x,\xi)=\sqrt{s(x)s(\xi)}\ G(x,\xi)$$

which possesses the symmetry as $G(x,\xi)$.

In the case of homogeneous equation

$$L[y]+\lambda s(x)y(x)=0 \tag{7.14.4}$$

we have the corresponding homogeneous integral equation

$$y(x)=\lambda\int_a^b G(x,\xi)s(\xi)y(\xi)\,d\xi \tag{7.14.5}$$

With the use of the transformation (7.14.3), Eq. (7.14.5) takes the form

$$v(x)=\lambda\int_a^b K(x,\xi)v(\xi)\,d\xi \tag{7.14.6}$$

where $K(x,\xi)$ is the symmetric kernel of Eq. (7.14.6). Thus, the theorems on symmetric kernels (see Courant and Hilbert [62]) can be applied to obtain the solution of the differential equation (7.14.1).

We can now state the following theorem regarding the determination of the solution of the nonhomogeneous differential equation

$$\frac{d}{dx}[p(x)y']+[q(x)+\lambda s(x)]y(x)=g(x)$$

$$\begin{aligned} a_1y(a)+a_2y'(a)&=0 \\ b_1y(b)+b_2y'(b)&=0 \end{aligned} \tag{7.14.7}$$

Theorem 7.14.1 *Let $\lambda_1,\lambda_2,\lambda_3,\ldots$ be the eigenvalues corresponding to the eigenfunctions $\phi_1,\phi_2,\phi_3,\ldots$ of the associated homogeneous problem.*

1. *If $\lambda\neq\lambda_i$, that is, λ is not the eigenvalue, then problem* (7.14.7) *has a unique solution for arbitrary $g(x)$.*
2. *If $\lambda=\lambda_i$, then problem* (7.14.7) *has a solution if, and only if,*

$$\int_a^b s\phi_i g\,dx=0$$

Moreover, if a function $y(x)$ which has a continuous first and a piecewise continuous second derivative satisfies the boundary conditions, then $y(x)$ may be expanded in an absolutely and uniformly convergent series of eigenfunctions

$$y(x)=\sum_{k=1}^{\infty}c_k\phi_k(x)$$

where

$$c_k=\int_a^b sy\phi_k\,dx$$

For proof, see Courant and Hilbert [62].

Exercises

1. Determine the eigenvalues and eigenfunctions of the following regular Sturm-Liouville systems:
   a. $y''+\lambda y=0$
      $y(0)=0 \quad y(\pi)=0$
   b. $y''+\lambda y=0$
      $y(0)=0 \quad y'(1)=0$
   c. $y''+\lambda y=0$
      $y'(0)=0 \quad y'(\pi)=0$
   d. $y''+\lambda y=0$
      $y(1)=0 \quad y(0)+y'(0)=0$

2. Find the eigenvalues and eigenfunctions of the following periodic Sturm-Liouville systems:
   a. $y''+\lambda y=0$
      $y(-1)=y(1) \quad y'(-1)=y'(1)$
   b. $y''+\lambda y=0$
      $y(0)=y(2\pi) \quad y'(0)=y'(2\pi)$
   c. $y''+\lambda y=0$
      $y(0)=y(\pi) \quad y'(0)=y'(\pi)$

3. Obtain the eigenvalues and eigenfunctions of the following Sturm-Liouville systems:
   a. $y''+y'+(1+\lambda)y=0$
      $y(0)=0 \quad y(1)=0$
   b. $y''+2y'+(1-\lambda)y=0$
      $y(0)=0 \quad y'(1)=0$
   c. $y''-3y'+3(1+\lambda)y=0$
      $y'(0)=0 \quad y'(\pi)=0$

4. Show that if ϕ_1 and ϕ_2 are two linearly independent solutions of the Sturm-Liouville system

$$\frac{d}{dx}[p(x)y']+[q(x)+\lambda s(x)]y=0$$

$$a_1y(a)+a_2y(b)+a_3y'(a)+a_4y'(b)=0$$

$$b_1y(a)+b_2y(b)+b_3y'(a)+b_4y'(b)=0$$

   then they are orthogonal with respect to the weight function $s(x)$ in $[a,b]$.

5. Find the eigenvalues and eigenfunctions of the following regular Sturm-Liouville systems:
   a. $x^2y''+3xy'+\lambda y=0, \quad 1\leqslant x\leqslant e$
      $y(1)=0, \quad y(e)=0$
   b. $\dfrac{d}{dx}[(2+x)^2y']+\lambda y=0, \quad -1\leqslant x\leqslant 1$
      $y(-1)=0, \quad y(1)=0$
   c. $(1+x)^2y''+2(1+x)y'+3\lambda y=0, \quad 0\leqslant x\leqslant 1$
      $y(0)=0, \quad y(1)=0$

6. Determine all eigenvalues and eigenfunctions of the singular Sturm-Liouville systems

a. $x^2y''+xy'+\lambda y=0$

$y(1)=0 \quad y,y'$ bounded at $x=0$

b. $y''+\lambda y=0$

$y(0)=0 \quad y,y'$ bounded at infinity

7. Expand the function

$$f(x)=\sin x \quad 0\leqslant x\leqslant \pi$$

in terms of the eigenfunctions of the Sturm-Liouville problems

$$y''+\lambda y=0$$
$$y(0)=0 \quad y(\pi)+y'(\pi)=0$$

8. Find the expansion of

$$f(x)=x \quad 0\leqslant x\leqslant \pi$$

in a series of eigenfunctions of the Sturm-Liouville system

$$y''+\lambda y=0$$
$$y'(0)=0 \quad y'(\pi)=0$$

9. Transform each of the following equations into the equivalent self-adjoint form:

a. The Leguerre equation

$$xy''+(1-x)y'+ny=0 \quad n=0,1,2,\ldots$$

b. The Hermite equation

$$y''-2xy'+2ny=0 \quad n=0,1,2,\ldots$$

c. The Tchebycheff equation

$$(1-x^2)y''-xy'+n^2y=0 \quad n=0,1,2,\ldots$$

10. If $q(x)$ and $s(x)$ are continuous and $p(x)$ is twice continuously differentiable in $[a,b]$, then the solutions of the fourth-order Sturm-Liouville system

$$[p(x)y'']''+[q(x)+\lambda s(x)]y=0$$
$$[a_1y+a_2(py'')']_{x=a}=0 \quad [b_1y+b_2(py'')']_{x=b}=0$$
$$[c_1y'+c_2(py'')]_{x=a}=0 \quad [d_1y'+d_2(py'')]_{x=b}=0$$

with $a_1^2+a_2^2\neq 0$, $b_1^2+b_2^2\neq 0$, $c_1^2+c_2^2\neq 0$, $d_1^2+d_2^2\neq 0$ are orthogonal with respect to $s(x)$ in $[a,b]$.

11. If the eigenfunctions of the problem

$$\frac{1}{r}\frac{d}{dr}(ry')+\lambda y=0 \quad 0<r<a$$
$$c_1y(a)+c_2y'(a)=0$$
$$\lim_{r\to 0+} y(r)<\infty$$

satisfy

$$\lim_{r\to 0+} ry'(r)=0$$

show that all the eigenvalues are real for real c_1 and c_2.

12. Prove the Schwarz inequality for square-integrable functions $\phi(x)$ and $\psi(x)$:

$$\left|\int_I \phi(x)\psi(x)\rho(x)\,dx\right|^2 \leqslant \int_I \phi^2(x)\rho(x)\,dx \int_I \psi^2(x)\rho(x)\,dx.$$

13. Find the Green's functions for the following problems:

a. $L[y]=y''=0$
   $y(0)=0 \quad y'(1)=0$

b. $L[y]=(1-x^2)y''-2xy'=0$
   $y(0)=0 \quad y'(1)=0$

c. $L[y]=y''+a^2y=0 \quad a=\text{constant}$
   $y(0)=0 \quad y(1)=0$

14. Determine the solution of each of the following boundary-value problems:

a. $y''+y=1$
   $y(0)=0 \quad y(1)=0$

b. $y''+4y=e^x$
   $y(0)=0, \quad y'(1)=0$

c. $y''=\sin x$
   $y(0)=0 \quad y(1)+2y'(1)=0$

d. $y''+4y=-2$
   $y(0)=0, \quad y[\frac{\pi}{4}]=0$

e. $y''=-x$
   $y(0)=2 \quad y(1)+y'(1)=4$

f. $y''=-x^2$
   $y(0)+y'(0)=4 \quad y'(1)=2$

g. $y''=-x$
   $y(0)=1 \quad y'(1)=2$

15. Determine the solution of the following boundary-value problems:

a. $y''=-f(x), \quad y(0)=0, \quad y'(1)=0$

b. $y''=-f(x), \quad y(-1)=0, \quad y(1)=0$

16. Find the solution of the following boundary-value problems:

a. $y''-y=-f(x), \quad y(0)=y(1)=0$

b. $y''-y=-f(x), \quad y'(0)=y'(1)=0$

17. If the boundary condition at $x=0$ is replaced by a finiteness condition, explain why the discussion for some boundary-value problems with homogeneous boundary conditions and with $p(x)=x$ is still valid.

18. Determine the Green's function for the boundary-value problem

$$xy''+y'=-f(x)$$
$$y(1)=0$$
$$\lim_{x\to 0}|y(x)|<\infty$$

19. Determine the Green's function for the boundary-value problem

$$xy'' + y' - \frac{n^2}{x}y = -f(x)$$
$$y(1) = 0$$
$$\lim_{x\to 0} |y(x)| < \infty$$

20. Determine the Green's function for the boundary-value problem

$$[(1-x^2)y']' - \frac{h^2}{1-x^2}y = -f(x) \quad h = 1,2,3\ldots$$
$$\lim_{r\to \pm 1} |y(x)| < \infty$$

21. Prove the uniqueness of the Green's function for the boundary-value problem

$$L[y] = -f(x)$$
$$a_1 y(a) + a_2 y'(a) = 0$$
$$b_1 y(b) + b_2 y'(b) = 0$$

22. Find the Green's function for the boundary-value problem

$$L[y] = y^{\text{iv}} = -f(x)$$
$$y(0) = y(1) = y'(0) = y'(1) = 0$$

23. Consider the nonhomogeneous boundary-value problem with continuous coefficients.

$$L[y] = y'' + p(x)y' + q(x)y = g(x)$$
$$U_1[y] = a_1 y(a) + a_2 y'(a) + b_1 y(b) + b_2 y'(b) = \alpha$$
$$U_2[y] = a_3 y(a) + a_4 y'(a) + b_3 y(b) + b_4 y'(b) = \beta$$

If the associated homogeneous boundary-value problem

$$L[y] = 0$$
$$U_j[y] = 0 \quad j = 1,2$$

has a trivial solution only, prove that the nonhomogeneous problem has a unique solution.

24. Determine the Green's function for the boundary-value problem

$$y'' = -f(x)$$
$$y(-1) = y(1)$$
$$y'(-1) = y'(1)$$

25. Consider the nonself-adjoint boundary-value problem

$$L[y] = y'' + 3y' + 2y = -f(x)$$
$$2y(0) - y(1) = 0$$
$$y'(1) = 2$$

By direct integration of $GL[y]$ from 0 to 1, show that

$$y(x) = -2G(1,x) - \int_0^1 G(x,\xi)f(\xi)\,d\xi$$

is the solution of the boundary-value problem if G satisfies

$$G_{\xi\xi} - 3G_{\xi} + 2G = 0, \ \xi \neq x$$

$$G(0,x) = 0$$

$$6G(1,x) - 2G_{\xi}(1,x) + G_{\xi}(0,x) = 0$$

Find $G(x,\xi)$.

26. Show that

$$\left.\frac{dG(x,\xi)}{dx}\right|_{\xi=x-}^{\xi=x+} = \frac{1}{p(x)}$$

is equivalent to

$$\left.\frac{dG(x,\xi)}{dx}\right|_{x=\xi-}^{x=\xi+} = -\frac{1}{p(\xi)}$$

CHAPTER EIGHT

Boundary Value Problems

8.1 Boundary Value Problems

In the preceding chapters, we have treated the initial value and initial-boundary value problems. In this chapter, we shall be concerned with boundary value problems. Mathematically, a boundary value problem is finding a function which satisfies a given partial differential equation and particular boundary conditions. Physically speaking, the problem is independent of time, involving only space coordinates. Just as initial value problems are associated with hyberbolic partial differential equations, boundary value problems are associated with partial differential equations of elliptic type. In marked contrast to initial value problems, boundary value problems are considerably more difficult to solve. This is due to physical requirement that solutions must attain in the large unlike the case of initial value problems where solutions in the small, say over a short interval of time, may still be of physical interest.

Second-order partial differential equation of elliptic type in n independent variables $x_1, x_2, \ldots, x_n$ is of the form[9]

$$\nabla^2 u = F(x_1, x_2, \ldots, x_n, u_{x_1}, u_{x_2}, \ldots, u_{x_n}) \tag{8.11}$$

Some well-known elliptic equations are the

A. Laplace equation

$$\nabla^2 u = 0 \tag{8.1.2}$$

B. Poisson equation

$$\nabla^2 u = g(x) \tag{8.1.3}$$

[9] $\nabla^2 u = \sum_{i=1}^{n} u_{x_i x_i}$

where

$$g(x)=g(x_1,x_2,\ldots,x_n)$$

C. Helmoltz equation

$$\nabla^2 u+\lambda u=0 \tag{8.1.4}$$

where λ is a positive constant.

D. Schrödinger equation (time independent)

$$\nabla^2 u+[\lambda-q(x)]u=0 \tag{8.1.5}$$

We shall not attempt to treat general elliptic partial differential equations. Instead, we shall begin by presenting the simplest boundary value problems for the Laplace equation in two dimensions.

Let us first define a harmonic function. A function is said to be *harmonic* in a domain D if it satisfies the Laplace equation and if it and its first two derivatives are continuous in D.

We may note here that, since the Laplace equation is linear and homogeneous, a linear combination of harmonic functions is harmonic.

1. The First Boundary Value Problem

(The Dirichlet Problem): Find a function $u(x,y)$, harmonic in D, which satisfies

$$u=f(s) \qquad \text{on } B \tag{8.1.6}$$

where $f(s)$ is a prescribed continuous function on the boundary B of the domain D. D is the interior of a simple closed piecewise smooth curve B.

We may interpret physically the solution u of the Dirichlet problem as the steady-state temperature distribution in a body containing no sources or sinks of heat, with the temperature prescribed at all points on the boundary.

2. The Second Boundary Value Problem

(The Neumann Problem): Find a function $u(x,y)$, harmonic in D, which satisfies

$$\frac{\partial u}{\partial n}=f(s) \qquad \text{on } B \tag{8.1.7}$$

with

$$\int_B f(s)\,ds=0 \tag{8.1.8}$$

The symbol $\partial u/\partial n$ denotes the directional derivative of u along the outward normal to the boundary B. The last condition (8.1.8) is known as *the compatibility condition*, since it is a consequence of (8.1.7) and the equation $\nabla^2 u = 0$. Here the solution u may be interpreted as the steady-state temperature distribution in a body containing no heat sources or heat sinks when the heat flux across the boundary is prescribed.

The compatibility condition, in this case, may be interpreted physically as the heat requirement that the net heat flux across the boundary be zero.

3. The Third Boundary Value Problem

Find a function $u(x,y)$ harmonic in D which satisfies

$$\frac{\partial u}{\partial n} + h(s)u = f(s) \qquad \text{on } B \tag{8.1.9}$$

where h and f are given continuous functions. In this problem, the solution u may be interpreted as the steady-state temperature distribution in a body, from the boundary of which the heat radiates freely into the surrounding medium of prescribed temperature.

4. The Fourth Boundary Value Problem

(The Robin Problem): Find a function $u(x,y)$, harmonic in D, which satisfies boundary conditions of different types on different portions of the boundary B. An example involving such boundary conditions is

$$\begin{aligned} u &= f_1(s) \qquad \text{on} \qquad B_1 \\ \frac{\partial u}{\partial n} &= f_2(s) \qquad \text{on} \qquad B_2 \end{aligned} \tag{8.1.10}$$

where $B = B_1 + B_2$.

Problems 1 through 4 are called *interior boundary value problems*. These differ from *exterior boundary value problems* in two respects:

i. For problems of the latter variety, part of the boundary is at infinity.
ii. Solutions of exterior problems must satisfy an additional requirement, namely, that of boundedness at infinity.

8.2 Maximum and Minimum Principles

Before we prove the uniqueness and continuity theorems for the interior Dirichlet problem for the two-dimensional Laplace equation, we first prove the maximum and minimum principles.

Theorem 8.2.1 (The Maximum Principle) *Suppose that $u(x,y)$ is harmonic in a bounded domain D and continuous in $\mathbf{D} = D + B$. Then u attains its maximum on the boundary B of D.*

Physically, we may interpret this as the temperature of a body which has neither a source or a sink of heat acquires its largest (and smallest) values on the surface of the body and the electrostatic potential in a region which does not contain any free charge attains its maximum (and minimum) values on the boundary of the region.

PROOF. Let the maximum of u on B be M. Let us now suppose that the maximum of u in $\mathbf{D}$ is not attained at any point of B. Then it must be attained at some point $P_0(x_0,y_0)$ in D. If $M_0=u(x_0,y_0)$ denote the maximum of u in D, then M_0 must also be the maximum of u in $\mathbf{D}$.

Consider the function

$$v(x,y)=u(x,y)+\frac{M_0-M}{4R^2}\left[(x-x_0)^2+(y-y_0)^2\right] \tag{8.2.1}$$

where the point $P(x,y)$ is in D and where R is the radius of a circle containing D. Note that

$$v(x_0,y_0)=u(x_0,y_0)=M_0.$$

We have $v(x,y)\leqslant M+(M_0-M)/2=\frac{1}{2}(M+M_0)<M_0$ on B. Thus, $v(x,y)$ like $u(x,y)$ must attain its maximum at a point in D. It follows from the definition of v that

$$v_{xx}+v_{yy}=u_{xx}+u_{yy}+\frac{(M_0-M)}{R^2}=\frac{(M_0-M)}{R^2}>0 \tag{8.2.2}$$

But for v to be a maximum in D,

$$v_{xx}\leqslant 0, v_{yy}\leqslant 0$$

Thus,

$$v_{xx}+v_{yy}\leqslant 0$$

which contradicts Eq. (8.2.2). Hence the maximum of u must be attained on B. □

Theorem 8.2.2 (The Minimum Principle) *If $u(x,y)$ is harmonic in a bounded domain* D *and continuous in $\mathbf{D}=D+B$, then u attains its minimum on the boundary B of D.*

PROOF. The proof follows directly by applying the preceding theorem to the harmonic function $-u(x,y)$. □

As a result of the above theorems, we see that $u=$constant which is evidently harmonic attains the same value in the domain D as on the boundary B.

8.3 Uniqueness and Continuity Theorems

Theorem 8.3.1 (Uniqueness Theorem) *The solution of the Dirichlet problem, if it exists, is unique.*

PROOF. Let $u_1(x,y)$ and $u_2(x,y)$ be two solutions of the Dirichlet problem. Then u_1 and u_2 satisfy

$$\nabla^2 u_1 = 0 \quad \nabla^2 u_2 = 0 \qquad \text{in } D$$
$$u_1 = f \qquad u_2 = f \qquad \text{on } B$$

Since u_1 and u_2 are harmonic in D, $u_1 - u_2$ is also harmonic in D. But

$$u_1 - u_2 = 0 \qquad \text{on } B$$

By the maximum-minimum principles

$$u_1 - u_2 = 0$$

at all interior points of D. Thus, we have

$$u_1 = u_2$$

Therefore, the solution is unique. □

Theorem 8.3.2 (Continuity Theorem) *The solution of the Dirichlet problem depends continuously on the boundary data.*

PROOF. Let u_1 and u_2 be the solutions of

$$\nabla^2 u_1 = 0 \qquad \text{in } D$$
$$u_1 = f_1 \qquad \text{on } B$$

and

$$\nabla^2 u_2 = 0 \qquad \text{in } D$$
$$u_2 = f_2 \qquad \text{on } B$$

If $v = u_1 - u_2$, then v satisfies

$$\nabla^2 v = 0 \qquad \text{in } D$$
$$v = f_1 - f_2 \quad \text{on } B$$

By the maximum and minimum principles, $f_1 - f_2$ attains the maximum and minimum of v on B. Thus, if $|f_1 - f_2| < \varepsilon$, then

$$-\varepsilon < v_{\min} \leqslant v_{\max} < \varepsilon \qquad \text{on } B$$

Thus, at any interior points in D, we have

$$-\varepsilon < v_{\min} \leqslant v \leqslant v_{\max} < \varepsilon$$

Therefore $|v| < \varepsilon$ in D. Hence

$$|u_1 - u_2| < \varepsilon$$

□

Theorem 8.3.3 *Let $\{u_n\}$ be a sequence of functions harmonic in D and continuous in $\mathbf{D}$. Let f_i be the values of u_i on B. If $\{u_n\}$ converges uniformly on B, then it converges uniformly in $\mathbf{D}$.*

PROOF. By hypothesis $\{f_n\}$ converges uniformly on B. Thus, for $\varepsilon>0$, there exists an integer N such that everywhere on B

$$|f_n-f_m|<\varepsilon \quad \text{for } n,m>N$$

It follows from the continuity theorem that for all $n,m>N$

$$|u_n-u_m|<\varepsilon$$

in D, and hence the theorem is proved. □

8.4 Dirichlet Problem for a Circle

1. Interior Problem

We shall now establish the existence of the solution of the Dirichlet problem for a circle.

The Dirichlet problem is

$$\nabla^2 u = u_{rr}+\frac{1}{r}u_r+\frac{1}{r^2}u_{\theta\theta}=0, \quad 0\leqslant r<a \tag{8.4.1}$$

$$u(a,\theta)=f(\theta) \tag{8.4.2}$$

By the method of separation of variables, we seek a solution in the form

$$u(r,\theta)=R(r)\Theta(\theta)$$

Substitution of this in Eq. (8.4.1) yields

$$r^2\frac{R''}{R}+r\frac{R'}{R}=-\frac{\Theta''}{\Theta}=\lambda$$

Hence

$$r^2R''+rR'-\lambda R=0 \tag{8.4.3}$$

$$\Theta''+\lambda\Theta=0 \tag{8.4.4}$$

Because of the periodicity conditions $\Theta(0)=\Theta(2\pi)$ and $\Theta'(0)=\Theta'(2\pi)$ which ensure that the function Θ is single-valued, the case $\lambda<0$ does not yield an acceptable solution. When $\lambda=0$, we have

$$u(r,\theta)=(A+B\log r)(C\theta+D)$$

Since $\log r\to-\infty$ as $r\to 0+$ (note that $r=0$ is a singular point of Eq. (8.4.1)), B must vanish in order for u to be finite at $r=0$. C must also vanish in order for u to be periodic with period 2π. Hence the solution for $\lambda=0$ is $u=$constant. When $\lambda>0$, the solution of Eq. (8.4.4) is

$$\Theta(\theta)=A\cos\sqrt{\lambda}\,\theta+B\sin\sqrt{\lambda}\,\theta$$

The periodicity conditions imply

$$\sqrt{\lambda} = n \text{ for } n=1, 2, 3, \ldots .$$

Equation (8.4.3) is the Euler equation and therefore the general solution is

$$R(r) = Cr^{\sqrt{\lambda}} + Dr^{-\sqrt{\lambda}}$$

Since $r^{-\sqrt{\lambda}} \to \infty$ as $r \to 0$, D must vanish for u to be continuous at $r=0$. Thus, the solution is

$$u(r,\theta) = Cr^{\sqrt{\lambda}}(A\cos\sqrt{\lambda}\,\theta + B\sin\sqrt{\lambda}\,\theta) \quad \text{for } \sqrt{\lambda} = 1, 2, \ldots .$$

Hence the general solution of Eq. (8.4.1) may be written in the form

$$u(r,\theta) = \frac{a_0}{2} + \sum_{n=1}^{\infty} \left(\frac{r}{a}\right)^n (a_n \cos n\theta + b_n \sin n\theta) \tag{8.4.5}$$

where the constant term $a_0/2$ represents the solution for $\lambda=0$, and where a_n and b_n are constants. Letting $\rho = r/a$, we have

$$u(\rho,\theta) = \frac{a_0}{2} + \sum_{n=1}^{\infty} \rho^n (a_n \cos n\theta + b_n \sin n\theta). \tag{8.4.6}$$

Our next task is to show that $u(r,\theta)$ is harmonic in $0 \leqslant r < a$ and continuous in $0 \leqslant r \leqslant a$. We must also show that u satisfies the boundary condition (8.4.2).

We first assume that a_n and b_n are the Fourier coefficients of $f(\theta)$, that is,

$$a_n = \frac{1}{\pi}\int_0^{2\pi} f(\theta)\cos n\theta \, d\theta \quad n = 0, 1, 2, 3, \ldots$$

$$b_n = \frac{1}{\pi}\int_0^{2\pi} f(\theta)\sin n\theta \, d\theta \quad n = 1, 2, 3, \ldots \tag{8.4.7}$$

Thus, from their very definitions, a_n and b_n are bounded, that is, there exists some number $M > 0$ such that

$$|a_0| < M \quad |a_n| < M \quad |b_n| < M, \quad n = 1, 2, 3, \ldots$$

Thus, if we consider the sequence of functions $\{u_n\}$ defined by

$$u_n(\rho,\theta) = \rho^n (a_n \cos n\theta + b_n \sin n\theta) \tag{8.4.8}$$

we see that

$$|u_n| < 2\rho_0^n M \quad 0 \leqslant \rho \leqslant \rho_0 < 1$$

Hence in any closed circular region, series (8.4.6) converges uniformly.

Next, differentiate u_n with respect to r. Then for $0 \leqslant \rho \leqslant \rho_0 < 1$

$$\left|\frac{\partial u_n}{\partial r}\right| = \left|\frac{n}{a}\rho^{n-1}(a_n \cos n\theta + b_n \sin n\theta)\right| < 2\frac{n}{a}\rho_0^{n-1} M$$

Thus, the series obtained by differentiating series (8.4.6) term by term with respect to r converges uniformly. In a similar manner, we can prove that the series obtained by twice differentiating series (8.4.6) term by term with respect to r and θ converge uniformly. Consequently,

$$\begin{aligned}\nabla^2 u &= u_{rr} + \frac{1}{r}u_r + \frac{1}{r^2}u_{\theta\theta}\\ &= \sum_{n=1}^{\infty} \frac{\rho^{n-2}}{a^2}(a_n\cos n\theta + b_n \sin n\theta)\left[n(n-1)+n-n^2\right]\\ &= 0, \quad 0 \leqslant \rho \leqslant \rho_0 < 1\end{aligned}$$

Since each term of series (8.4.6) is a harmonic function, and since the series converges uniformly, $u(r,\theta)$ is harmonic at any interior point of the region $0 \leqslant \rho < 1$. It now remains to show that u satisfies the boundary data $f(\theta)$.

Substitution of the Fourier coefficients a_n and b_n into Eq. (8.4.6) yields

$$\begin{aligned}u(\rho,\theta) &= \frac{1}{2\pi}\int_0^{2\pi} f(\theta)\,d\theta + \frac{1}{\pi}\sum_{n=1}^{\infty}\rho^n\int_0^{2\pi} f(\tau)\left[\cos n\tau\cos n\theta + \sin n\tau \sin n\theta\right]d\tau\\ &= \frac{1}{2\pi}\int_0^{2\pi}\left[1 + 2\sum_{n=1}^{\infty}\rho^n \cos n(\theta-\tau)\right]f(\tau)\,d\tau \qquad (8.4.9)\end{aligned}$$

The interchange of summation and integration is permitted due to the uniform convergence of the series. For $0 \leqslant \rho < 1$

$$\begin{aligned}1+2\sum_{n=1}^{\infty}\left[\rho^n\cos n(\theta-\tau)\right] &= 1+\sum_{n=1}^{\infty}\left[\rho^n e^{in(\theta-\tau)} + \rho^n e^{-in(\theta-\tau)}\right]\\ &= 1 + \frac{\rho e^{i(\theta-\tau)}}{1-\rho e^{i(\theta-\tau)}} + \frac{\rho e^{-i(\theta-\tau)}}{1-\rho e^{-i(\theta-\tau)}}\\ &= \frac{1-\rho^2}{1-\rho e^{i(\theta-\tau)} - \rho e^{-i(\theta-\tau)} + \rho^2}\\ &= \frac{1-\rho^2}{1-2\rho\cos(\theta-\tau)+\rho^2}\end{aligned}$$

Hence

$$u(\rho,\theta) = \frac{1}{2\pi}\int_0^{2\pi}\frac{1-\rho^2}{1-2\rho\cos(\theta-\tau)+\rho^2}f(\tau)\,d\tau \qquad (8.4.10)$$

The integral on the right side of the preceding equation is called the *Poisson integral for a circle*.

Now if $f(\theta)=1$, then according to series (8.4.9), $u(r,\theta)=1$ for $0 \leqslant \rho < 1$. Thus, Eq. (8.4.10) gives

$$1 = \frac{1}{2\pi}\int_0^{2\pi}\frac{1-\rho^2}{1-2\rho\cos(\theta-\tau)+\rho^2}\,d\tau.$$

Hence,

$$f(\theta)=\frac{1}{2\pi}\int_0^{2\pi}\frac{1-\rho^2}{1-2\rho\cos(\theta-\tau)+\rho^2}f(\theta)\,d\tau,\quad 0\leqslant\rho<1.$$

Therefore,

$$u(\rho,\theta)-f(\theta)=\frac{1}{2\pi}\int_0^{2\pi}\frac{(1-\rho^2)[f(\tau)-f(\theta)]}{1-2\rho\cos(\theta-\tau)+\rho^2}\,d\tau. \tag{8.4.11}$$

Since $f(\theta)$ is uniformly continuous on $[0,2\pi]$, for given $\varepsilon>0$, there exists a positive number $\delta(\varepsilon)$ such that $|\theta-\tau|<\delta$ implies $|f(\theta)-f(\tau)|<\varepsilon$. If $|\theta-\tau|\geqslant\delta$ so that $\theta-\tau\neq 2n\pi$ for $n=0,1,2,\ldots,$ then

$$\lim_{\rho\to 1-}\frac{1-\rho^2}{1-2\rho\cos(\theta-\tau)+\rho^2}=0$$

In other words, there exists ρ_0 such that if $|\theta-\tau|\geqslant\delta$, then

$$\frac{1-\rho^2}{1-2\rho\cos(\theta-\tau)+\rho^2}<\varepsilon$$

for $0\leqslant\rho\leqslant\rho_0<1$. Hence Eq. (8.4.10) yields

$$\begin{aligned}|u(r,\theta)-f(\theta)|&\leqslant\frac{1}{2\pi}\int_{|\theta-\tau|\geqslant\delta}^{2\pi}\frac{(1-\rho^2)|f(\theta)-f(\tau)|}{1-2\rho\cos(\theta-\tau)+\rho^2}\,d\tau\\&\quad+\frac{1}{2\pi}\int_{|\theta-\tau|<\delta}^{2\pi}\frac{(1-\rho^2)|f(\theta)-f(\tau)|}{1-2\rho\cos(\theta-\tau)+\rho^2}\,d\tau\\&\leqslant\frac{1}{2\pi}2\pi\varepsilon\Big[2\max_{0\leqslant\theta\leqslant 2\pi}|f(\theta)|\Big]+\frac{\varepsilon}{2\pi}\cdot 2\pi\\&\leqslant\varepsilon\Big[1+2\Big(\max_{0\leqslant\theta\leqslant 2\pi}|f(\theta)|\Big)\Big]\end{aligned}$$

which implies that

$$\lim_{\rho\to 1-}u(r,\theta)=f(\theta)$$

uniformly in θ. Therefore we state the following theorem.

Theorem 8.4.1 *There exists one and only one harmonic function $u(r,\theta)$ which satisfies the continuous boundary data $f(\theta)$. This function is either given by*

$$u(r,\theta)=\frac{1}{2\pi}\int_0^{2\pi}\frac{a^2-r^2}{a^2-2ar\cos(\theta-\tau)+r^2}f(\tau)\,d\tau \tag{8.4.12}$$

or

$$u(r,\theta)=\frac{a_0}{2}+\sum_{n=1}^{\infty}\frac{r^n}{a^n}(a_n\cos n\theta+b_n\sin n\theta) \tag{8.4.13}$$

where a_n and b_n are the Fourier coefficients of $f(\theta)$.

For $\rho=0$, the Poisson integral formula (8.4.10) becomes

$$u(0,\theta)=u(0)=\frac{1}{2\pi}\int_0^{2\pi} f(\tau)\,dt \tag{8.4.14}$$

This result may be stated as follows.

Theorem 8.4.2 (Mean Value Theorem) *If u is harmonic in a circle, then the value of u at the center is equal to the mean value of u on the boundary of the circle.*

As in the preceding section, the *exterior Dirichlet problem* for a circle can readily be solved. For the exterior problem u must be bounded as $r\to\infty$. The general solution, therefore, is

$$u(r,\theta)=\frac{a_0}{2}+\sum_{n=1}^{\infty}\left(\frac{r^{-n}}{a^{-n}}\right)(a_n\cos n\theta+b_n\sin n\theta) \tag{8.4.15}$$

Applying the boundary condition $u(a,\theta)=f(\theta)$, we obtain

$$f(\theta)=\frac{a_0}{2}+\sum_{n=1}^{\infty}(a_n\cos n\theta+b_n\sin n\theta)$$

Hence, we find

$$a_n=\frac{1}{\pi}\int_0^{2\pi} f(\tau)\cos n\tau\,d\tau \qquad n=0,1,2,\ldots \tag{8.4.16}$$

$$b_n=\frac{1}{\pi}\int_0^{2\pi} f(\tau)\sin n\tau\,d\tau \qquad n=1,2,3,\ldots \tag{8.4.17}$$

Substitution of a_n and b_n into Eq. (8.4.15) yields

$$u(r,\theta)=\frac{1}{2\pi}\int_0^{2\pi}\left[1+2\sum_{n=1}^{\infty}\left(\frac{a}{r}\right)^n\cos n(\theta-\tau)\right]f(\tau)\,d\tau$$

Comparing with Eq. (8.4.9), we see that the only difference between the exterior and interior problem is ρ^n is replaced by ρ^{-n}. Therefore, the final result takes the form

$$u(\rho,\theta)=\frac{1}{2\pi}\int_0^{2\pi}\frac{\rho^2-1}{1-2\rho\cos(\theta-\tau)+\rho^2}f(\tau)\,d\tau \tag{8.4.18}$$

for $\rho>1$.

8.5 Dirichlet Problem for a Circular Annulus

The natural extension of the Dirichlet problem for a circle is the Dirichlet problem for a circular annulus, that is

$$\nabla^2 u=0 \qquad r_2<r<r_1 \tag{8.5.1}$$

$$u(r_1,\theta)=f(\theta) \tag{8.5.2}$$

$$u(r_2,\theta)=g(\theta) \tag{8.5.3}$$

In addition $u(r,\theta)$ must satisfy the periodicity condition. Accordingly, $f(\theta)$ and $g(\theta)$ must also be periodic with period 2π.

Proceeding as in the case of the Dirichlet problem for a circle, we obtain for $\lambda=0$

$$u(r,\theta)=(A+B\log r)(C\theta+D)$$

The periodicity condition on u requires that $C=0$. Then $u(r,\theta)$ becomes

$$u(r,\theta)=\frac{a_0}{2}+\frac{b_0}{2}\log r$$

where $a_0=2AD$ and $b_0=2BD$.

The solution for the case $\lambda>0$ is

$$u(r,\theta)=(Cr^{\sqrt{\lambda}}+Dr^{-\sqrt{\lambda}})(A\cos\sqrt{\lambda}\,\theta+B\sin\sqrt{\lambda}\,\theta)$$

for $\sqrt{\lambda}=n=1,2,3,\ldots$. Thus, the general solution is

$$u(r,\theta)=\frac{1}{2}(a_0+b_0\log r)+\sum_{n=1}^{\infty}\left[(a_nr^n+b_nr^{-n})\cos n\theta\right.$$
$$\left.+(c_nr^n+d_nr^{-n})\sin n\theta\right] \tag{8.5.4}$$

where a_n, b_n, c_n, and d_n are constants.

Applying the boundary conditions (8.5.2) and (8.5.3), we find that the coefficients are given by

$$a_0+b_0\log r_1=\frac{1}{\pi}\int_0^{2\pi}f(\tau)\,d\tau$$
$$a_nr_1^n+b_nr_1^{-n}=\frac{1}{\pi}\int_0^{2\pi}f(\tau)\cos n\tau\,d\tau$$
$$c_nr_1^n+d_nr_1^{-n}=\frac{1}{\pi}\int_0^{2\pi}f(\tau)\sin n\tau\,d\tau$$

and

$$a_0+b_0\log r_2=\frac{1}{\pi}\int_0^{2\pi}g(\tau)\,d\tau$$
$$a_nr_2^n+b_nr_2^{-n}=\frac{1}{\pi}\int_0^{2\pi}g(\tau)\cos n\tau\,d\tau$$
$$c_nr_2^n+d_nr_2^{-n}=\frac{1}{\pi}\int_0^{2\pi}g(\tau)\sin n\tau\,d\tau$$

The constants $a_0, b_0, a_n, b_n, c_n, d_n$ for $n=1,2,3,\ldots$ can then be determined. Hence the solution of the Dirichlet problem for an annulus is given by (8.5.4).

8.6 Neumann Problem for a Circle

Let u be a solution of the Neumann problem

$$\nabla^2 u = 0 \quad \text{in } D$$
$$\frac{\partial u}{\partial n} = f \quad \text{on } B$$

It is evident that u+constant is also a solution. Thus, we see that the solution of the Neumann problem is not unique, and it differs from another by a constant.

Consider the interior Neumann problem

$$\nabla^2 u = 0 \qquad r < R \tag{8.6.1}$$
$$\frac{\partial u}{\partial n} = \frac{\partial u}{\partial r} = f(\theta) \quad r = R \tag{8.6.2}$$

Before we determine a solution of the Neumann problem, a necessary condition for the existence of solution will be established.

In Green's second formula

$$\int_D\int (v\nabla^2 u - u\nabla^2 v)\,dS = \int_B \left(v\frac{\partial u}{\partial n} - u\frac{\partial v}{\partial n}\right)ds, \tag{8.6.3}$$

we put $v=1$, so that $\nabla^2 v = 0$ in D and $\partial v/\partial n = 0$ on B. Thus, the result is

$$\int_D\int \nabla^2 u\,dS = \int_B \frac{\partial u}{\partial n}\,ds \tag{8.6.4}$$

Substitution of Eqs. (8.6.1) and (8.6.2) into Eq. (8.6.4) yields

$$\int_B f ds = 0 \tag{8.6.5}$$

which may also be written in the form

$$R\int_0^{2\pi} f(\theta)\,d\theta = 0 \tag{8.6.6}$$

As in the case of the interior Dirichlet problem for a circle, the solution of the Laplace equation is

$$u(r,\theta) = \frac{a_0}{2} + \sum_{k=1}^{\infty} r^k (a_k \cos k\theta + b_k \sin k\theta) \tag{8.6.7}$$

Differentiating this with respect to r and applying the boundary condition (8.6.2), we obtain

$$\frac{\partial u}{\partial r}(R,\theta) = \sum_{k=1}^{\infty} kR^{k-1}(a_k \cos k\theta + b_k \sin k\theta) = f(\theta) \tag{8.6.8}$$

Hence the coefficients are given by

$$a_k = \frac{1}{k\pi R^{k-1}} \int_0^{2\pi} f(\tau)\cos k\tau \, d\tau \quad k=1,2,3,\ldots$$

$$b_k = \frac{1}{k\pi R^{k-1}} \int_0^{2\pi} f(\tau)\sin k\tau \, dt \quad k=1,2,3,\ldots \tag{8.6.9}$$

Note that the expansion of $f(\theta)$ in a series of the form (8.6.8) is possible only by virtue of the compatibility condition (8.6.6) since

$$a_0 = \frac{1}{\pi} \int_0^{2\pi} f(\tau)\, d\tau = 0.$$

Inserting a_k and b_k in Eq. (8.6.7), we obtain

$$u(r,\theta) = \frac{a_0}{2} + \frac{R}{\pi} \int_0^{2\pi} \left[\sum_{k=1}^{\infty} \left(\frac{r}{R}\right)^k \cos k(\theta-\tau) \right] f(\tau)\, d\tau$$

Using the identity

$$-\frac{1}{2}\log\left[1+\rho^2-2\rho\cos(\theta-\tau)\right] = \sum_{k=1}^{\infty} \frac{1}{k}\rho^k \cos k(\theta-\tau)$$

with $\rho = r/R$, we find that

$$u(r,\theta) = \frac{a_0}{2} - \frac{R}{2\pi} \int_0^{2\pi} \log\left[R^2 - 2rR\cos(\theta-\tau) + r^2\right] f(\tau)\, d\tau \tag{8.6.10}$$

in which a constant factor R^2 in the argument of the logarithm was eliminated by virtue of Eq. (8.6.6).

In a similar manner, for the exterior Neumann problem, we can readily find that

$$u(r,\theta) = \frac{a_0}{2} + \frac{R}{2\pi} \int_0^{2\pi} \log\left[R^2 - 2rR\cos(\theta-\tau) + r^2\right] f(\tau)\, d\tau \tag{8.6.11}$$

8.7 Dirichlet Problem for a Rectangle

Let us first consider the problem

$$\nabla^2 u = u_{xx} + u_{yy} = 0, \qquad 0<x<a,\ 0<y<b \tag{8.7.1}$$

$$u(x,0) = f(x), \qquad 0 \leqslant x \leqslant a \tag{8.7.2}$$

$$u(x,b) = 0 \tag{8.7.3}$$

$$u(0,y) = 0 \tag{8.7.4}$$

$$u(a,y) = 0 \tag{8.7.5}$$

We seek a solution in the form

$$u(x,y) = X(x)\,Y(y)$$

Substituting $u(x,y)$ in the Laplace equation, we obtain

$$X'' - \lambda X = 0 \tag{8.7.6}$$

$$X'' + \lambda X = 0 \tag{8.7.7}$$

where λ is a separation constant. Since the boundary conditions are homogeneous on $x=0$ and $x=a$, we choose $\lambda=-\alpha^2$ with $\alpha>0$ in order to obtain nontrivial solutions of the eigenvalue problem

$$x''+\alpha^2x=0$$
$$x(0)=x(a)=0$$

It is easily found that the eigenvalues are

$$\alpha=\frac{n\pi}{a}\qquad n=1,2,3,\ldots$$

and the corresponding eigenfunctions are sin $n\pi x/a$. Hence

$$X_n(x)=B_n\sin\frac{n\pi x}{a}$$

The solution of Eq. (8.7.7) is $Y(y)=C\cosh\alpha y+D\sinh\alpha y$ which may also be written in the form

$$Y(y)=E\sinh\alpha(y+F)$$

where $E=(D^2-C^2)^{1/2}$ and $F=1/\alpha\tanh^{-1}(C/D)$. Applying the remaining homogeneous boundary condition

$$u(x,b)=X(x)Y(b)=0,$$

we obtain

$$Y(b)=E\sinh\alpha(b+F)=0$$

and hence

$$F=-b,\quad E\neq 0$$

for a nontrivial solution $u(x,y)$. Thus, we have

$$Y_n(y)=E_n\sinh\frac{n\pi}{a}(y-b)$$

Because of linearity, the solution is

$$u(x,y)=\sum_{n=1}^{\infty}a_n\sin\frac{n\pi x}{a}\sinh\frac{n\pi}{a}(y-b)$$

where $a_n=B_nE_n$. Now, we apply the nonhomogeneous boundary condition to obtain

$$u(x,0)=f(x)=\sum_{n=1}^{\infty}a_n\sinh\left(\frac{-n\pi b}{a}\right)\sin\frac{n\pi x}{a}$$

This is a Fourier sine series and hence

$$a_n=\frac{-2}{a\sinh\left(\dfrac{n\pi b}{a}\right)}\int_0^a f(x)\sin\frac{n\pi x}{a}\,dx$$

Thus, the formal solution is given by

$$u(x,y)=\sum_{n=1}^{\infty} a_n^* \frac{\sinh \frac{n\pi}{a}(b-y)}{\sinh \frac{n\pi b}{a}} \sin \frac{n\pi x}{a} \tag{8.7.8}$$

where

$$a_n^* = \frac{2}{a}\int_0^a f(x)\sin\frac{n\pi x}{a}\,dx$$

To prove the existence of solution (8.7.8), we first note that

$$\frac{\sinh \frac{n\pi}{a}(b-y)}{\sinh \frac{n\pi b}{a}} = e^{\frac{-n\pi y}{a}}\left[\frac{1-e^{-\frac{2n\pi}{a}(b-y)}}{1-e^{\frac{-2n\pi b}{a}}}\right]$$

$$\leqslant C_1 e^{-n\pi y/a}$$

where C_1 is a constant. Since $f(x)$ is bounded, we have

$$|a_n^*| \leqslant \frac{2}{a}\int_0^a |f(x)|\,dx = C_2$$

Thus, the series for $u(x,y)$ is dominated by the series

$$\sum_{n=1}^{\infty} Me^{-n\pi y_0/a} \quad \text{for } y \geqslant y_0 > 0, M=\text{constant}$$

and hence $u(x,y)$ converges uniformly in x and y whenever $0 \leqslant x \leqslant a, y \geqslant y_0 > 0$. Consequently, $u(x,y)$ is continuous in this region and satisfies the boundary values $u(0,y)=u(a,y)=u(x,b)=0$.

Now differentiating u twice with respect to x, we obtain

$$u_{xx}(x,y)=\sum_{n=1}^{\infty} -a_n^*\left(\frac{n\pi}{a}\right)^2 \frac{\sinh \frac{n\pi}{a}(b-y)}{\sinh \frac{n\pi b}{a}} \sin \frac{n\pi x}{a}$$

and differentiating u twice with respect to y, we obtain

$$u_{yy}(x,y)=\sum_{n=1}^{\infty} a_n^*\left(\frac{n\pi}{a}\right)^2 \frac{\sinh \frac{n\pi}{a}(b-y)}{\sinh \frac{n\pi b}{a}} \sin \frac{n\pi x}{a}$$

It is evident that the series for u_{xx} and u_{yy} are both dominated by

$$\sum_{n=1}^{\infty} M^* n^2 e^{-n\pi y_0/a}$$

and hence converges uniformly for any $0<y_0<b$. It follows that u_{xx} and u_{yy} exist, and u satisfies the Laplace equation.

It now remains to be shown that $u(x,0)=f(x)$. Let $f(x)$ be a continuous function and let $f'(x)$ be piecewise continuous on $[0,a]$. If, in addition, $f(0)=f(a)=0$, then the Fourier series for $f(x)$ converges uniformly. Putting $y=0$ in the series for $u(x,y)$, we obtain

$$u(x,0)=\sum_{n=1}^{\infty} a_n^* \sin\frac{n\pi x}{a}$$

Since $u(x,0)$ converges uniformly to $f(x)$, we write for $\varepsilon>0$

$$|s_m(x,0)-s_n(x,0)|<\varepsilon \quad \text{for} \quad m,n>N_\varepsilon$$

where

$$s_m(x,y)=\sum_{n=1}^{m} a_n^* \sin\frac{n\pi x}{a}$$

We also know that $s_m(x,y)-s_n(x,y)$ satisfies the Laplace equation and the boundary conditions on $x=0, x=a$ and $y=b$. Then by the maximum principle,

$$|s_m(x,y)-s_n(x,y)|<\varepsilon \quad \text{for} \quad m,n>N_\varepsilon$$

in the region $0\leqslant x\leqslant a, 0\leqslant y\leqslant b$. Thus, the series for $u(x,y)$ converges uniformly, and as a consequence, $u(x,y)$ is continuous in the region $0\leqslant x\leqslant a, 0\leqslant y\leqslant b$. Hence, we obtain

$$u(x,0)=\sum_{n=1}^{\infty} a_n^* \sin\frac{n\pi x}{a}=f(x)$$

Thus the solution (8.7.8) is established.

The general Dirichlet problem

$$\nabla^2 u=0 \qquad 0<x<a \qquad 0<y<b$$

$$u(x,0)=f_1(x)$$

$$u(x,a)=f_2(x)$$

$$u(0,y)=f_3(y)$$

$$u(b,y)=f_4(y)$$

can be solved by separating it into four problems each of which has one nonhomogeneous boundary condition and the rest zero. Thus, determining each solution as in the preceding problem and then by adding the four solutions, the solution of the Dirichlet problem for a rectangle is obtained.

8.8 Dirichlet Problem Involving Poisson Equation

The solution of the Dirichlet problem involving the Poisson equation can be obtained for simple regions when the solution of the corresponding Dirichlet problem for the Laplace equation is known.

Consider the Poisson equation

$$\nabla^2 u = u_{xx} + u_{yy} = f(x,y) \quad \text{in } D$$

with the condition

$$u = g(x,y) \quad \text{on } B$$

Assume that the solution can be written in the form

$$u = v + w$$

where v is a particular solution of the Poisson equation and w is the solution of the associated homogeneous equation, that is,

$$\nabla^2 v = f$$

$$\nabla^2 w = 0$$

As soon as v is ascertained, the solution of the Dirichlet problem

$$\begin{aligned} \nabla^2 w &= 0 && \text{in } D \\ w &= -v + g(x,y) && \text{on } B \end{aligned}$$

can be determined. The usual method of finding a particular solution for the case in which $f(x,y)$ is a polynomial of degree n is to seek a solution in the form of a polynomial of degree $(n+2)$ with undetermined coefficients.

As an example, consider the torsion problem

$$\nabla^2 u = -2 \qquad 0<x<a \qquad 0<y<b$$

$$u(0,y) = 0$$

$$u(a,y) = 0$$

$$u(x,0) = 0$$

$$u(x,b) = 0$$

We let $u = v + w$. Now assume v to be of the form

$$v(x,y) = A + Bx + Cy + Dx^2 + Exy + Fy^2$$

Substituting this in the Poisson equation, we obtain

$$2D + 2F = -2.$$

The simplest way of satisfying this equation is to choose

$$D = -1 \text{ and } F = 0.$$

The remaining coefficients are arbitrary. Thus we take

$$v(x,y) = ax - x^2$$

so that v reduces to zero on the sides $x=0$ and $x=a$. Next, we find w from

$$\nabla^2 w = 0 \qquad 0<x<a, \qquad 0<y<b$$

$$w(0,y) = -v(0,y) = 0$$

$$w(a,y) = -v(a,y) = 0$$

$$w(x,0) = -v(x,0) = -(ax - x^2)$$

$$w(x,b) = -v(x,b) = -(ax - x^2)$$

As in the Dirichlet problem (Sec. 8.7), the solution is found to be

$$w(x,y)=\sum_{n=1}^{\infty}\left(a_n\cosh\frac{n\pi y}{a}+b_n\sinh\frac{n\pi y}{a}\right)\sin\frac{n\pi x}{a}$$

Application of the nonhomogeneous boundary conditions yield

$$w(x,0)=-(ax-x^2)=\sum_{n=1}^{\infty}a_n\sin\frac{n\pi x}{a}$$

$$w(x,b)=-(ax-x^2)=\sum_{n=1}^{\infty}\left(a_n\cosh\frac{n\pi b}{a}+b_n\sinh\frac{n\pi b}{a}\right)\sin\frac{n\pi x}{a}$$

from which we find

$$\begin{aligned} a_n &= \frac{2}{a}\int_0^a (x^2-ax)\sin\frac{n\pi x}{a}\,dx \\ &= \begin{cases} 0 & \text{if } n \text{ is even} \\ \dfrac{-8a^2}{\pi^3 n^3} & \text{if } n \text{ is odd} \end{cases} \end{aligned}$$

and

$$\left(a_n\cosh\frac{n\pi b}{a}+b_n\sinh\frac{n\pi b}{a}\right)=\frac{2}{a}\int_0^a(x^2-ax)\sin\frac{n\pi x}{a}\,dx$$

Thus, we have

$$b_n=\frac{\left(1-\cosh\dfrac{n\pi b}{a}\right)a_n}{\sinh\dfrac{n\pi b}{a}}$$

Hence the solution of the Dirichlet problem for the Poisson equation is given by

$$u(x,y)=(a-x)x$$
$$-\frac{8a^2}{\pi^3}\sum_{n=1}^{\infty}\frac{\left[\sinh(2n-1)\dfrac{\pi(b-y)}{a}+\sinh(2n-1)\dfrac{\pi y}{a}\right]}{\sinh(2n-1)\dfrac{\pi b}{a}}\,\frac{\sin(2n-1)\dfrac{\pi x}{a}}{(2n-1)^3}$$

8.9 Neumann Problem for a Rectangle

Consider the Neumann problem

$$\nabla^2 u=0 \qquad 0<x<a \qquad 0<y<b \tag{8.9.1}$$

$$u_x(0,y)=f_1(y) \tag{8.9.2}$$

$$u_x(a,y)=f_2(y) \tag{8.9.3}$$

$$u_y(x,0)=g_1(x) \tag{8.9.4}$$

$$u_y(x,b)=g_2(x) \tag{8.9.5}$$

The compatibility condition that must be fulfilled in this case is

$$\int_0^a [g_1(x)-g_2(x)]\,dx+\int_0^b [f_1(y)-f_2(y)]\,dy=0 \tag{8.9.6}$$

We assume a solution in the form

$$u(x,y)=u_1(x,y)+u_2(x,y) \tag{8.9.7}$$

where $u_1(x,y)$ is a solution of

$$\begin{aligned}
&\nabla^2 u_1=0\\
&\frac{\partial u_1}{\partial x}(0,y)=0\\
&\frac{\partial u_1}{\partial x}(a,y)=0\\
&\frac{\partial u_1}{\partial y}(x,0)=g_1(x)\\
&\frac{\partial u_1}{\partial y}(x,b)=g_2(x)
\end{aligned} \tag{8.9.8}$$

and where g_1 and g_2 satisfy the compatibility condition

$$\int_0^a [g_1(x)-g_2(x)]\,dx=0 \tag{8.9.9}$$

The function $u_2(x,y)$ is a solution of

$$\begin{aligned}
&\nabla^2 u=0\\
&\frac{\partial u_2}{\partial x}(0,y)=f_1(y)\\
&\frac{\partial u_2}{\partial x}(a,y)=f_2(y)\\
&\frac{\partial u_2}{\partial y}(x,0)=0\\
&\frac{\partial u_2}{\partial y}(x,b)=0
\end{aligned} \tag{8.9.10}$$

where f_1 and f_2 satisfy the compatibility condition

$$\int_0^b [f_1(y)-f_2(y)]\,dy=0 \tag{8.9.11}$$

$u_1(x,y)$ and $u_2(x,y)$ can be determined. Conditions (8.9.9) and (8.9.11) ensure that condition (8.9.6) is fulfilled. Thus the problem is solved.

However, the solution obtained in this manner is rather restrictive. In general condition (8.9.6) does not imply conditions (8.9.9) and (8.9.11). Thus, generally speaking, it is not possible to obtain a solution of the Neumann problem for a rectangle by the method described above.

To obtain a general solution, Grunberg [17] proposed the following method. Suppose we assume a solution in the form

$$u(x,y)=\frac{Y_0}{2}(y)+\sum_{n=1}^{\infty} X_n(x)\,Y_n(y) \tag{8.9.12}$$

where $X_n(x)=\cos n\pi x/a$ is an eigenfunction of the eigenvalue problem

$$X''+\lambda X=0$$
$$X'(0)=X'(a)=0$$

corresponding to the eigenvalue $\lambda_n=(n\pi/a)^2$. Then from Eq. (8.9.12), we see that

$$\begin{aligned} Y_n(y)&=\frac{2}{a}\int_0^a u(x,y)X_n(x)\,dx \\ &=\frac{2}{a}\int_0^a u(x,y)\cos\frac{n\pi x}{a}\,dx \end{aligned} \tag{8.9.13}$$

Multiplying both sides of Eq. (8.9.1) by $2\cos(n\pi x/a)$ and integrating with respect to x from 0 to a, we obtain

$$\frac{2}{a}\int_0^a (u_{xx}+u_{yy})\cos\frac{n\pi x}{a}\,dx=0$$

or

$$Y_n''+\frac{2}{a}\int_0^a u_{xx}\cos\frac{n\pi x}{a}\,dx=0$$

Integrating the second term by parts and applying the boundary conditions (8.9.2) and (8.9.3), we obtain

$$Y_n''-\left(\frac{n\pi}{a}\right)^2 Y_n=F_n(y) \tag{8.9.14}$$

where $F_n(y)=2[f_1(y)-(-1)^n f_2(y)]/a$. This is an ordinary differential equation whose solution may be written in the form

$$Y_n(y)=A_n\cosh\frac{n\pi y}{a}+B_n\sinh\frac{n\pi y}{a}+\frac{2}{\pi n}\int_0^y F_n(\tau)\sinh\frac{n\pi}{a}(y-\tau)\,d\tau \tag{8.9.15}$$

The coefficients A_n and B_n are determined from the boundary conditions

$$\begin{aligned} Y_n'(0)&=\frac{2}{a}\int_0^a u_y(x,0)\cos\frac{n\pi x}{a}\,dx \\ &=\frac{2}{a}\int_0^a g_1(x)\cos\frac{n\pi x}{a}\,dx \end{aligned} \tag{8.9.16}$$

and

$$Y_n'(b)=\frac{2}{a}\int_0^a g_2(x)\cos\frac{n\pi x}{a}\,dx \tag{8.9.17}$$

For $n=0$, Eq. (8.9.14) takes the form

$$Y_0'' = \frac{2}{a}[f_1(y) - f_2(y)]$$

and hence

$$Y_0' = \frac{2}{a}\int_0^y [f_1(\tau) - f_2(\tau)]\,d\tau + C$$

where C is an integration constant. Employing the condition (8.9.16) for $n=0$, we find

$$C = \frac{2}{a}\int_0^a g_1(x)\,dx$$

Thus, we have

$$Y_0'(y) = \frac{2}{a}\left\{\int_0^y [f_1(\tau) - f_2(\tau)]\,d\tau + \int_0^a g_1(x)\,dx\right\}$$

Consequently,

$$Y_0'(b) = \frac{2}{a}\left\{\int_0^b [f_1(\tau) - f_2(\tau)]\,d\tau + \int_0^a g_1(x)\,dx\right\}$$

Also from Eq. (8.9.16), we have

$$Y_0'(b) = \frac{2}{a}\int_0^a g_2(x)\,dx$$

It follows from these two expressions for $Y_0'(b)$ that

$$\int_0^b [f_1(y) - f_2(y)]\,dy + \int_0^a [g_1(x) - g_2(x)]\,dx = 0$$

which is the necessary condition for the existence of a solution to the Neumann problem for a rectangle.

Exercises

1. Reduce the Neumann problem to the Dirichlet problem in the two-dimensional case.

2. Reduce the wave equation

$$u_{tt} = c^2(u_{xx} + u_{yy} + u_{zz})$$

to the Laplace equation

$$u_{xx} + u_{yy} + u_{zz} + u_{\tau\tau} = 0$$

by letting $\tau = ict$ where $i = \sqrt{-1}$. Obtain the solution of the wave equation in cylindrical coordinates via the solution of the Laplace equation. Assume that $u(r,\theta,z,\tau)$ is independent of z.

3. Prove that, if $u(x,t)$ satisfies

$$u_t = ku_{xx}$$

for $0 \leqslant x \leqslant l, 0 \leqslant t \leqslant t_0$, then the maximum value of u is attained either at $t=0$ or at the end points $x=0$ or $x=l$ for $0 \leqslant t \leqslant t_0$. This is called the maximum principle for the heat equation.

4. Prove that a function which is harmonic everywhere on a plane and is bounded either above or below is a constant. This is called the Liouville theorem.

5. Show that the compatibility condition for the Neumann problem

$$\nabla^2 u = f \quad \text{in } D$$

$$\frac{\partial u}{\partial n} = g \quad \text{on } B$$

is

$$\int_D f dS + \int_B g\,ds = 0$$

where B is the boundary of domain D.

6. Show that the second degree polynomial

$$P = Ax^2 + Bxy + Cy^2 + Dyz + Fz^2 + Fxz$$

is harmonic if

$$E = -(A+C)$$

and obtain

$$P = A(x^2 - z^2) + Bxy + C(y^2 - z^2) + Dyz + Fxz$$

7. Prove that a solution of the Neumann problem

$$\nabla^2 u = f \quad \text{in } D$$

$$u = g \quad \text{on } B$$

differs from another solution by at most a constant.

8. Determine the solution of each of the following problems:

a. $\nabla^2 u = 0, \quad 1 < r < 2 \quad 0 < \theta < \pi$

$u(1,\theta) = \sin\theta, \quad 0 \leqslant \theta \leqslant \pi$

$u(2,\theta) = 0$

$u(r,0) = 0$

$u(r,\pi) = 0$

b. $$\nabla^2 u=0, \quad 1<r<2, \quad 0<\theta<\pi$$
$$u(1,\theta)=0$$
$$u(2,\theta)=\theta(\theta-\pi), \quad 0\leqslant\theta\leqslant\pi$$
$$u(r,0)=0$$
$$u(r,\pi)=0$$

c. $$\nabla^2 u=0 \quad 1<r<3 \quad 0<\theta<\pi/2$$
$$u(1,\theta)=0$$
$$u(3,\theta)=0$$
$$u(r,0)=(r-1)(r-3), \quad 1\leqslant r\leqslant 3$$
$$u\left(r,\frac{\pi}{2}\right)=0$$

d. $$\nabla^2 u=0 \quad 1<r<3 \quad 0<\theta<\pi/2$$
$$u(1,\theta)=0$$
$$u(3,\theta)=0$$
$$u(r,0)=0$$
$$u\left(r,\frac{\pi}{2}\right)=f(r), \quad 1\leqslant r\leqslant 3$$

9. Solve

$$\nabla^2 u=0 \quad a<r<b, \quad 0<\theta<\alpha$$
$$u(a,\theta)=f(\theta) \quad 0\leqslant\theta\leqslant\alpha$$
$$u(b,\theta)=0$$
$$u(r,0)=f(r) \quad a\leqslant r\leqslant b$$
$$u(r,\alpha)=0$$

10. Verify directly that the Poisson integral is a solution of the Laplace equation.

11. Solve

$$\nabla^2 u=0 \quad 0<r<a \quad 0<\theta<\pi$$
$$u(r,0)=0$$
$$u(r,\pi)=0$$
$$u(a,\theta)=\theta(\pi-\theta), \quad 0\leqslant\theta\leqslant\pi$$
$$u(0,\theta) \text{ is bounded}$$

12. Solve

$$\begin{aligned} &\nabla^2 u + u = 0 \qquad 0<r<a \qquad 0<\theta<\alpha \\ &u(r,0)=0 \\ &u(r,\alpha)=0 \\ &u(a,\theta)=f(\theta) \quad 0\leqslant\theta\leqslant\alpha \\ &u(0,\theta) \text{ is bounded} \end{aligned}$$

13. Find the solution of the Dirichlet problem

$$\begin{aligned} &\nabla^2 u = -2, \qquad r<a, \qquad 0<\theta<2\pi \\ &u(a,\theta)=0 \end{aligned}$$

14. Solve the following problems

a.
$$\begin{aligned} &\nabla^2 u = 0 \qquad 1<r<2 \qquad 0<\theta<2\pi \\ &u_r(1,\theta)=\sin\theta \quad 0\leqslant\theta\leqslant 2\pi \\ &u_r(2,\theta)=0 \end{aligned}$$

b.
$$\begin{aligned} &\nabla^2 u = 0 \qquad 1<r<2 \qquad 0<\theta<2\pi \\ &u_r(1,\theta)=0 \\ &u_r(2,\theta)=\theta-\pi \quad 0\leqslant\theta\leqslant 2\pi \end{aligned}$$

15. Solve

$$\begin{aligned} &\nabla^2 u = 0 \qquad a<r<b \qquad 0<\theta<2\pi \\ &u_r(a,\theta)=f(\theta) \quad 0\leqslant\theta\leqslant 2\pi \\ &u_r(b,\theta)=g(\theta) \quad 0\leqslant\theta\leqslant 2\pi \end{aligned}$$

where

$$\int_{r=a} f\,ds + \int_{r=b} g\,ds = 0$$

16. Solve the Robin problem for a semicircular disk

$$\begin{aligned} &\nabla^2 u = 0 \qquad r<R \qquad 0<\theta<\pi \\ &u_r(R,\theta)=\sin\theta, \quad 0\leqslant\theta\leqslant\pi \\ &u(r,0)=0 \\ &u(r,\pi)=0 \end{aligned}$$

17. Solve

$$\begin{aligned} \nabla^2 u &= 0 \qquad a<r<b \qquad 0<\theta<\alpha \\ u_r(a,\theta) &= 0 \\ u_r(b,\theta) &= f(\theta) \quad 0\leqslant\theta\leqslant\alpha \\ u(r,0) &= 0 \\ u(r,\alpha) &= 0 \end{aligned}$$

18. Determine the solution of the mixed boundary value problem

$$\nabla^2 u = 0, \qquad r<R, \qquad 0<\theta<2\pi$$
$$u_r(R,\theta) + hu(R,\theta) = f(\theta), \quad h = \text{constant}$$

19. Solve

$$\begin{aligned} \nabla^2 u &= 0 \qquad a<r<b, \qquad 0<\theta<2\pi \\ u_r(a,\theta) + hu(a,\theta) &= f(\theta) \\ u_r(b,\theta) + hu(b,\theta) &= g(\theta) \end{aligned}$$

20. Find a solution of the Neumann problem

$$\begin{aligned} \nabla^2 u &= -r^2 \sin 2\theta, \qquad r_1<r<r_2, \qquad 0<\theta<2\pi \\ u_r(r_1,\theta) &= 0 \\ u_r(r_2,\theta) &= 0 \end{aligned}$$

21. Solve the Robin problem

$$\begin{aligned} \nabla^2 u &= -r^2 \sin 2\theta \\ u(r_1,\theta) &= 0 \\ u_r(r_2,\theta) &= 0 \end{aligned}$$

22. Solve the following Dirichlet problems:

a.
$$\begin{aligned} \nabla^2 u &= 0 \qquad 0<x<1 \qquad 0<y<1 \\ u(x,0) &= x(x-1) \quad 0\leqslant x\leqslant 1 \\ u(x,1) &= 0 \\ u(0,y) &= 0 \\ u(1,y) &= 0 \end{aligned}$$

b.
$$\begin{aligned} \nabla^2 u &= 0 \qquad 0<x<1 \qquad 0<y<1 \\ u(x,0) &= 0 \\ u(x,1) &= \sin \pi x \quad 0\leqslant x\leqslant 1 \\ u(0,y) &= 0 \\ u(1,y) &= 0 \end{aligned}$$

c. $\nabla^2 u = 0 \qquad 0<x<1 \qquad 0<y<1$

$u(x,0)=0$

$u(x,1)=0$

$u(0,y)=\left(\cos\frac{\pi y}{2}-1\right)\cos\frac{\pi y}{2}, \quad 0\leqslant y\leqslant 1$

$u(1,y)=0$

d. $\nabla^2 u = 0 \qquad 0<x<1 \qquad 0<y<1$

$u(x,0)=0$

$u(x,1)=0$

$u(0,y)=0$

$u(1,y)=\sin\pi y\cos\pi y \quad 0\leqslant y\leqslant 1$

23. Solve the following Neumann problems:

a. $\nabla^2 u = 0 \qquad 0<x<\pi \qquad 0<y<\pi$

$u_x(0,y)=\left(y-\frac{\pi}{2}\right) \quad 0\leqslant y\leqslant \pi$

$u_x(\pi,y)=0$

$u_y(x,0)=x$

$u_y(x,\pi)=x$

b. $\nabla^2 u = 0 \qquad 0<x<\pi \qquad 0<y<\pi$

$u_x(0,y)=0$

$u_x(\pi,y)=2\cos y \quad 0\leqslant y\leqslant \pi$

$u_y(x,0)=0$

$u_y(x,\pi)=0$

c. $\nabla^2 u = 0 \qquad 0<x<\pi \qquad 0<y<\pi$

$u_x(0,y)=0$

$u_x(\pi,y)=0$

$u_y(x,0)=\cos x \quad 0\leqslant x\leqslant \pi$

$u_y(x,\pi)=0$

d. $\nabla^2 u = 0 \qquad 0<x<\pi \qquad 0<y<\pi$

$u_x(0,y) = y$

$u_x(\pi,y) = y$

$u_y(x,0) = x$

$u_y(x,\pi) = x$

24. The steady state temperature distribution in a rectangular plate of length a and width b is described by

$$\nabla^2 u = 0 \qquad 0<x<a \qquad 0<y<b$$

At $x=0$, the temperature is kept at zero degree, while at $x=a$, the plate is insulated. The temperature is prescribed at $y=0$. At $y=b$, heat is allowed to radiate freely into the surrounding medium of zero degree temperature. That is,

$$u(0,y) = 0$$
$$u_x(a,y) = 0$$
$$u(x,0) = f(x) \qquad 0 \leqslant x \leqslant a$$
$$u_y(x,b) + hu(x,b) = 0$$

Determine the temperature distribution.

25. Solve the Dirichlet problem

$$\nabla^2 u = -2y \qquad 0<x<1 \qquad 0<y<1$$
$$u(0,y) = 0$$
$$u(1,y) = 0$$
$$u(x,0) = 0$$
$$u(x,1) = 0$$

26. Find the harmonic function which vanishes on the hypotenuse and has prescribed values on the other two sides of an isosceles right-angled triangle formed by $x=0, y=0$ and $y=a-x$ where $a=$constant.

27. Find a solution of the Neumann problem

$$\nabla^2 u = x^2 - y^2 \qquad 0<x<a \qquad 0<y<a$$
$$u_x(0,y) = 0$$
$$u_x(a,y) = 0$$
$$u_y(x,0) = 0$$
$$u_y(x,a) = 0$$

28. Solve the third boundary value problem

$$\nabla^2 u = 0 \qquad 0<x<1 \qquad 0<y<1$$

$$u_x(0,y) + hu(0,y) = 0 \qquad h = \text{constant}$$

$$u_x(1,y) + hu(1,y) = 0$$

$$u_y(x,0) + hu(x,0) = 0$$

$$u_y(x,1) + hu(x,1) = f(x) \quad 0 \leqslant x \leqslant 1$$

29. Determine the solution of

$$\nabla^2 u = 1 \qquad 0<x<\pi \qquad 0<y<\pi$$

$$u(0,y) = 0$$

$$u_x(\pi,y) = 0$$

$$u_y(x,0) = 0$$

$$u_y(x,\pi) + hu(x,\pi) = f(x) \quad 0 \leqslant x \leqslant \pi$$

30. Obtain the integral representation of the Neumann problem

$$\nabla^2 u = f \quad \text{in } D$$

$$\frac{\partial u}{\partial n} = g \quad \text{on } B$$

31. Find the solution in terms of Green's function of

$$\nabla^2 u = f \quad \text{in } D$$

$$\frac{\partial u}{\partial n} + hu = g \quad \text{on } B$$

CHAPTER NINE

Higher Dimensional Problems

The treatment of problems in more than two space variables is much more involved than problems in two space variables. Here a number of multidimensional problems with various boundary conditions will be presented.

9.1 Dirichlet Problem for a Cube

The steady state temperature distribution in a cube is described by the Laplace equation

$$\nabla^2 u = u_{xx} + u_{yy} + u_{zz} = 0$$

for $0<x<\pi, 0<y<\pi, 0<z<\pi$. The faces are kept at zero degree temperature except for the face $z=0$, that is,

$$u(0,y,z) = u(\pi,y,z) = 0$$
$$u(x,0,z) = u(x,\pi,z) = 0$$
$$u(x,y,\pi) = 0$$
$$u(x,y,0) = f(x,y), \qquad 0 \leqslant x \leqslant \pi, \qquad 0 \leqslant y \leqslant \pi$$

By the method of separation of variables, we assume a solution in the form

$$u(x,y,z) = X(x)\,Y(y)Z(z)$$

Substituting this in the Laplace equation, we obtain

$$X''YZ + XY''Z + XYZ'' = 0$$

Division by XYZ yields

$$\frac{X''}{X} + \frac{Y''}{Y} = -\frac{Z''}{Z}$$

Since the right side depends only on z and the left side is independent of z, both terms must be equal to a constant. Thus we have

$$\frac{X''}{X}+\frac{Y''}{Y}=-\frac{Z''}{Z}=\lambda$$

By the same reasoning, we have

$$\frac{X''}{X}=\lambda-\frac{Y''}{Y}=\mu$$

Hence we obtain the equations

$$X''-\mu X=0$$
$$Y''-(\lambda-\mu)Y=0$$
$$Z''+\lambda Z=0$$

With use of the boundary conditions, the eigenvalue problem for X

$$X''-\mu X=0$$
$$X(0)=X(\pi)=0$$

yields the eigenvalues $\mu=-m^2$ for $m=1,2,3,\ldots$ and the corresponding eigenfunctions $\sin mx$.

Similarly, the eigenvalue problem for Y

$$Y''-(\lambda-\mu)Y=0$$
$$Y(0)=Y(\pi)=0$$

gives the eigenvalues $\lambda-\mu=-n^2$ where $n=1,2,3,\ldots$ and the corresponding eigenfunctions $\sin ny$.

Since λ is given by $-(m^2+n^2)$, it follows that the solution of $Z''+\lambda Z=0$ satisfying the condition $Z(\pi)=0$ is

$$Z(z)=C\sinh\sqrt{m^2+n^2}\,(\pi-z)$$

Thus, the solution of the Laplace equation satisfying the homogeneous boundary conditions takes the form

$$u(x,y,z)=\sum_{m=1}^{\infty}\sum_{n=1}^{\infty}a_{mn}\sinh\sqrt{m^2+n^2}\,(\pi-z)\sin mx\sin ny$$

Applying the nonhomogeneous boundary condition, we formally obtain

$$f(x,y)=\sum_{m=1}^{\infty}\sum_{n=1}^{\infty}a_{mn}\sinh(\sqrt{m^2+n^2}\,\pi)\sin mx\sin ny$$

The coefficient of the double Fourier series is thus given by

$$a_{mn}\sinh\sqrt{m^2+n^2}\,\pi=\frac{4}{\pi^2}\int_0^{\pi}\int_0^{\pi}f(x,y)\sin mx\sin ny\,dx\,dy$$

Therefore, the formal solution to the Dirichlet problem for a cube may be written in the form

$$u(x,y,z)=\sum_{m=1}^{\infty}\sum_{n=1}^{\infty} b_{mn}\frac{\sinh\sqrt{m^2+n^2}\,(\pi-z)}{\sinh\sqrt{m^2+n^2}\,\pi}\sin mx\sin ny$$

where

$$b_{mn}=a_{mn}\sinh\sqrt{m^2+n^2}\,\pi$$

9.2 Dirichlet Problem for a Cylinder

EXAMPLE 9.2.1. We consider the problem of determining the electric potential u inside a charge-free cylinder. The potential u satisfies the Laplace equation

$$\nabla^2 u=u_{rr}+\frac{1}{r}u_r+\frac{1}{r^2}u_{\theta\theta}+u_{zz}=0 \tag{9.2.1}$$

for $0\leqslant r<a, 0<z<l$. Let the lateral surface $r=a$ and the top $z=l$ be grounded, that is, be at zero potential. Let the potential on the base $z=0$ be given by

$$u(r,\theta,0)=f(r,\theta) \tag{9.2.2}$$

where $f(a,\theta)=0$.

We assume a solution in the form

$$u(r,\theta,z)=R(r)\Theta(\theta)Z(z)$$

Substitution of this in the Laplace equation yields

$$\frac{R''+\frac{1}{r}R'}{R}+\frac{1}{r^2}\frac{\Theta''}{\Theta}=-\frac{Z''}{Z}=\lambda$$

It follows that

$$\frac{r^2R''+rR'}{R}-r^2\lambda=-\frac{\Theta''}{\Theta}=\mu$$

Thus, we obtain the equations

$$r^2R''+rR'-(\lambda r^2+\mu)R=0 \tag{9.2.3}$$

$$\Theta''+\mu\Theta=0 \tag{9.2.4}$$

$$Z''+\lambda Z=0 \tag{9.2.5}$$

Using the periodicity conditions, the eigenvalue problem for $\Theta(\theta)$

$$\Theta''+\mu\Theta=0$$

$$\Theta(0)=\Theta(2\pi)$$

$$\Theta'(0)=\Theta'(2\pi)$$

yields the eigenvalues $\mu = n^2$ for $n = 0, 1, 2, \ldots$ with the corresponding eigenfunctions $\sin n\theta, \cos n\theta$. Thus

$$\Theta(\theta) = A\cos n\theta + B\sin n\theta \tag{9.2.6}$$

Suppose λ is real and negative and let $\lambda = -\beta^2$ where $\beta > 0$. If the condition $Z(l) = 0$ is imposed, then the solution of Eq. (9.2.5) can be written in the form

$$Z(z) = C\sinh\beta(l - z) \tag{9.2.7}$$

Next we introduce the new independent variable $\xi = \beta r$. Equation (9.2.3) transforms into

$$\xi^2 \frac{d^2R}{d\xi^2} + \xi\frac{dR}{d\xi} + (\xi^2 - n^2)R = 0$$

which is the Bessel equation of order n. The general solution is

$$R_n(\xi) = DJ_n(\xi) + EY_n(\xi)$$

where J_n and Y_n are the Bessel functions of the first and second kind respectively. In terms of the original variable, we have

$$R_n(r) = DJ_n(\beta r) + EY_n(\beta r)$$

Since $Y_n(\beta r)$ is unbounded at $r = 0$, we choose $E = 0$. The condition $R(a) = 0$ requires that

$$J_n(\beta a) = 0$$

For each $n \geqslant 0$, there exist positive zeros. Arranging these in an infinite increasing sequence, we have

$$0 < \alpha_{n1} < \alpha_{n2} < \cdots < \alpha_{nm} < \cdots$$

Thus, we obtain

$$\beta_{nm} = \alpha_{nm}/a$$

Consequently

$$R_n(r) = DJ_n(\alpha_{nm}r/a)$$

The solution u then finally takes the form

$$u(r,\theta,z) = \sum_{n=0}^{\infty}\sum_{m=1}^{\infty} J_n\left(\alpha_{nm}\frac{r}{a}\right)(a_{nm}\cos n\theta + b_{nm}\sin n\theta)\sinh\alpha_{nm}\frac{(l-z)}{a}$$

To satisfy the nonhomogeneous boundary condition, it is required that

$$f(r,\theta) = \sum_{n=0}^{\infty}\sum_{m=1}^{\infty} J_n\left(\alpha_{nm}\frac{r}{a}\right)(a_{nm}\cos n\theta + b_{nm}\sin n\theta)\sinh\alpha_{nm}\frac{l}{a}$$

The coefficients a_{nm} and b_{nm} are given by

$$a_{0m} = \frac{1}{\pi a^2 \sinh\left(\alpha_{0m}\frac{1}{a}\right)[J_1(\alpha_{0m})]^2} \int_0^a \int_0^{2\pi} f(r,\theta) J_0\left(\alpha_{0m}\frac{r}{a}\right) r\, dr\, d\theta$$

$$a_{nm} = \frac{2}{\pi a^2 \sinh\left(\alpha_{nm}\frac{1}{a}\right)[J_{n+1}(\alpha_{nm})]^2} \int_0^a \int_0^{2\pi} f(r,\theta) J_n\left(\alpha_{nm}\frac{r}{a}\right) \cos n\theta\, r\, dr\, d\theta$$

$$b_{nm} = \frac{2}{\pi a^2 \sinh\left(\alpha_{nm}\frac{1}{a}\right)[J_{n+1}(\alpha_{nm})]^2} \int_0^a \int_0^{2\pi} f(r,\theta) J_n\left(\alpha_{nm}\frac{r}{a}\right) \sin n\theta\, r\, dr\, d\theta$$

EXAMPLE 9.2.2. We shall illustrate the same problem with different boundary conditions. Consider the problem

$$\nabla^2 u = 0, \qquad 0 \leqslant r < a, \qquad 0 < z < \pi$$
$$u(r,\theta,0) = 0$$
$$u(r,\theta,\pi) = 0$$
$$u(a,\theta,z) = f(\theta,z)$$

As before, by the separation of variables we obtain

$$r^2 R'' + rR' - (\lambda r^2 + \mu)R = 0$$
$$\Theta'' + \mu\Theta = 0$$
$$Z'' + \lambda Z = 0$$

By the periodicity conditions, again as in the previous example, Θ equation yields the eigenvalues $\mu = n^2$ with $n = 0, 1, 2, \ldots$; the corresponding eigenfunctions are $\sin n\theta, \cos n\theta$. Thus, we have

$$\Theta(\theta) = A_n \cos n\theta + B_n \sin n\theta$$

Now let $\lambda = \beta^2$ with $\beta > 0$. Then the problem

$$Z'' + \beta^2 Z = 0$$
$$Z(0) = 0 \qquad Z(\pi) = 0$$

has the solution

$$Z(z) = C_m \sin mz, \quad m = 1, 2, 3 \ldots$$

Finally, we have

$$r^2 R'' + rR' - (m^2 r^2 + n^2)R = 0$$

or

$$R'' + \frac{1}{r}R' - \left(m^2 + \frac{n^2}{r^2}\right)R = 0$$

the general solution of which is

$$R(r)=DI_n(mr)+EK_n(mr)$$

where I_n and K_n are the modified Bessel functions of the first and second kind, respectively.

Since R must remain finite at $r=0$, we set $E=0$. Then R appears in the form

$$R(r)=DI_n(mr)$$

Applying the nonhomogeneous condition, we find the solution to be

$$u(r,\theta,z)=\sum_{m=1}^{\infty}\frac{a_{m0}}{2}\frac{I_0(mr)}{I_0(ma)}\sin mz$$

$$+\sum_{m=1}^{\infty}\sum_{n=1}^{\infty}(a_{mn}\cos n\theta+b_{mn}\sin n\theta)\frac{I_n(mr)}{I_n(ma)}\sin mz$$

where

$$a_{mn}=\frac{2}{\pi^2}\int_0^{\pi}\int_0^{2\pi}f(\theta,z)\sin mz\cos n\theta\,d\theta\,dz$$

$$b_{mn}=\frac{2}{\pi^2}\int_0^{\pi}\int_0^{2\pi}f(\theta,z)\sin mz\sin n\theta\,d\theta\,dz$$

9.3 Dirichlet Problem for a Sphere

EXAMPLE 9.3.1. To determine the potential in a sphere, we transform the Laplace equation in spherical coordinates. It has the form

$$\nabla^2 u=u_{rr}+\frac{2}{r}u_r+\frac{1}{r^2}u_{\theta\theta}+\frac{\cot\theta}{r^2}u_\theta+\frac{1}{r^2\sin^2\theta}u_{\varphi\varphi}=0 \tag{9.3.1}$$

$0\leqslant r<a, 0<\theta<\pi, 0<\varphi<2\pi$.

Let the prescribed potential on the sphere be

$$u(a,\theta,\varphi)=f(\theta,\varphi) \tag{9.3.2}$$

By the method of separation of variables, we assume a solution in the form

$$u(r,\theta,\varphi)=R(r)\Theta(\theta)\Phi(\varphi)$$

Substitution of u in the Laplace equation yields

$$r^2R''+2rR'-\lambda R=0 \tag{9.3.3}$$

$$\sin^2\theta\,\Theta''+\sin\theta\cos\theta\,\Theta'+(\lambda\sin^2\theta-\mu)\Theta=0 \tag{9.3.4}$$

$$\Phi''+\mu\Phi=0 \tag{9.3.5}$$

The general solution of Eq. (9.3.5) is

$$\Phi(\varphi)=A\cos\sqrt{\mu}\,\varphi+B\sin\sqrt{\mu}\,\varphi \tag{9.3.6}$$

The periodicity condition requires that

$$\sqrt{\mu} = m, \qquad m = 0, 1, 2, \ldots$$

Since Eq. (9.3.3) is of the Euler type, the solution is of the form

$$R(r) = r^{\beta}$$

Inserting this in Eq. (9.3.3), we obtain

$$\beta^2 + \beta - \lambda = 0$$

The roots are $\beta = (-1 + \sqrt{1+4\lambda})/2$ and $-(1+\beta)$. Hence the general solution of Eq. (9.3.3) is

$$R(r) = Cr^{\beta} + Dr^{-(1+\beta)} \tag{9.3.7}$$

The variable $\xi = \cos\theta$ transforms Eq. (9.3.4) into

$$(1-\xi^2)\Theta'' - 2\xi\Theta' + \left[\beta(\beta+1) - \frac{m^2}{1-\xi^2}\right] = 0 \tag{9.3.8}$$

which is the Legendre's associated equation. The general solution with $\beta = n$ for $n = 0, 1, 2, \ldots$ is

$$\Theta(\theta) = EP_n^m(\cos\theta) + FQ_n^m(\cos\theta)$$

Continuity of $\Theta(\theta)$ at $\theta = 0$, π corresponds to continuity of $\Theta(\xi)$ at $\xi = \pm 1$. Since $Q_n^m(\xi)$ has a logarithmic singularity at $\xi = 1$, we choose $F = 0$. Thus, the solution of Eq. (9.3.8) becomes

$$\Theta(\theta) = EP_n^m(\cos\theta)$$

Consequently, the solution of the Laplace equation in spherical coordinates is

$$u(r,\theta,\varphi) = \sum_{n=0}^{\infty} \sum_{m=0}^{n} r^n P_n^m(\cos\theta)[a_{nm}\cos m\varphi + b_{nm}\sin m\varphi]$$

In order for u to satisfy the prescribed function on the boundary, it is necessary that

$$f(\theta,\varphi) = \sum_{n=0}^{\infty} \sum_{m=0}^{n} a^n P_n^m(\cos\theta)[a_{nm}\cos m\varphi + b_{nm}\sin m\varphi]$$

for $0 \leqslant \theta \leqslant \pi, 0 \leqslant \varphi \leqslant 2\pi$. By the orthogonal properties of the functions $P_n^m(\cos\theta)\cos m\varphi$ and $P_n^m(\cos\theta)\sin m\varphi$, the coefficients are given by

$$a_{nm} = \frac{(2n+1)}{2\pi a^n} \frac{(n-m)!}{(n+m)!} \int_0^{2\pi} \int_0^{\pi} f(\theta,\varphi) P_n^m(\cos\theta)\cos m\varphi \sin\theta \, d\theta \, d\varphi$$

$$b_{nm} = \frac{(2n+1)}{2\pi a^n} \frac{(n-m)!}{(n+m)!} \int_0^{2\pi} \int_0^{\pi} f(\theta,\varphi) P_n^m(\cos\theta)\sin m\varphi \sin\theta \, d\theta \, d\varphi$$

for $m = 1, 2, \ldots$ and $n = 1, 2, \ldots$, and

$$a_{n0} = \frac{2n+1}{4\pi a^n} \int_0^{2\pi} \int_0^{\pi} f(\theta,\varphi) P_n(\cos\theta) \sin\theta \, d\theta \, d\varphi$$

for $n = 0, 1, 2, \ldots$.

EXAMPLE 9.3.2. To determine the potential of a grounded conducting sphere in a uniform field, we are required to solve

$$\nabla^2 u=0, \qquad 0\leq r<a,\, 0<\theta<\pi,\, 0<\phi<2\pi$$
$$u(a,\theta)=0$$
$$u\to -E_0 r\cos\theta \qquad \text{as } r\to\infty$$

Let the field be in the z direction so that the potential u will be independent of ϕ. Then the Laplace equation takes the form

$$u_{rr}+\frac{2}{r}u_r+\frac{1}{r^2}u_{\theta\theta}+\frac{\cot\theta}{r^2}u_\theta=0$$

We assume a solution in the form

$$u(r,\theta)=R(r)\Theta(\theta)$$

Substitution of this in the Laplace equation yields

$$r^2R''+2rR'-\lambda R=0$$
$$\sin^2\theta\,\Theta''+\sin\theta\cos\theta\,\Theta'+\lambda\sin^2\theta\,\Theta=0$$

If we set $\lambda=n(n+1)$, with $n=0,1,2,\ldots$, then the second equation is the Legendre equation. The general solution of this equation is

$$\Theta(\theta)=A_nP_n(\cos\theta)+B_nQ_n(\cos\theta)$$

where P_n and Q_n are the Legendre functions of the first and second kind, respectively. In order for the solution not to be singular at $\theta=0$ and $\theta=\pi$, we set $B_n=0$. Thus, $\Theta(\theta)$ becomes

$$\Theta(\theta)=A_nP_n(\cos\theta)$$

The solution of the R-equation is obtained in the form

$$R(r)=C_nr^n+D_nr^{-(n+1)}$$

Thus, the potential function is

$$u(r,\theta)=\sum_{n=0}^{\infty}\left(a_nr^n+b_nr^{-(n+1)}\right)P_n(\cos\theta)$$

To satisfy the condition at infinity, we must have

$$a_1=-E_0 \qquad \text{and } a_n=0 \qquad \text{for } n\geq 2$$

and hence

$$u(r,\theta)=-E_0r\cos\theta+\sum_{n=1}^{\infty}\frac{b_n}{r^{n+1}}P_n(\cos\theta)$$

The condition $u(a,\theta)=0$ yields

$$0=-E_0a\cos\theta+\sum_{n=1}^{\infty}\frac{b_n}{a^{n+1}}P_n(\cos\theta)$$

Using the orthogonality of the Legendre functions, we find that b_n are given by

$$b_n = \frac{2n+1}{2} E_0 a^{n+2} \int_{-\pi}^{\pi} \cos\theta P_n(\cos\theta)\,d(\cos\theta)$$
$$= E_0 a^3 \delta_{n1}$$

since the integral vanishes for all n except $n=1$. Hence the potential function is given by

$$u(r,\theta) = -E_0 r\cos\theta + E_0 \frac{a^3}{r^2}\cos\theta$$

EXAMPLE 9.3.3. A dielectric sphere of radius a is placed in a uniform electric field E_0. Determine the potentials inside and outside the sphere.

The problem is

$$\nabla^2 u_1 = \nabla^2 u_2 = 0$$

$$K\frac{\partial u_1}{\partial r} = \frac{\partial u_2}{\partial r} \qquad \text{on } r=a$$

$$u_1 = u_2 \qquad \text{on } r=a$$

$$u_2 \to -E_0 r\cos\theta \text{ as } r\to\infty$$

where u_1 and u_2 are the potentials inside and outside the sphere respectively, and K is the dielectric constant.

As in the preceding example, the potential function is

$$u(r,\theta) = \sum_{n=0}^{\infty} \left(a_n r^n + b_n r^{-(n+1)}\right) P_n(\cos\theta) \tag{9.3.9}$$

Since u_1 must be finite at the origin, we take

$$u_1(r,\theta) = \sum_{n=0}^{\infty} a_n r^n P_n(\cos\theta) \qquad \text{for } r \leqslant a \tag{9.3.10}$$

For u_2, which must approach infinity in the prescribed manner, we choose

$$u_2(r,\theta) = -E_0 r\cos\theta + \sum_{n=0}^{\infty} b_n r^{-(n+1)} P_n(\cos\theta) \tag{9.3.11}$$

From the two continuity conditions at $r=a$, we obtain

$$a_1 = -E_0 + \frac{b_1}{a^3}$$

$$Ka_1 = -E_0 - \frac{2b_1}{a^3}$$

$$a_n = b_n = 0, \qquad n \geqslant 2$$

The coefficients a_1 and b_1 are then found to be

$$a_1 = -\frac{3E_0}{K+2} \qquad b_1 = E_0 a^3 \frac{(K-1)}{(K+2)}$$

Hence the potential for $r \leqslant a$ is given by

$$u_1(r,\theta) = -\frac{3E_0}{K+2} r\cos\theta$$

and the potential for $r \geqslant a$ is given by

$$u_2(r,\theta) = -E_0 r\cos\theta + E_0 a^3 \frac{(K-1)}{(K+2)} r^{-2}\cos\theta$$

EXAMPLE 9.3.4. Determine the potential between concentric spheres held at different constant potentials.

Here we need to solve

$$\nabla^2 u = 0 \qquad a<r<b$$
$$u = A \text{ on } r = a$$
$$u = B \text{ on } r = b$$

In this case, the potential depends only on the radial distance. Hence we have

$$\frac{1}{r^2}\frac{\partial}{\partial r}\left(r^2 \frac{\partial u}{\partial r}\right) = 0$$

By elementary integration, we obtain

$$u(r) = c_1 + \frac{c_2}{r}$$

Applying the boundary conditions, we obtain

$$c_1 = \frac{Bb - Aa}{b-a}$$

$$c_2 = (A-B)\frac{ab}{b-a}$$

Thus, the solution is

$$u(r) = \frac{Bb-Aa}{(b-a)} + \frac{(A-B)ab}{(b-a)r}$$
$$= \frac{Bb}{r}\frac{r-a}{b-a} + \frac{Aa}{r}\frac{b-r}{b-a}$$

9.4 Wave and Heat Equations

The wave equation in three space variables may be written

$$u_{tt} = c^2 \nabla^2 u \tag{9.4.1}$$

where ∇^2 is the Laplace operator.

By the method of separation of variables, we assume a solution in the form

$$u(x,y,z,t)=U(x,y,z)T(t)$$

Substituting this in Eq. (9.4.1) we obtain

$$T''+\lambda c^2 T=0 \tag{9.4.2}$$

$$\nabla^2 U+\lambda U=0 \tag{9.4.3}$$

where $-\lambda$ is a separation constant. The variables are separated and the solutions of Eqs. (9.4.2) and (9.4.3) are to be determined.

Next we consider the heat equation

$$u_t=k\nabla^2 u \tag{9.4.4}$$

As before, we seek a solution in the form

$$u(x,y,z,t)=U(x,y,z)T(t)$$

Substitution of this in Eq. (9.4.4) yields

$$T'+\lambda kT=0$$

$$\nabla^2 U+\lambda U=0$$

Thus, we see that the problem here, as in the previous case, is essentially that of solving the Helmholtz equation

$$\nabla^2 U+\lambda U=0$$

9.5 Vibrating Membrane

As a specific example of the higher dimensional wave equation, let us determine the solution of the problem of the vibrating membrane of length a and width b. The initial-boundary value problem for the displacement function $u(x,y,t)$ is

$$u_{tt}=c^2(u_{xx}+u_{yy}), \quad 0<x<a, \quad 0<y<b, \quad t>0 \tag{9.5.1}$$

$$u(x,y,0)=f(x,y), \quad 0\leqslant x\leqslant a, \quad 0\leqslant y\leqslant b \tag{9.5.2}$$

$$u_t(x,y,0)=g(x,y), \quad 0\leqslant x\leqslant a, \quad 0\leqslant y\leqslant b \tag{9.5.3}$$

$$u(0,y,t)=0 \tag{9.5.4}$$

$$u(a,y,t)=0 \tag{9.5.5}$$

$$u(x,0,t)=0 \tag{9.5.6}$$

$$u(x,b,t)=0 \tag{9.5.7}$$

We have just shown that the separated equations for the wave equation are

$$T''+\lambda c^2 T=0 \tag{9.5.8}$$

$$\nabla^2 U+\lambda U=0 \tag{9.5.9}$$

where, in this case, $\nabla^2 U = U_{xx} + U_{yy}$. Let $\lambda = \alpha^2$. Then the solution of Eq. (9.5.8) is

$$T(t) = A\cos\alpha ct + B\sin\alpha ct$$

Now we look for a solution of Eq. (9.5.9) in the form

$$U(x,y) = X(x)\,Y(y)$$

Substitution of this into Eq. (9.5.9) yields

$$X'' - \mu X = 0$$

$$Y'' + (\lambda + \mu)\,Y = 0$$

If we let $\mu = -\beta^2$, then the solutions of these equations take the form

$$X(x) = C\cos\beta x + D\sin\beta x$$

$$Y(y) = E\cos\gamma y + F\sin\gamma y$$

where

$$\gamma^2 = (\lambda + \mu) = \alpha^2 - \beta^2.$$

The homogeneous boundary conditions in x require that $C = 0$ and

$$D\sin\beta a = 0$$

which implies that $\beta = m\pi/a$ with $D \neq 0$. Similarly, the homogeneous boundary conditions in y require that $E = 0$ and

$$F\sin\gamma b = 0$$

which implies that $\gamma = n\pi/b$ with $F \neq 0$. Noting that m and n are independent integers, we obtain the displacement function in the form

$$u(x,y,t) = \sum_{m=1}^{\infty}\sum_{n=1}^{\infty}(a_{mn}\cos\alpha_{mn}ct + b_{mn}\sin\alpha_{mn}ct)\sin\frac{m\pi x}{a}\sin\frac{n\pi y}{b} \tag{9.5.10}$$

where $\alpha_{mn} = m^2\pi^2/a^2 + n^2\pi^2/b^2$, a_{mn} and b_{mn} are constants.

Now applying the nonhomogeneous initial conditions, we have

$$u(x,y,0) = f(x,y) = \sum_{m=1}^{\infty}\sum_{n=1}^{\infty} a_{mn}\sin\frac{m\pi x}{a}\sin\frac{n\pi y}{b}$$

and thus,

$$a_{mn} = \frac{4}{ab}\int_0^a\int_0^b f(x,y)\sin\frac{m\pi x}{a}\sin\frac{n\pi y}{b}\,dx\,dy$$

In a similar manner, the initial condition on u_t implies

$$u_t(x,y,0) = g(x,y) = \sum_{m=1}^{\infty}\sum_{n=1}^{\infty} b_{mn}\alpha_{mn}c\sin\frac{m\pi x}{a}\sin\frac{n\pi y}{b}$$

from which it follows that

$$b_{mn}=\frac{4}{\alpha_{mn}abc}\int_0^a\int_0^b g(x,y)\sin\frac{m\pi x}{a}\sin\frac{n\pi y}{b}\,dx\,dy$$

The solution of the rectangular membrane problem is, therefore, given by Eq. (9.5.10).

9.6 Heat Flow in a Rectangular Plate

Another example of two-dimensional problems is the conduction of heat in a thin rectangular plate. Let the plate of length a and width b be perfectly insulated at the faces $x=0$ and $x=a$. Let the two other sides be maintained at zero temperature. Let the initial temperature distribution be $f(x,y)$. Then we seek the solution of the initial-boundary value problem

$$u_t=k\nabla^2 u,\quad 0<x<a,\quad 0<y<b,\quad t>0 \tag{9.6.1}$$

$$u(x,y,0)=f(x,y),\quad 0\leqslant x\leqslant a,\quad 0\leqslant y\leqslant b \tag{9.6.2}$$

$$u_x(0,y,t)=0 \tag{9.6.3}$$

$$u_x(a,y,t)=0 \tag{9.6.4}$$

$$u(x,0,t)=0 \tag{9.6.5}$$

$$u(x,b,t)=0 \tag{9.6.6}$$

As shown earlier, the separated equations for this problem are found to be

$$T'+\lambda kT=0 \tag{9.6.7}$$

$$\nabla^2 U+\lambda U=0 \tag{9.6.8}$$

We assume a solution in the form

$$U(x,y)=X(x)\,Y(y)$$

Inserting this in Eq. (9.6.8), we obtain

$$X''-\mu X=0 \tag{9.6.9}$$

$$Y''+(\lambda+\mu)\,Y=0 \tag{9.6.10}$$

Because the conditions in x are homogeneous, we choose $\mu=-\alpha^2$ so that

$$X(x)=A\cos\alpha x+B\sin\alpha x$$

Since $X'(0)=0, B=0$ and since $X'(a)=0$

$$\sin\alpha a=0,\qquad A\neq 0$$

which gives

$$\alpha = m\pi/a \qquad m = 1, 2, 3, \ldots$$

We note that $\mu = 0$ is also an eigenvalue. Consequently,

$$X_m(x) = A_m \cos(m\pi x/a) \qquad m = 0, 1, 2, \ldots$$

Similarly, for nontrivial solution Y, we select $\beta^2 = \lambda + \mu = \lambda - \alpha^2$ so that the solution of Eq. (9.6.10) is

$$Y(y) = C\cos\beta y + D\sin\beta y$$

Applying the homogeneous conditions, we find $C = 0$ and

$$\sin\beta b = 0 \qquad D \neq 0$$

Thus, we obtain

$$\beta = n\pi/b \qquad n = 1, 2, 3, \ldots$$

and

$$Y_n(y) = D_n \sin(n\pi y/b)$$

Recalling that $\lambda = \alpha^2 + \beta^2$, the solution of Eq. (9.6.7) may be written in the form

$$T_{mn}(t) = E_{mn} e^{-\left(\frac{m^2}{a^2} + \frac{n^2}{b^2}\right)\pi^2 kt}$$

Thus, the solution of the heat equation satisfying the prescribed boundary conditions may be written as

$$u(x,y,t) = \sum_{m=0}^{\infty} \sum_{n=1}^{\infty} a_{mn} e^{-\left(\frac{m^2}{a^2} + \frac{n^2}{b^2}\right)\pi^2 kt} \cos\frac{m\pi x}{a} \sin\frac{n\pi y}{b} \tag{9.6.11}$$

where $a_{mn} = A_m D_m E_{mn}$ is an arbitrary constant.

Applying the initial condition, we obtain

$$u(x,y,0) = f(x,y) = \sum_{m=0}^{\infty} \sum_{n=1}^{\infty} a_{mn} \cos\frac{m\pi x}{a} \sin\frac{n\pi y}{b}$$

This is a double Fourier series and the coefficient is given by

$$a_{0n} = \frac{2}{ab} \int_0^a \int_0^b f(x,y) \sin\frac{n\pi y}{b}\, dx\, dy$$

and for $m \geqslant 1$

$$a_{mn} = \frac{4}{ab} \int_0^a \int_0^b f(x,y) \cos\frac{m\pi x}{a} \sin\frac{n\pi y}{b}\, dx\, dy$$

The solution of the heat equation is thus given by Eq. (9.6.11).

9.7 Waves in Three Dimensions

The propagation of waves due to an initial disturbance in a rectangular volume is best described by the solution of the initial-boundary value problem

$$u_{tt}=c^2\nabla^2 u,\qquad 0<x<a,\ 0<y<b,\ 0<z<d,\ t>0 \tag{9.7.1}$$

$$u(x,y,z,0)=f(x,y,z),\ 0\leqslant x\leqslant a,\ 0\leqslant y\leqslant b,\ 0\leqslant z\leqslant d \tag{9.7.2}$$

$$u_t(x,y,z,0)=g(x,y,z),\ 0\leqslant x\leqslant a,\ 0\leqslant y\leqslant b,\ 0\leqslant z\leqslant d \tag{9.7.3}$$

$$u(0,y,z,t)=0 \tag{9.7.4}$$

$$u(a,y,z,t)=0 \tag{9.7.5}$$

$$u(x,0,z,t)=0 \tag{9.7.6}$$

$$u(x,b,z,t)=0 \tag{9.7.7}$$

$$u(x,y,0,t)=0 \tag{9.7.8}$$

$$u(x,y,d,t)=0 \tag{9.7.9}$$

We assume a solution be in the form

$$u(x,y,z,t)=U(x,y,z)T(t)$$

The separated equations are

$$T''+\lambda c^2T=0 \tag{9.7.10}$$

$$\nabla^2U+\lambda U=0 \tag{9.7.11}$$

Assume U of the form

$$U(x,y,z)=X(x)Y(y)Z(z)$$

Substitution of this in Eq. (9.7.11) yields

$$X''-\mu X=0 \tag{9.7.12}$$

$$Y''-\nu Y=0 \tag{9.7.13}$$

$$Z''+(\lambda+\mu+\nu)Z=0 \tag{9.7.14}$$

Because of the homogeneous conditions in x, we let $\mu=-\alpha^2$ so that

$$X(x)=A\cos\alpha x+B\sin\alpha x$$

As in the preceding examples, we obtain

$$X_l(x)=B_l\sin\frac{l\pi x}{a}\qquad l=1,2,3,\ldots$$

In a similar manner, we let $\nu=-\beta^2$ to obtain

$$Y(y)=C\cos\beta y+D\sin\beta y$$

and accordingly

$$Y_m(y)=D_m\sin\frac{m\pi y}{b}\qquad m=1,2,3,\ldots$$

We again choose $\gamma^2=\lambda+\mu+\nu=\lambda-\alpha^2-\beta^2$ so that

$$Z(z)=E\cos\gamma z+F\sin\gamma z$$

Applying the homogeneous conditions in z, we obtain

$$Z_n(z) = F_n \sin\frac{n\pi z}{d}$$

Since the solution of Eq. (9.7.10) is

$$T(t) = G\cos\sqrt{\lambda}\;ct + H\sin\sqrt{\lambda}\;ct$$

the solution of the wave equation is

$$u(x,y,z,t) = \sum_{l=1}^{\infty}\sum_{m=1}^{\infty}\sum_{n=1}^{\infty}\left[a_{lmn}\cos\sqrt{\lambda}\;ct + b_{lmn}\sin\sqrt{\lambda}\;ct\right]\sin\frac{l\pi x}{a}\sin\frac{m\pi y}{b}\sin\frac{n\pi z}{d}$$

where a_{lmn} and b_{lmn} are arbitrary constants. The coefficient a_{lmn} is determined from the initial condition $u(x,y,z,0)=f(x,y,z)$ and found to be

$$a_{lmn} = \frac{8}{abd}\int_0^a\int_0^b\int_0^d f(x,y,z)\sin\frac{l\pi x}{a}\sin\frac{m\pi y}{b}\sin\frac{n\pi z}{d}\,dx\,dy\,dz$$

whereas b_{lmn} is determined from the initial condition $u_t(x,y,z,0)=g(x,y,z)$ and found to be

$$b_{lmn} = \frac{8}{\sqrt{\lambda}\;cabd}\int_0^a\int_0^b\int_0^d g(x,y,z)\sin\frac{l\pi x}{a}\sin\frac{m\pi y}{b}\sin\frac{n\pi z}{d}\,dx\,dy\,dz$$

where

$$\lambda = \left(\frac{l^2}{a^2}+\frac{m^2}{b^2}+\frac{n^2}{d^2}\right)\pi^2$$

9.8 Heat Conduction in a Rectangular Volume

As in the case of the wave equation, the solution of the heat equation in three space variables can be determined. Consider the problem of heat distribution in a rectangular volume. The faces are maintained at zero degree temperature. The solid is initially heated so that the problem may be written as

$$u_t = k\nabla^2 u, \qquad 0<x<a, \quad 0<y<b, \quad 0<z<d, \quad t>0$$

$$u(x,y,z,0) = f(x,y,z), \quad 0\leqslant x\leqslant a, \quad 0\leqslant y\leqslant b, \quad 0\leqslant z\leqslant d$$

$$u(0,y,z,t) = 0$$

$$u(a,y,z,t) = 0$$

$$u(x,0,z,t) = 0$$

$$u(x,b,z,t) = 0$$

$$u(x,y,0,t) = 0$$

$$u(x,y,d,t) = 0$$

As before, the separated equations are

$$T' + \lambda k T = 0 \tag{9.8.1}$$

$$\nabla^2 U + \lambda U = 0 \tag{9.8.2}$$

If we assume U of the form

$$U(x,y,z) = X(x)Y(y)Z(z),$$

then the solution of the Helmholtz equation is

$$U_{lmn}(x,y,z) = B_l D_m F_n \sin\frac{l\pi x}{a} \sin\frac{m\pi y}{b} \sin\frac{n\pi z}{d}$$

Since the solution of Eq. (9.8.1) is

$$T(t) = Ge^{-\lambda kt}$$

the solution of the heat equation takes the form

$$u(x,y,z,t) = \sum_{l=1}^{\infty} \sum_{m=1}^{\infty} \sum_{n=1}^{\infty} a_{lmn} e^{-\lambda kt} \sin\frac{l\pi x}{a} \sin\frac{m\pi y}{b} \sin\frac{n\pi z}{d}$$

where $\lambda = [(l^2/a^2) + (m^2/b^2) + (n^2/d^2)]\pi^2$ and a_{lmn} is a constant. Application of the initial condition yields

$$a_{lmn} = \frac{8}{abd} \int_0^a \int_0^b \int_0^d f(x,y,z) \sin\frac{l\pi x}{a} \sin\frac{m\pi y}{b} \sin\frac{n\pi z}{d}\, dx\, dy\, dz$$

9.9 The Hydrogen Atom

We shall now solve one of the fundamental problems of quantum mechanics, the hydrogen atom. The hydrogen atom is a system consisting of an electron of mass m^* and charge $-e$ attracted by a massive nucleus of charge e. The potential energy is given by

$$V = -\frac{e^2}{r} \tag{9.9.1}$$

where r is the distance of the electron from the center of the nucleus. The time dependent wave function ψ for a single particle moving in three dimensional space is described by the *Schrodinger equation*

$$\frac{ih}{2\pi}\psi_t = \frac{-h^2}{8\pi^2 m^*}\nabla^2\psi + V\psi \tag{9.9.2}$$

where h is Planck's constant and $i = \sqrt{-1}$.

We assume a solution in the form

$$\psi(x,y,z,t) = F(x,y,z)T(t)$$

Substitution of this in Eq. (9.9.2) gives

$$T' + \frac{2\pi i\lambda}{h} T = 0 \tag{9.9.3}$$

$$\nabla^2 F - \frac{8\pi^2 m^*}{h}(V-\lambda)F = 0 \tag{9.9.4}$$

where λ is a separation constant. Integration of the equation for T gives

$$T(t) = Ae^{-\frac{2\pi i\lambda t}{h}}$$

where A is an arbitrary constant.

It is required that the probability of finding the particle somewhere is unity so that

$$\int\int\int |\psi|^2 d\tau = 1 \tag{9.9.5}$$

where the integral extends over all space.

Thus, the solution of the Schrodinger equation is

$$\psi(x,y,z,t) = F(x,y,z)e^{-\frac{2\pi i\lambda t}{h}}$$

in which A was taken to be unity for normalization. This is permissible since Eq. (9.9.4) is homogeneous in F.

To determine the function F, we introduce spherical coordinates for convenience. Equation (9.9.4) in spherical coordinates then becomes

$$\frac{1}{r^2}\frac{\partial}{\partial r}\left(r^2\frac{\partial F}{\partial r}\right) + \frac{1}{r^2\sin\varphi}\frac{\partial}{\partial\varphi}\left(\sin\varphi\frac{\partial F}{\partial\theta}\right) + \frac{1}{r^2\sin^2\varphi}\frac{\partial^2 F}{\partial\theta^2} + \left(a + \frac{b}{r}\right)F = 0 \tag{9.9.7}$$

where $a = 8\pi^2 m^*\lambda/h$ and $b = 8\pi^2 m^* e^2/h$.

Again by the method of separation of variables, we assume

$$F(r,\theta,\varphi) = R(r)\Theta(\theta)\Phi(\varphi)$$

Then after separation, the following equations are obtained:

$$r^2R'' + 2rR' + \left[\left(a + \frac{b}{r}\right)r^2 - \mu\right]R = 0 \tag{9.9.8}$$

$$\Theta'' - \nu\Theta = 0 \tag{9.9.9}$$

$$\sin^2\varphi\Phi'' + \sin\varphi\cos\varphi\Phi' + (\mu\sin^2\varphi + \nu)\Phi = 0 \tag{9.9.10}$$

where μ and ν are separation constants.

In order for $\Theta(\theta)$ to be continuous and single-valued, ν must be zero or negative. Let $\nu = -m^2$ for $m = 1,2,3,\ldots$. Thus the fundamental solutions are $e^{-im\theta}, 1, e^{im\theta}$. Thus, the solution may be written in the form

$$\Theta_m(\theta) = \frac{1}{\sqrt{2\pi}} e^{im\theta}$$

for $m=0, \pm 1, \pm 2, \ldots$. The factor $1/\sqrt{2\pi}$ is introduced in order to normalize $\Theta_m(\theta)$. The constant m (not to be confused with the notation m^* for the mass used earlier) is called the *magnetic quantum number.*

Consider next the solution of

$$\sin^2\varphi\Phi'' + \sin\varphi\cos\varphi\Phi' + (\mu\sin^2\varphi - m^2)\Phi = 0$$

Now if we introduce

$$z = \cos\varphi$$

which varies between -1 and 1, the equation for $\Phi(z)$ takes the form

$$\frac{d}{dz}\left[(1-z^2)\frac{d\Phi}{dz}\right] + \left(\mu - \frac{m^2}{1-z^2}\right)\Phi = 0 \tag{9.9.11}$$

For certain values of μ, the only solution that exists is the series solution which diverges at $z=\pm 1$. In order for Φ to be continuous at the end points the solution must be necessarily bounded at those points.

If $\mu = l(l+1)$ where l is a positive integer or zero, Eq. (9.9.11) is the associated Legendre equation, the fundamental solutions of which are $P_l^m(z)$ and $Q_l^m(z)$. Since $Q_l^m(z)$ approaches infinity at $z=\pm 1$, the admissible solution for the present problem is

$$\Phi_{lm}(z) = \sqrt{\frac{(2l+1)}{2}\frac{(l-m)!}{(l+m)!}}\; P_l^m(z)$$

Here the constant is so chosen that Φ be normalized. l is called the *azimuthal quantum number.*

Note that when $m=0$, Eq. (9.9.11) is the Legendre equation of which $P_l(z)$ is the solution. The Legendre polynomials and the associated Legendre polynomials are related by

$$P_l^m(z) = (1-z^2)^{m/2}\frac{d^m P_l(z)}{dz^m}$$

From this, we see that $P_l^m(z)$ vanishes when $m>l$. As for the negative integral values of m, it can be readily shown that

$$P_l^{-m}(z) = (-1)^m\frac{(l-m)!}{(l+m)!}P_l^m(z)$$

Thus, the function $P_l^{-m}(z)$ differs from the function $P_l^m(z)$ by a constant factor, and as a consequence, m is restricted to a positive integer or zero.

To find $R(r)$, we divide Eq. (9.9.8) by r^2. We then obtain

$$R'' + \frac{2R'}{r} + \left[a + \frac{b}{r} - \frac{l(l+1)}{r^2}\right]R = 0$$

Introducing the new constant

$$n = b/2\sqrt{-a}$$

and the new variable

$$\zeta = 2\sqrt{-a}\ r$$

we are led to the equation

$$R_{\zeta\zeta} + \frac{2}{\zeta} R_{\zeta} + \left[-\frac{1}{4} + \frac{n}{\zeta} - \frac{l(l+1)}{\zeta^2} \right] R = 0$$

In order that $R(\zeta)$ be continuous at $\zeta = 0$ and remain finite at $\zeta \to \infty$ for real n, the integral values of n must be greater than l. n is called the *total quantum number*. The constant is chosen to be the normalization integral for the *associated Laguerre polynomial*

$$\int_0^{\infty} e^{-\zeta} \zeta^{2l} \left[L_{n+l}^{2l+1}(\zeta) \right]^2 \zeta^2 \, d\zeta = \frac{2n\left[(n+l)!\right]^3}{(n-l-1)!}$$

in which the factor ζ^2 arises from the volume element in spherical coordinates. The solution, therefore, is

$$R_{ln}(\zeta) = -\sqrt{\left(\frac{2}{na_0}\right)^3 \frac{(n-l-1)!}{2n\left[(n+l)!\right]^3}}\ e^{-\zeta/2} \zeta^l L_{n+l}^{2l+1}(\zeta)$$

where $a_0 = h^2/4\pi^2 m^* e^2$. The negative sign is chosen for convenience so that $R(\zeta)$ takes on positive values for small values of r. Here l assumes all integral values from 0 to $n - l$.

Thus the wave function of the hydrogen atom is

$$\psi_{lmn}(r,\theta,\varphi) = R_{ln}(r)\Theta_m(\theta)\Phi_{lm}(\varphi)$$

with

$$R_{ln}(r) = -\left[\left(\frac{2}{na_0}\right)^3 \frac{(n-l-1)!}{2n\{(n+l)!\}^3} \right]^{\frac{1}{2}} e^{-\zeta/2} \zeta^l L_{n+l}^{2l+1}(\zeta)$$

$$\Phi_{lm}(\varphi) = \left[\frac{(2l+1)(l-m)!}{2(l+m)!} \right]^{\frac{1}{2}} P_l^m(\cos\varphi)$$

$$\Theta_m(\theta) = \frac{1}{\sqrt{2\pi}} e^{im\theta}$$

in which

$$\zeta = 2\sqrt{-a}\ r$$

9.10 Method of Eigenfunctions

Consider the nonhomogeneous initial-boundary value problem

$$L[u] = \rho u_{tt} - G \qquad \text{in } D \tag{9.10.1}$$

with prescribed homogeneous boundary conditions on the boundary B of

D, and the initial conditions

$$u(x_1,x_2,\ldots,x_n,0)=f(x_1,x_2,\ldots,x_n) \tag{9.10.2}$$

$$u_t(x_1,x_2,\ldots,x_n,0)=g(x_1,x_2,\ldots,x_n) \tag{9.10.3}$$

Here $\rho(x_1,x_2,\ldots,x_n)$ is a real-valued positive continuous function and $G(x_1,x_2,\ldots,x_n)$ is a real-valued continuous function.

We assume the only solution of the associated homogeneous problem

$$L[u]=\rho u_{tt} \tag{9.10.4}$$

with the prescribed boundary conditions is the trivial solution. Then, if there exists a solution of the given problem (9.10.1)–(9.10.3), it can be represented by a series of eigenfunctions of the associated eigenvalue problem

$$L[\varphi]+\lambda\rho\varphi=0 \tag{9.10.5}$$

with φ satisfying the boundary conditions given for u. For problems with one space variable, see Sec. 6.7.

9.11 Forced Vibration of Membrane

As a specific example, we shall determine the solution of the problem of forced vibration of a rectangular membrane of length a and width b. The problem is

$$u_{tt}-c^2\nabla^2 u=F(x,y,t) \quad \text{in } D \tag{9.11.1}$$

$$u(x,y,0)=f(x,y), \quad 0\leqslant x\leqslant a, \quad 0\leqslant y\leqslant b \tag{9.11.2}$$

$$u_t(x,y,0)=g(x,y), \quad 0\leqslant x\leqslant a, \quad 0\leqslant y\leqslant b \tag{9.11.3}$$

$$u(0,y,t)=0 \tag{9.11.4}$$

$$u(a,y,t)=0 \tag{9.11.5}$$

$$u(x,0,t)=0 \tag{9.11.6}$$

$$u(x,b,t)=0 \tag{9.11.7}$$

The associated eigenvalue problem is

$$\nabla^2\varphi+\lambda\varphi=0 \quad \text{in } D$$

$$\varphi=0 \quad \text{on } B$$

The eigenvalues for this problem according to Sec. 9.5 are

$$\alpha_{mn}=\frac{m^2\pi^2}{a^2}+\frac{n^2\pi^2}{b^2} \qquad m,n=1,2,3,\ldots$$

and the corresponding eigenfunctions are

$$\varphi_{mn}(x,y)=\sin\frac{m\pi x}{a}\sin\frac{n\pi y}{b}$$

Thus, we assume the solution

$$u(x,y,t) = \sum_{m=1}^{\infty} \sum_{n=1}^{\infty} u_{mn}(t) \sin\frac{m\pi x}{a} \sin\frac{n\pi y}{b}$$

and the forcing function

$$F(x,y,t) = \sum_{m=1}^{\infty} \sum_{n=1}^{\infty} F_{mn}(t) \sin\frac{m\pi x}{a} \sin\frac{n\pi y}{b}$$

Here $F_{mn}(t)$ are given by

$$F_{mn}(t) = \frac{4}{ab} \int_0^a \int_0^b F(x,y,t) \sin\frac{m\pi x}{a} \sin\frac{n\pi y}{b}\, dx\, dy$$

Note that u automatically satisfies the homogeneous boundary conditions. Now inserting $u(x,y,t)$ and $F(x,y,t)$ in Eq. (9.11.1), we obtain

$$u''_{mn} + c^2\alpha_{mn}^2 u_{mn} = F_{mn}$$

where $\alpha_{mn}^2 = (m\pi/a)^2 + (n\pi/b)^2$. We have assumed that u is twice continuously differentiable with respect to t. Thus the solution of the preceding ordinary differential equation takes the form

$$u_{mn}(t) = A_{mn} \cos\alpha_{mn}ct + B_{mn} \sin\alpha_{mn}ct$$
$$+ \frac{1}{\alpha_{mn}c} \int_0^t F_{mn}(\tau) \sin\alpha_{mn}c(t-\tau)\, d\tau$$

The first initial condition gives

$$u(x,y,0) = f(x,y) = \sum_{m=1}^{\infty} \sum_{n=1}^{\infty} A_{mn} \sin\frac{m\pi x}{a} \sin\frac{n\pi y}{b}$$

Assuming that $f(x,y)$ is continuous in x and y, the coefficient A_{mn} of the double Fourier series is given by

$$A_{mn} = \frac{4}{ab} \int_0^a \int_0^b f(x,y) \sin\frac{m\pi x}{a} \sin\frac{n\pi y}{b}\, dx\, dy$$

Similarly, from the remaining initial condition, we have

$$u_t(x,y,0) = g(x,y) = \sum_{m=1}^{\infty} \sum_{n=1}^{\infty} B_{mn}\alpha_{mn}c \sin\frac{m\pi x}{a} \sin\frac{n\pi y}{b}$$

and hence for continuous $g(x,y)$

$$B_{mn} = \frac{4}{ab\alpha_{mn}c} \int_0^a \int_0^b g(x,y) \sin\frac{m\pi x}{a} \sin\frac{n\pi y}{b}\, dx\, dy$$

The solution of the given initial-boundary value problem is therefore given by

$$u(x,y,t) = \sum_{m=1}^{\infty} \sum_{n=1}^{\infty} u_{mn}(t) \sin\frac{m\pi x}{a} \sin\frac{n\pi y}{b},$$

provided the series for u and its first and second derivatives converge uniformly.

If $F(x,y,t)=e^{x+y}\cos\omega t$, then

$$F_{mn}(t)=\frac{4mn\pi^2}{(m^2\pi^2+a^2)(n^2\pi^2+b^2)}\left[1+(-1)^{m+1}e^a\right]\left[1+(-1)^{n+1}e^b\right]\cos\omega t$$

$$=C_{mn}\cos\omega t$$

Hence we have

$$u_{mn}(t)=\frac{1}{\alpha_{mn}c}\int_0^t C_{mn}\cos\omega\tau\sin\alpha_{mn}c(t-\tau)dr$$

$$=\frac{C_{mn}}{(\alpha_{mn}^2c^2-\omega^2)}(\cos\omega t-\cos\alpha_{mn}ct)$$

provided $\omega\neq\alpha_{mn}c$. Thus, the solution may be written in the form

$$u(x,y,t)=\sum_{m=1}^{\infty}\sum_{n=1}^{\infty}\frac{C_{mn}}{(\alpha_{mn}^2c^2-\omega^2)}(\cos\omega t-\cos\alpha_{mn}ct)\sin\frac{m\pi x}{a}\sin\frac{n\pi y}{b}$$

9.12 Time-dependent Boundary Conditions

The preceding chapters have been devoted to problems with homogeneous boundary conditions. Due to occurrence of problems with time dependent boundary conditions in practice, we consider the forced vibration of rectangular membrane with moving boundaries. The problem here is to determine the displacement function u of

$$u_{tt}-c^2\nabla^2u=F(x,y,t)\quad 0<x<a\quad 0<y<b \tag{9.12.1}$$

$$u(x,y,0)=f(x,y)\quad 0\leqslant x\leqslant a\quad 0\leqslant y\leqslant b \tag{9.12.2}$$

$$u_t(x,y,0)=g(x,y)\quad 0\leqslant x\leqslant a\quad 0\leqslant y\leqslant b \tag{9.12.3}$$

$$u(0,y,t)=p_1(y,t)\quad 0\leqslant y\leqslant b\quad t\geqslant 0 \tag{9.12.4}$$

$$u(a,y,t)=p_2(y,t)\quad 0\leqslant y\leqslant b\quad t\geqslant 0 \tag{9.12.5}$$

$$u(x,0,t)=q_1(x,t)\quad 0\leqslant x\leqslant a\quad t\geqslant 0 \tag{9.12.6}$$

$$u(x,b,t)=q_2(x,t)\quad 0\leqslant x\leqslant a\quad t\geqslant 0 \tag{9.12.7}$$

For such problems, we seek a solution in the form

$$u(x,y,t)=U(x,y,t)+v(x,y,t) \tag{9.12.8}$$

where v is the new dependent variable to be determined. Before finding v, we must first determine U. If we substitute Eq. (9.12.8) in Eqs. (9.12.1)–(9.12.7), we repectively obtain

$$v_{tt}-c^2(v_{xx}+v_{yy})=F-U_{tt}+c^2(U_{xx}+U_{yy})$$

$$=\tilde{F}(x,y,t)$$

and

$$\begin{aligned}
v(x,y,0)&=f(x,y)-U(x,y,0)=\tilde{f}(x,y)\\
v_t(x,y,0)&=g(x,y)-U_t(x,y,0)=\tilde{g}(x,y)\\
v(0,y,t)&=p_1(y,t)-U(0,y,t)=\tilde{p}_1(y,t)\\
v(a,y,t)&=p_2(y,t)-U(a,y,t)=\tilde{p}_2(y,t)\\
v(x,0,t)&=q_1(x,t)-U(x,0,t)=\tilde{q}_1(x,t)\\
v(x,b,t)&=q_2(x,t)-U(x,b,t)=\tilde{q}_2(x,t)
\end{aligned}$$

In order to make the conditions on v homogeneous, we set

$$\tilde{p}_1=\tilde{p}_2=\tilde{q}_1=\tilde{q}_2=0$$

so that

$$\begin{aligned}
U(0,y,t)&=p_1(y,t)\\
U(a,y,t)&=p_2(y,t)\\
U(x,0,t)&=q_1(x,t)\\
U(x,b,t)&=q_2(x,t)
\end{aligned}\tag{9.12.9}$$

In order that the boundary conditions be compatible, we assume the prescribed functions take the forms

$$\begin{aligned}
p_1(y,t)&=\varphi(y)p_1^*(y,t)\\
p_2(y,t)&=\varphi(y)p_2^*(y,t)\\
q_1(x,t)&=\psi(x)q_1^*(x,t)\\
q_2(x,t)&=\psi(x)q_2^*(x,t)
\end{aligned}$$

where the function φ must vanish at the end points $y=0, y=b$ and the function ψ must vanish at $x=0, x=a$. Thus, $U(x,y,t)$ which satisfies Eq. (9.12.9) takes the form

$$U(x,y,t)=\varphi(y)\left[p_1^*+\frac{x}{a}(p_2^*-p_1^*)\right]+\psi(x)\left[q_1^*+\frac{y}{b}(q_2^*-q_1^*)\right]$$

The problem then is to find the function $v(x,y,t)$ of

$$\begin{aligned}
v_{tt}-c^2(v_{xx}+v_{yy})&=\tilde{F}(x,y,t)\\
v(x,y,0)&=\tilde{f}(x,y)\\
v_t(x,y,0)&=\tilde{g}(x,y)\\
v(0,y,t)&=0\\
v(a,y,t)&=0\\
v(x,0,t)&=0\\
v(x,b,t)&=0
\end{aligned}$$

This is an initial-boundary value problem with homogeneous boundary conditions which has already been solved.

As a particular case, consider the following problem

$$u_{tt} - c^2(u_{xx} + u_{yy}) = 0$$

$$u(x,y,0) = 0$$

$$u_t(x,y,0) = \frac{y}{b}\sin\frac{\pi x}{a}$$

$$u(0,y,t) = 0$$

$$u(a,y,t) = 0$$

$$u(x,0,t) = 0$$

$$u(x,b,t) = \sin\frac{\pi x}{a}\sin t$$

We assume a solution in the form

$$u(x,y,t) = v(x,y,t) + U(x,y,t)$$

The function which satisfies

$$U(0,y,t) = 0$$

$$U(a,y,t) = 0$$

$$U(x,0,t) = 0$$

$$U(x,b,t) = \sin\frac{\pi x}{a}\sin t$$

is

$$U(x,y,t) = \sin\frac{\pi x}{a}\left(\frac{y}{b}\sin t\right)$$

Thus the new problem to be solved is

$$v_{tt} - c^2(v_{xx} + v_{yy}) = \left(1 - \frac{c^2\pi^2}{a^2}\right)\frac{y}{b}\sin\frac{\pi x}{a}\sin t$$

$$v(x,y,0) = 0$$

$$v_t(x,y,0) = 0$$

$$v(0,y,t) = 0$$

$$v(a,y,t) = 0$$

$$v(x,0,t) = 0$$

$$v(x,b,t) = 0$$

Then we find F_{mn} from

$$F_{mn}(t) = \frac{4}{ab}\int_0^a\int_0^b F(x,y,t)\sin\frac{m\pi x}{a}\sin\frac{n\pi y}{b}\,dx\,dy$$

where

$$F(x,y,t)=\left(1-\frac{c^2\pi^2}{a^2}\right)\frac{y}{b}\sin\frac{\pi x}{a}\sin t$$

and obtain

$$F_{mn}(t)=\frac{2(-1)^{n+1}}{an}\left(1-\frac{c^2\pi^2}{a^2}\right)\sin t$$

Now we determine $v_{mn}(t)$ which is given by

$$v_{mn}(t)=A_{mn}\cos\alpha_{mn}ct+B_{mn}\sin\alpha_{mn}ct$$
$$+\frac{1}{\alpha_{mn}c}\int_0^t F_{mn}(\tau)\sin\alpha_{mn}c(t-\tau)\,d\tau$$

Since $v(x,y,0)=0, A_{mn}=0$, but

$$B_{mn}=\frac{4}{ab\alpha_{mn}c}\int_0^a\int_0^b\left(-\frac{y}{b}\sin\frac{\pi x}{a}\right)\sin\frac{m\pi x}{a}\sin\frac{n\pi y}{b}\,dx\,dy$$
$$=\frac{2(-1)^n}{\alpha_{mn}can}$$

Thus, we have

$$v_{mn}(t)=\frac{2(-1)^n}{\alpha_{mn}can}\sin\alpha_{mn}ct$$
$$+\frac{2(-1)^n}{\alpha_{mn}ca^3n(1-\alpha^2c^2)}(a^2-c^2\pi^2)(\sin\alpha_{mn}ct-\alpha c\sin t)$$

The solution is therefore given by

$$u(x,y,t)=\frac{y}{b}\sin\frac{\pi x}{a}\sin t+\sum_{m=1}^{\infty}\sum_{n=1}^{\infty}v_{mn}(t)\sin\frac{m\pi x}{a}\sin\frac{n\pi y}{b}$$

Exercises

1. Solve the Dirichlet problem

$$\nabla^2u=0 \qquad 0<x<a \qquad 0<y<b \qquad 0<z<c$$
$$u(0,y,z)=\sin\frac{\pi y}{b}\sin\frac{\pi z}{c}$$
$$u(a,y,z)=0$$
$$u(x,0,z)=0$$
$$u(x,b,z)=0$$
$$u(x,y,0)=0$$
$$u(x,y,c)=0$$

2. Solve the Neumann problem

$$\nabla^2 u=0 \qquad 0<x<1 \qquad 0<y<1 \qquad 0<z<1$$
$$u_x(0,y,z)=0$$
$$u_x(1,y,z)=0$$
$$u_y(x,0,z)=0$$
$$u_y(x,1,z)=0$$
$$u_z(x,y,0)=\cos\pi x\cos\pi y$$
$$u_z(x,y,1)=0$$

3. Solve the Robin problem

$$\nabla^2 u=0 \qquad 0<x<\pi \qquad 0<y<\pi \qquad 0<z<\pi$$
$$u(0,y,z)=f(y,z)$$
$$u(\pi,y,z)=0$$
$$u_y(x,0,z)=0$$
$$u_y(x,\pi,z)=0$$
$$u_z(x,y,0)+hu(x,y,0)=0 \qquad h=\text{constant}$$
$$u_z(x,y,\pi)+hu(x,y,\pi)=0$$

4. Determine the solution of each of the following problems for a cylinder:

a.

$$\nabla^2 u=0 \qquad r<a \qquad 0<\theta<2\pi \qquad 0<z<l$$
$$u(a,\theta,z)=0$$
$$u(r,\theta,l)=0$$
$$u(r,\theta,0)=f(r,\theta)$$

b.

$$\nabla^2 u=0 \qquad r<a \qquad 0<\theta<2\pi \qquad 0<z<l$$
$$u(a,\theta,z)=f(\theta,z)$$
$$u_z(r,\theta,0)=0$$
$$u_z(r,\theta,l)=0$$

5. Find the solution of the Dirichlet problem for a sphere

$$\nabla^2 u=0 \qquad r<a \qquad 0<\theta<\pi \qquad 0<\varphi<2\pi$$
$$u(a,\theta,\varphi)=\cos^2\theta$$

6. Solve the Dirichlet problem for a concentric sphere

$$\nabla^2 u=0 \qquad a<r<b \qquad 0<\theta<\pi \qquad 0<\phi<2\pi$$
$$u(a,\theta,\phi)=f(\theta,\phi)$$
$$u(b,\theta,\phi)=g(\theta,\phi)$$

7. Find the steady state temperature distribution in a cylinder of radius a if a constant flow of heat T is supplied at the end $z=0$, and the surface $r=a$ and the end $z=l$ are maintained at zero temperature.

8. Find the potential of the electrostatic field inside a cylinder of length l and radius a if both ends of the cylinder are each earthed, and the surface is charged to a potential u_0.

9. Determine the potential of the electric field inside a sphere of radius a if the upper half of the sphere is charged to a potential u_1 and the lower half to a potential u_2.

10. Solve the Dirichlet problem for a half cylinder

$$\begin{aligned}
\nabla^2 u &= 0 \qquad r<1 \qquad 0<\theta<\pi \qquad 0<z<1\\
u(1,\theta,z) &= 0\\
u(r,0,z) &= 0\\
u(r,\pi,z) &= 0\\
u(r,\theta,0) &= 0\\
u(r,\theta,1) &= f(r,\theta)
\end{aligned}$$

11. Solve the Neumann problem for a sphere

$$\begin{aligned}
\nabla^2 u &= 0 \qquad r<1 \qquad 0<\theta<\pi \qquad 0<\varphi<2\pi\\
u_r(1,\theta,\varphi) &= f(\theta,\varphi)
\end{aligned}$$

where

$$\int_0^{2\pi}\int_0^{\pi} f(\theta,\varphi)\sin\theta\, d\theta\, d f\varphi = 0$$

12. Find the solution of the initial-boundary value problem

$$\begin{aligned}
u_{tt} &= c^2\nabla^2 u \qquad 0<x<1 \qquad 0<y<1 \qquad t>0\\
u(x,y,0) &= \sin^2\pi x \sin\pi y, \qquad 0\leqslant x\leqslant 1, \qquad 0\leqslant y\leqslant 1\\
u_t(x,y,0) &= 0\\
u(0,y,t) &= 0\\
u(1,y,t) &= 0\\
u(x,0,t) &= 0\\
u(x,1,t) &= 0
\end{aligned}$$

13. Obtain the solution of

$$\begin{aligned}
u_{tt} &= c^2\nabla^2 u \qquad r<a \qquad 0<\theta<2\pi \qquad t>0\\
u(r,\theta,0) &= f(r,\theta)\\
u_t(r,\theta,0) &= g(r,\theta)\\
u(a,\theta,t) &= 0
\end{aligned}$$

14. Determine the temperature distribution in a rectangular plate with radiation from its surface. The temperature distribution is described by

$$u_t = k(u_{xx} + u_{yy}) - h(u - u_0) \qquad 0<x<a \qquad 0<y<b \qquad t>0$$

$$u(x,y,0) = f(x,y)$$

$$u(0,y,t) = 0$$

$$u(a,y,t) = 0$$

$$u(x,0,t) = 0$$

$$u(x,b,t) = 0$$

where u_0 is a constant.

15. Solve the heat conduction problem in a circular plate

$$u_t = k\left(u_{rr} + \frac{1}{r}u_r + \frac{1}{r^2}u_{\theta\theta}\right) \qquad r<1 \qquad 0<\theta<2\pi \qquad t>0$$

$$u(r,\theta,0) = f(r,\theta)$$

$$u(1,\theta,t) = 0$$

16. Solve the initial-boundary value problem

$$u_{tt} = c^2\nabla^2 u \qquad 0<x<1 \qquad 0<y<1 \qquad 0<z<1 \qquad t>0$$

$$u(x,y,z,0) = \sin \pi x \sin \pi y \sin \pi z$$

$$u_t(x,y,z,0) = 0$$

$$u(0,y,z,t) = u(1,y,z,t) = 0$$

$$u(x,0,z,t) = u(x,1,z,t) = 0$$

$$u(x,y,0,t) = u(x,y,1,t) = 0$$

17. Solve

$$u_{tt} + ku_t = c^2\nabla^2 u \qquad 0<x<a \qquad 0<y<b \qquad 0<z<d \qquad t>0$$

$$u(x,y,z,0) = f(x,y,z)$$

$$u_t(x,y,z,0) = g(x,y,z)$$

$$u(0,y,z,t) = u(a,y,z,t) = 0$$

$$u(x,0,z,t) = u(x,b,z,t) = 0$$

$$u(x,y,0,t) = u(x,y,d,t) = 0$$

18. Obtain the solution of

$$u_{tt} = c^2\left(u_{rr} + \frac{1}{r}u_r + \frac{1}{r^2}u_{\theta\theta} + u_{zz}\right) \qquad r<a \qquad 0<z<l \qquad t>0$$

$$u(r,\theta,z,0) = f(r,\theta,z)$$

$$u_t(r,\theta,z,0) = q(r,\theta,z)$$

$$u(a,\theta,z,t) = 0$$

$$u(r,\theta,0,t) = u(r,\theta,l,t) = 0$$

19. Determine the solution of the heat conduction problem

$$u_t = k\nabla^2 u \qquad 0<x<a \qquad 0<y<b \qquad 0<z<c \qquad t>0$$
$$u(x,y,z,0) = f(x,y,z)$$
$$u_x(0,y,z,t) = u_x(a,y,z,t) = 0$$
$$u_y(x,0,z,t) = u_y(x,b,z,t) = 0$$
$$u_z(x,y,0,t) = u_z(x,y,c,t) = 0$$

20. Solve

$$u_t = k\nabla^2 u \qquad r<a \qquad 0<z<l \qquad t>0$$
$$u(r,\theta,z,0) = f(r,\theta,z)$$
$$u_r(a,\theta,z,t) = 0$$
$$u(r,\theta,0,t) = u(r,\theta,l,t) = 0$$

21. Find the temperature distribution in the section of a sphere cut out by the cone $\theta = \theta_0$. The surface temperature is zero while the initial temperature is given by $f(r,\theta,\varphi)$.

22. Solve the initial-boundary value problem

$$u_{tt} = c^2\nabla^2 u + F(x,y,t) \qquad 0<x<a \qquad 0<y<b$$
$$u(x,y,0) = f(x,y)$$
$$u_t(x,y,0) = g(x,y)$$
$$u_x(0,y,t) = u_x(a,y,t) = 0$$
$$u_y(x,0,t) = u_y(x,b,t) = 0$$

23. Solve

$$u_{tt} = c^2\nabla^2 u + xy\sin t \qquad 0<x<\pi \qquad 0<y<\pi \qquad t>0$$
$$u(x,y,0) = 0$$
$$u_t(x,y,0) = 0$$
$$u(0,y,t) = u(\pi,y,t) = 0$$
$$u(x,0,t) = u(x,\pi,t) = 0$$

24. Solve

$$u_t = k\nabla^2 u + F(x,y,z,t) \qquad 0<x<a \qquad 0<y<b \qquad 0<z<c$$
$$u(x,y,z,0) = f(x,y,z)$$
$$u(0,y,z,t) = u(a,y,z,t) = 0$$
$$u(x,0,z,t) = u(x,b,z,t) = 0$$
$$u_z(x,y,0,t) = u_z(x,y,c,t) = 0$$

25. Solve

$$u_t = k\nabla^2 u + A \qquad 0<x<\pi \qquad 0<y<\pi \qquad t>0$$
$$u(x,y,0)=0$$
$$u(0,y,t)=u(\pi,y,t)=0$$
$$u_y(x,0,t)+u(x,0,t)=0$$
$$u_y(x,\pi,t)+u(x,\pi,t)=0$$

where A is a constant.

26. Find the temperature distribution of the composite cylinder consisting of an inner cylinder $0\leqslant r\leqslant r_0$ and an outer cylindrical tube $r_0\leqslant r\leqslant a$. The surface temperature is maintained at zero degree, and the initial temperature distribution is given by $f(r,\theta,z)$.

27. Solve the initial-boundary value problem

$$u_t - c^2\nabla^2 u=0 \qquad 0<x<\pi \qquad 0<y<\pi \qquad t>0$$
$$u(x,y,0)=0$$
$$u(0,y,t)=u(\pi,y,t)=0$$
$$u(x,0,t)=x(x-\pi)\sin t, \qquad 0\leqslant x\leqslant\pi, \qquad t\geqslant 0$$
$$u(x,\pi,t)=0$$

28. Solve

$$u_{tt}=c^2\nabla^2 u \qquad r<a \qquad t>0$$
$$u(r,\theta,0)=f(r,\theta)$$
$$u_t(r,\theta,0)=g(r,\theta)$$
$$u(a,\theta,t)=p(\theta,t)$$

29. Solve

$$u_t=c^2\nabla^2 u \qquad r<a \qquad t>0$$
$$u(r,\theta,0)=f(r,\theta)$$
$$u_r(a,\theta,t)=g(\theta,t)$$

30. Determine the solution of the biharmonic equation

$$\nabla^4 u=q/D$$

with the boundary conditions

$$u\left(-\frac{a}{2},y\right)=u\left(\frac{a}{2},y\right)=0$$
$$u_{xx}\left(-\frac{a}{2},y\right)=u_{xx}\left(\frac{a}{2},y\right)=0$$
$$u(x,0)=u(x,b)=0$$
$$u_{yy}(x,0)=u_{yy}(x,b)=0$$

where q is the load per unit area and D is the flexural rigidity of the plate. This is the problem of the deflection of a uniformly loaded plate, the sides of which are simply supported.

CHAPTER TEN

Green's Function

10.1 The Delta Function

The application of Green's functions to boundary-value problems in ordinary differential equations was described earlier in Chapter 7. The Green's function method is applied here to boundary value problems in partial differential equations. This method provides solutions in integral form and is applicable to a wide class of problems in mathematical physics.

Before developing the method of Green's function, we will first define the *Dirac delta function* $\delta(x-\xi,y-\eta)$ in two dimensions by

$$\text{a.}\quad \delta(x-\xi,y-\eta)=0, \qquad x\neq\xi,\quad y\neq\eta \tag{10.1.1}$$

$$\text{b.}\quad \iint_{R_\epsilon}\delta(x-\xi,y-\eta)\,dx\,dy=1, \qquad R_\epsilon:(x-\xi)^2+(y-\eta)^2<\epsilon^2 \tag{10.1.2}$$

$$\text{c.}\quad \iint_{R}F(x,y)\delta(x-\xi,y-\eta)\,dx\,dy=F(\xi,\eta) \tag{10.1.3}$$

for arbitrary continuous function F in the region R.

The delta function is not a function in the ordinary sense.[10] It is a symbolic function, and is often viewed as the limit of a distribution.

If $\delta(x-\xi)$ and $\delta(y-\nu)$ are one-dimensional delta funstions, we have

$$\iint_{R}F(x,y)\delta(x-\xi)\delta(y-\eta)\,dx\,dy=F(\xi,\eta) \tag{10.1.4}$$

[10]For an elegant treatment of the delta function as a generalized function, see L. Schwartz, *Theory of Distributions*.

Since (10.1.3) and (10.1.4) hold for an arbitrary continuous function F, we conclude that

$$\delta(x-\xi,y-\eta)=\delta(x-\xi)\delta(y-\eta) \tag{10.1.5}$$

Thus, we may state that the two-dimensional delta function is the product of one-dimensional delta functions.

Higher dimensional delta functions can be defined in a similar manner.

10.2 Green's Function

The solution of the Dirichlet problem

$$\begin{aligned} \nabla^2 u &= h(x,y) \quad &\text{in } D \\ u &= f(x,y) \quad &\text{on } B \end{aligned} \tag{10.2.1}$$

is given by[11]

$$u(x,y)=\int\int_D G(x,y;\xi,\eta)h(\xi,\eta)\,d\xi\,d\eta+\int_B f\frac{\partial G}{\partial n}\,ds \tag{10.2.2}$$

where G is the Green's function and n denotes the outward normal to the boundary B of the region D. It is rather obvious then that the solution $u(x,y)$ can be determined as soon as the Green's function G is ascertained. So the problem really in this technique is to find the Green's function.

First we shall define the Green's function for the Dirichlet problem involving the Laplace operator. Then, the Green's function for the Dirichlet problem involving the Helmholtz operator may be defined in a completely analogous manner.

The Green's function for the Dirichlet problem involving the Laplace operator is the function which satisfies

a. $\nabla^2 G=\delta(x-\xi,y-\eta)$[12] in D (10.2.3)
   $G=0$ on B (10.2.4)

b. G is symmetric, that is,

$$G(x,y;\xi,\eta)=G(\xi,\eta;x,y) \tag{10.2.5}$$

c. G is continuous in x,y,ξ,η, but $\partial G/\partial h$ has a discontinuity at the point (ξ,η) which is specified by the equation

$$\lim_{\epsilon\to 0}\int_{C_\epsilon}\frac{\partial G}{\partial n}\,ds=1 \tag{10.2.6}$$

where n is the outward normal to the circle

$$C_\epsilon:(x-\xi)^2+(y-\eta)^2=\epsilon^2$$

---

[11]The proof is given in Sec. 10.4.

[12]For other operators, see Greenberg [16].

The Green's function G may be interpreted as the response of the system at a field point (x,y) due to a δ function input at the source point (ξ,η). G is continuous everywhere in D, and its first and second derivatives are continuous in D except at (ξ,η). Thus property (a) essentially states that $\nabla^2 G=0$ everywhere except at the source point (ξ,η).

We will now prove property (b).

Theorem 10.2.1 *The Green's function is symmetric.*

PROOF. Applying Green's second formula[13]

$$\iint_D (\phi\nabla^2\psi-\psi\nabla^2\phi)\,dS=\int_B\left(\phi\frac{\partial\psi}{\partial n}-\psi\frac{\partial\phi}{\partial n}\right)ds \tag{10.2.7}$$

to the functions $\phi=G(x,y;\xi,\eta)$ and $\psi=G(x,y;\xi^*,\eta^*)$ we obtain

$$\iint_D\left[G(x,y;\xi,\eta)\nabla^2G(x,y;\xi^*,\eta^*)-G(x,y;\xi^*,\eta^*)\nabla^2G(x,y;\xi,\eta)\right]dx\,dy$$

$$=\int_B\left[G(x,y;\xi,\eta)\frac{\partial G}{\partial n}(x,y;\xi^*,\eta^*)-G(x,y;\xi^*,\eta^*)\frac{\partial G}{\partial n}(x,y;\xi,\eta)\right]ds$$

Since $G(x,y;\xi,\eta)$ and hence $G(x,y;\xi^*,\eta^*)$ must vanish on B, we have

$$\iint_D\left[G(x,y;\xi,\eta)\nabla^2G(x,y;\xi^*,\eta^*)-G(x,y;\xi^*,\eta^*)\nabla^2G(x,y;\xi,\eta)\right]dx\,dy=0$$

But

$$\nabla^2G(x,y;\xi,\eta)=\delta(x-\xi,y-\eta)$$

and

$$\nabla^2G(x,y;\xi^*\eta^*)=\delta(x-\xi^*,y-\eta^*)$$

Since

$$\iint_D G(x,y;\xi,\eta)\delta(x-\xi^*,y-\eta^*)\,dx\,dy=G(\xi^*,\eta^*;\xi,\eta)$$

and

$$\iint_D G(x,y;\xi^*,\eta^*)\delta(x-\xi,y-\eta)\,dx\,dy=G(\xi,\eta;\xi^*,\eta^*)$$

we obtain

$$G(\xi,\eta;\xi^*,\eta^*)=G(\xi^*,\eta^*;\xi,\eta)$$ □

[13]For the general second-order operator in two variables see No. 1, Exercise 10.

Theorem 10.2.2 *$\partial G/\partial n$ is discontinuous at (ξ,η); in particular,*

$$\lim_{\epsilon\to 0}\int_{C_\epsilon}\frac{\partial G}{\partial n}\,ds=1 \qquad C_\epsilon:(x-\xi)^2+(y-\eta)^2=\epsilon^2$$

PROOF. Let R_ϵ be the region bounded by C_ϵ. Then integrating both sides of Eq. (10.2.3) we obtain

$$\int\int_{R_\epsilon}\nabla^2 G\,dx\,dy=\int\int_{R_\epsilon}\delta(x-\xi,y-\eta)\,dx\,dy=1$$

It therefore follows that

$$\lim_{\epsilon\to 0}\int\int_{R_\epsilon}\nabla^2 G\,dx\,dy=1 \tag{10.2.8}$$

Thus, by the Divergence theorem,

$$\lim_{\epsilon\to 0}\int_{C_\epsilon}\frac{\partial G}{\partial n}\,ds=1$$

□

10.3 Method of Green's Function

It is often convenient to seek G as the sum of a particular integral of the nonhomogeneous equation and the solution of the associated homogeneous equation. That is, G may assume the form[14]

$$G(\xi,\eta;x,y)=F(\xi,\eta;x,y)+g(\xi,\eta;x,y) \tag{10.3.1}$$

where F, known as the free-space Green's function, satisfies

$$\nabla^2 F=\delta(\xi-x,\eta-y) \qquad \text{in } D \tag{10.3.2}$$

and g satisfies

$$\nabla^2 g=0 \qquad \text{in } D \tag{10.3.3}$$

so that by superposition $G=F+g$ satisfies Eq. (10.2.3). Also $G=0$ on B requires that

$$g=-F \text{ on } B \tag{10.3.4}$$

Note that F need not satisfy the boundary condition.

Before we determine the solution of a particular problem, let us first find F for the Laplace and the Helmholtz operators.

---

[14]Hereafter, (x,y) will denote the source point.

(1) Laplace Operator

In this case F must satisfy

$$\nabla^2 F = \delta(\xi - x, \eta - y) \text{ in } D$$

Then for $r = [(\xi - x)^2 + (\eta - y)^2]^{1/2} > 0$, that is, for $\xi \neq x, \eta \neq y$, we have by taking (x,y) as the center

$$\nabla^2 F = \frac{1}{r}\frac{\partial}{\partial r}\left(r\frac{\partial F}{\partial r}\right) = 0$$

since F is independent of θ. The solution, therefore, is

$$F = A + B\log r$$

Applying condition (10.2.6), we see that

$$\lim_{\epsilon \to 0}\int_{C_\epsilon}\frac{\partial F}{\partial n}\,ds = \lim_{\epsilon \to 0}\int_0^{2\pi}\frac{B}{r}\,r\,d\theta = 1^{15}$$

Thus $B = 1/2\pi$ and A is arbitrary. For simplicity we choose $A = 0$. Then F takes the form

$$F = \frac{1}{2\pi}\log r \tag{10.3.5}$$

(2) Helmholtz Operator

Here F is required to satisfy

$$\nabla^2 F + \kappa^2 F = \delta(x - \xi, y - \eta)$$

Again for $r > 0$, we find

$$\frac{1}{r}\frac{\partial}{\partial r}\left(r\frac{\partial F}{\partial r}\right) + \kappa^2 F = 0$$

or

$$r^2 F_{rr} + rF_r + \kappa^2 r^2 F = 0$$

This is the Bessel equation of order zero, the solution of which is

$$F(\kappa r) = AJ_0(\kappa r) + BY_0(\kappa r)$$

Since the behavior of J_0 at $r = 0$ is not singular, we set $A = 0$. Thus, we have

$$F(\kappa r) = BY_0(\kappa r)$$

But for very small r,

$$Y_0(\kappa r) \equiv \frac{2}{\pi}\log r$$

---

[15]This follows directly from Eq. (10.2.8) with $\nabla^2 g = 0$.

Applying condition (10.2.6), we obtain

$$\lim_{\epsilon\to 0}\int_{C_\epsilon}\frac{\partial F}{\partial n}\,ds=\lim_{\epsilon\to 0}\int_{C_\epsilon}B\frac{\partial Y_0}{\partial r}\,ds=1$$

and hence $B=1/4$. $F(\kappa r)$ thus becomes

$$F(\kappa r)=\frac{1}{4}Y_0(\kappa r) \tag{10.3.6}$$

We may point out that since

$$\nabla^2+\kappa^2 \text{ approaches } \nabla^2 \text{ as } \kappa\to 0$$

it should follow (and does follow) that

$$\frac{1}{4}Y_0(\kappa r)\to\frac{1}{2\pi}\log r \qquad \text{as } \kappa\to 0$$

10.4 Dirichlet Problem for the Laplace Operator

We are now in a position to determine the solution of the Dirichlet problem

$$\begin{aligned}\nabla^2 u&=h \qquad \text{in } D\\ u&=f \qquad \text{on } B\end{aligned} \tag{10.4.1}$$

by the method of Green's function.

By putting $\phi(\xi,\eta)=G(\xi,\eta;x,y)$ and $\psi(\xi,\eta)=u(\xi,\eta)$ in Eq. (10.2.7), we obtain

$$\int\int_D\left[G(\xi,\eta;x,y)\nabla^2 u-u(\xi,\eta)\nabla^2 G\right]d\xi\,d\eta$$
$$=\int_B\left[G(\xi,\eta;x,y)\frac{\partial u}{\partial n}-u(\xi,\eta)\frac{\partial G}{\partial n}\right]ds$$

But

$$\nabla^2 u=h(\xi,\eta)$$

and

$$\nabla^2 G=\delta(\xi-x,\eta-y)$$

in D. Thus, we have

$$\int\int_D\left[G(\xi,\eta;x,y)h(\xi,\eta)-u(\xi,\eta)\delta(\xi-x,\eta-y)\right]d\xi\,d\eta$$
$$=\int_B\left[G(\xi,\eta;x,y)\frac{\partial u}{\partial n}-u(\xi,\eta)\frac{\partial G}{\partial n}\right]ds \tag{10.4.2}$$

Since $G=0$ and $u=f$ on B, and noting that G is symmetric, it follows that

$$u(x,y)=\int\int_D G(x,y;\xi,\eta)h(\xi,\eta)\,d\xi\,d\eta+\int_B f\frac{\partial G}{\partial n}\,ds \qquad (10.4.3)$$

which is the solution given in Sec. 10.2.

As a specific example, consider the Dirichlet problem for a unit circle. Then

$$\begin{aligned}\nabla^2 g = g_{\xi\xi}+g_{\eta\eta}&=0 \qquad \text{in } D\\ g&=-F \qquad \text{on } B\end{aligned} \qquad (10.4.4)$$

But we already have from Eq. (10.3.5) that $F=(1/2\pi)\log r$.

If we introduce the polar coordinates (see Fig. 10.4.1) ρ,θ,σ,β by means of the equations

$$\begin{aligned}x&=\rho\cos\theta \qquad & \xi&=\sigma\cos\beta\\ y&=\rho\sin\theta \qquad & \eta&=\sigma\sin\beta,\end{aligned} \qquad (10.4.5)$$

then the solution of Eq. (10.4.4) is [see Sec. 8.4]

$$g=\frac{a_0}{2}+\sum_{n=1}^{\infty}\sigma^n(a_n\cos n\beta+b_n\sin n\beta)$$

where

$$g=-\frac{1}{4\pi}\log\left[1+\rho^2-2\rho\cos(\beta-\theta)\right] \qquad \text{on } B.$$

By using the relation

$$\log\left[1+\rho^2-2\rho\cos(\beta-\theta)\right]=-2\sum_{n=1}^{\infty}\frac{\rho^n\cos n(\beta-\theta)}{n}$$

and equating the coefficients of $\sin n\beta$ and $\cos n\beta$ to determine a_n and b_n, we find

$$a_n=\frac{\rho^n}{2\pi n}\cos n\theta$$

$$b_n=\frac{\rho^n}{2\pi n}\sin n\theta$$

It therefore follows that

$$\begin{aligned}g(\rho,\theta;\sigma,\beta)&=\frac{1}{2\pi}\sum_{n=1}^{\infty}\frac{(\sigma\rho)^n}{n}\cos n(\beta-\theta)\\ &=-\frac{1}{4\pi}\log\left[1+(\sigma\rho)^2-2(\sigma\rho)\cos(\beta-\theta)\right]\end{aligned}$$

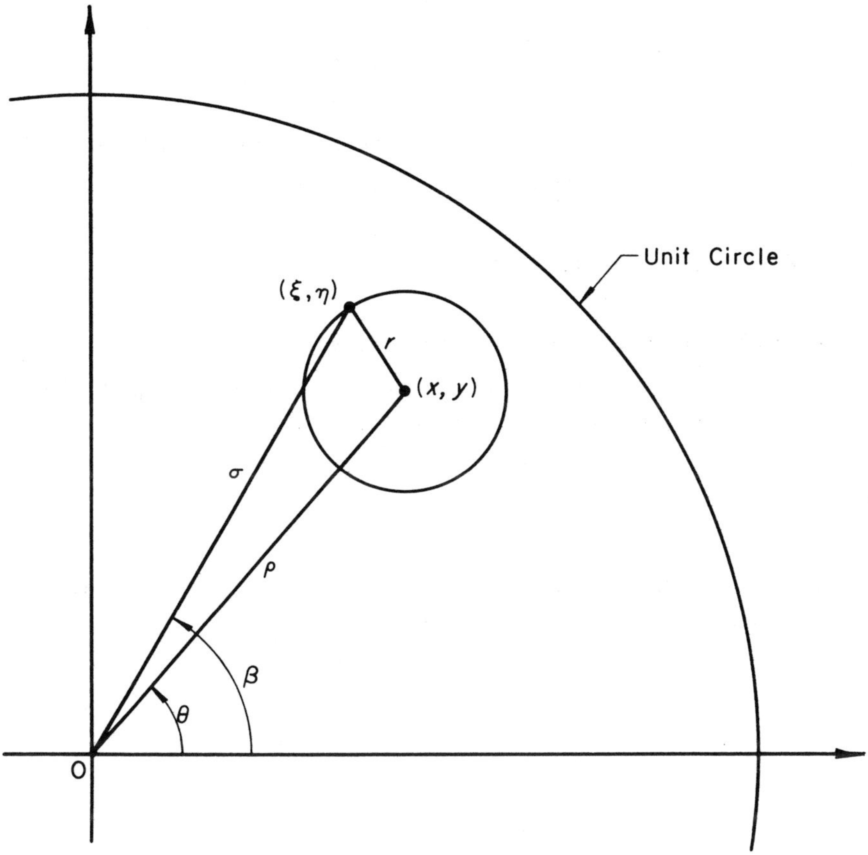

Figure 10.4.1

Hence the Green's function for the problem is

$$
\begin{aligned}
G(\rho,\theta;\sigma,\beta) &= \frac{1}{4\pi}\log\left[\sigma^2+\rho^2-2\sigma\rho\cos(\beta-\theta)\right] \\
&\quad -\frac{1}{4\pi}\log\left[1+(\sigma\rho)^2-2\sigma\rho\cos(\beta-\theta)\right] \\
&= \frac{1}{4\pi}\log\left[\sigma^2+\rho^2-2\sigma\rho\cos(\beta-\theta)\right] \\
&\quad -\frac{1}{4\pi}\log\left[\frac{1}{\sigma^2}+\rho^2-\frac{2\rho}{\sigma}\cos(\beta-\theta)\right]+\frac{1}{2\pi}\log\frac{1}{\sigma}
\end{aligned}
\tag{10.4.6}
$$

from which we find

$$
\left.\frac{\partial G}{\partial n}\right|_{\text{on } B} = \left(\frac{\partial G}{\partial \sigma}\right)_{\sigma=1} = \frac{1}{2\pi}\,\frac{1-\rho^2}{1+\rho^2-2\rho\cos(\beta-\theta)}
$$

If $h=0$, then the solution (10.4.3) becomes

$$u(\rho,\theta)=\frac{1}{2\pi}\int_0^{2\pi}\frac{1-\rho^2}{1+\rho^2-2\rho\cos(\beta-\theta)}f(\beta)\,d\beta$$

which is known as *Poisson's integral formula*. It is often written in the form

$$u(\rho,\theta)=\int_0^{2\pi}P(\rho,\beta-\theta)f(\beta)\,d\beta$$

where P is called the *Poission kernel*.

10.5 Dirichlet Problem for the Helmholtz Operator

We will now determine the Green's function solution of the Dirichlet problem involving the Helmholtz operator, namely,

$$\begin{aligned} \nabla^2 u+\kappa^2 u&=h \qquad \text{in } D\\ u&=f \qquad \text{on } B \end{aligned} \tag{10.5.1}$$

where D is a circular domain of unit radius with boundary B. Then the Green's function must satisfy

$$\begin{aligned} \nabla^2 G+\kappa^2 G&=\delta(\xi-x,\eta-y) \qquad \text{in } D\\ G&=0 \qquad \text{on } B \end{aligned} \tag{10.5.2}$$

Again, we seek the solution in the form

$$G(\xi,\eta;x,y)=F(\xi,\eta;x,y)+g(\xi,\eta;x,y)$$

From Eq. (10.3.6), we have

$$F=\frac{1}{4}Y_0(\kappa r) \tag{10.5.3}$$

where $r=[(\xi-x)^2+(\eta-y)^2]^{1/2}$. The function g must satisfy

$$\begin{aligned} \nabla^2 g+\kappa^2 g&=0 \qquad \text{in } D\\ g&=-\frac{1}{4}Y_0(\kappa r) \qquad \text{on } B \end{aligned} \tag{10.5.4}$$

the solution of which can be easily determined by the method of separation of variables. Thus, the solution in the polar coordinates defined by Eq. (10.4.5) may be written in the form

$$g(\rho,\theta;\sigma,\beta)=\sum_{n=0}^{\infty}J_n(\kappa\sigma)\left[a_n\cos n\beta+b_n\sin n\beta\right] \tag{10.5.5}$$

where

$$a_0 = -\frac{1}{8\pi J_0(\kappa)}\int_{-\pi}^{\pi} Y_0\left[\kappa\sqrt{1+\rho^2-2\rho\cos(\beta-\theta)}\,\right]d\beta$$

$$\left.\begin{aligned} a_n &= -\frac{1}{4\pi J_n(\kappa)}\int_{-\pi}^{\pi} Y_0\left[\kappa\sqrt{1+\rho^2-2\rho\cos(\beta-\theta)}\,\right]\cos n\beta\, d\beta \\ b_n &= -\frac{1}{4\pi J_n(\kappa)}\int_{-\pi}^{\pi} Y_0\left[\kappa\sqrt{1+\rho^2-2\rho\cos(\beta-\theta)}\,\right]\sin n\beta\, d\beta. \end{aligned}\right\}\ n=1,2,\ldots$$

To find the solution of the Dirichlet problem, we multiply both sides of the first equation of Eq. (10.5.1) by G and integrate. Thus, we have

$$\int\int_D (\nabla^2 u+\kappa^2 u)G(\xi,\eta;x,y)\,d\xi\,d\eta = \int\int_D h(\xi,\eta)G(\xi,\eta;x,y)\,d\xi\,d\eta$$

We then apply Green's theorem on the left side of the preceding equation and obtain

$$\int\int_D h(\xi,\eta)G(\xi,\eta;x,y)\,d\xi\,d\eta - \int\int_D u(\nabla^2 G+\kappa^2 G)\,d\xi\,d\eta$$

$$= \int_B (Gu_n - uG_n)\,ds$$

But $\nabla^2 G+\kappa^2 G=\delta(\xi-x,\eta-y)$ in D and $G=0$ on B. We therefore have

$$u(x,y) = \int\int_D h(\xi,\eta)G(\xi,\eta;x,y)\,d\xi\,d\eta + \int_B f(\xi,\eta)G_n\,ds \qquad (10.5.6)$$

where G is given by Eqs. (10.5.3) and (10.5.5).

10.6 Method of Images

We shall describe another method of obtaining Green's function. This method, called the method of images, is based essentially on the construction of Green's function for a finite domain from that of an infinite domain. The disadvantage of this method is that it can be applied only to problems with simple boundary geometries.

As an illustration, we consider the same Dirichlet problem solved in Sec. 10.4.

Let $P(\xi,\eta)$ be a point in the unit circle D, and let $Q(x,y)$ be the source point also in D. The distance between P and Q is r. Let Q' be the *image* which lies outside of D on the ray from the origin opposite to the source point Q (as shown in Fig. 10.6.1) such that $OQ/\sigma=\sigma/OQ'$ where σ is the radius of the circle passing through P centered at the origin.

Since the two triangles OPQ and OPQ' are similar by virtue of the hypothesis $(OQ)(OQ')=\sigma^2$ and by possessing a common angle at 0, we

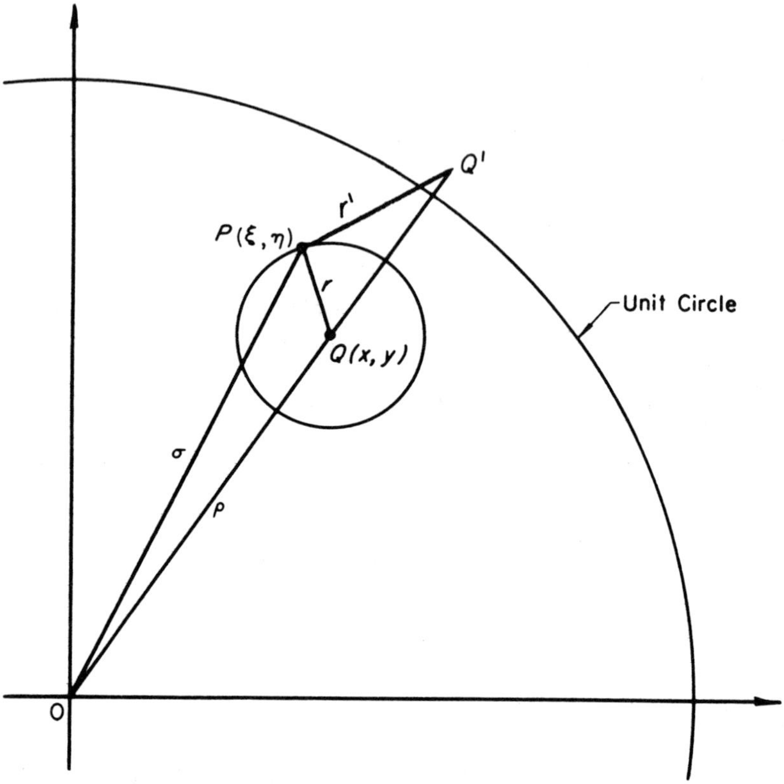

Figure 10.6.1

have

$$\frac{r'}{r} = \frac{\sigma}{\rho} \tag{10.6.1}$$

where $r' = PQ'$ and $\rho = OQ$.

If $\sigma = 1$, Eq. (10.6.1) becomes

$$\frac{r}{r'}\frac{1}{\rho} = 1$$

Then we can clearly see that the quantity

$$\frac{1}{2\pi}\log\left(\frac{r}{r'}\frac{1}{\rho}\right) = \frac{1}{2\pi}\log r - \frac{1}{2\pi}\log r' + \frac{1}{2\pi}\log\frac{1}{\rho} \tag{10.6.2}$$

which vanishes on the boundary $\sigma = 1$, is harmonic in D except at Q, and satisfies Eq. (10.2.3). [Note the $\log r'$ is harmonic everywhere except at Q' which is outside the domain D.] This suggests that we should choose the

Green's function

$$G=\frac{1}{2\pi}\log r-\frac{1}{2\pi}\log r'+\frac{1}{2\pi}\log\frac{1}{\rho} \tag{10.6.3}$$

Noting that Q' is at $(1/\rho,\theta)$, G in polar coordinates takes the form

$$G(\rho,\theta;\sigma,\beta)=\frac{1}{4\pi}\log\left[\sigma^2+\rho^2-2\sigma\rho\cos(\beta-\theta)\right]$$
$$-\frac{1}{4\pi}\log\left[\frac{1}{\sigma^2}+\rho^2-2\frac{\rho}{\sigma}\cos(\beta-\theta)\right]+\frac{1}{2\pi}\log\frac{1}{\sigma} \tag{10.6.4}$$

which is the same as G given by (10.4.6).

It is quite interesting to observe the physical interpretation of the Green's function (10.6.3) or (10.6.4). The first term represents the potential due to a unit line charge at the source point, whereas the second term represents the potential due to a negative unit charge at the image point. The third term represents a uniform potential. The sum of these potentials makes up the potential field.

EXAMPLE 10.6.1. To illustrate an obvious and simple case, consider the semiinfinite plane $\eta>0$. The problem is to solve

$$\nabla^2u=h \quad \text{in} \quad \eta>0$$
$$u=f \quad \text{on} \quad \eta=0$$

The image point should be obvious by inspection. Thus, if we construct

$$G=\frac{1}{4\pi}\log\left[(\xi-x)^2+(\eta-y)^2\right]-\frac{1}{4\pi}\log\left[(\xi-x)^2+(\eta+y)^2\right] \tag{10.6.5}$$

the condition that $G=0$ on $\eta=0$ is clearly satisfied. It is also evident that G is harmonic in $\eta>0$ except at the source point, and that G satisfies Eq. (10.2.3).

With $G_n|_B=[-G_\eta]_{\eta=0}$, the solution (10.4.3) is thus given by

$$u(x,y)=\frac{y}{\pi}\int_{-\infty}^{\infty}\frac{f(\xi)\,d\xi}{(\xi-x)^2+y^2}$$
$$+\frac{1}{4\pi}\int_0^{\infty}\int_{-\infty}^{\infty}\log\left[\frac{(\xi-x)^2+(\eta-y)^2}{(\xi-x)^2+(\eta+y)^2}\right]h(\xi,\eta)\,d\xi\,d\eta \tag{10.6.6}$$

EXAMPLE 10.6.2. Another example that illustrates the method of images well is the Robin's problem on the quarter infinite plane, namely.

$$\begin{aligned} \nabla^2u&=h(\xi,\eta) && \text{in} \quad \xi>0 \quad \eta>0 \\ u&=f(\eta) && \text{on} \quad \xi=0 \\ u_n&=g(\xi) && \text{on} \quad \eta=0 \end{aligned} \tag{10.6.7}$$

This is illustrated in Fig. 10.6.2.

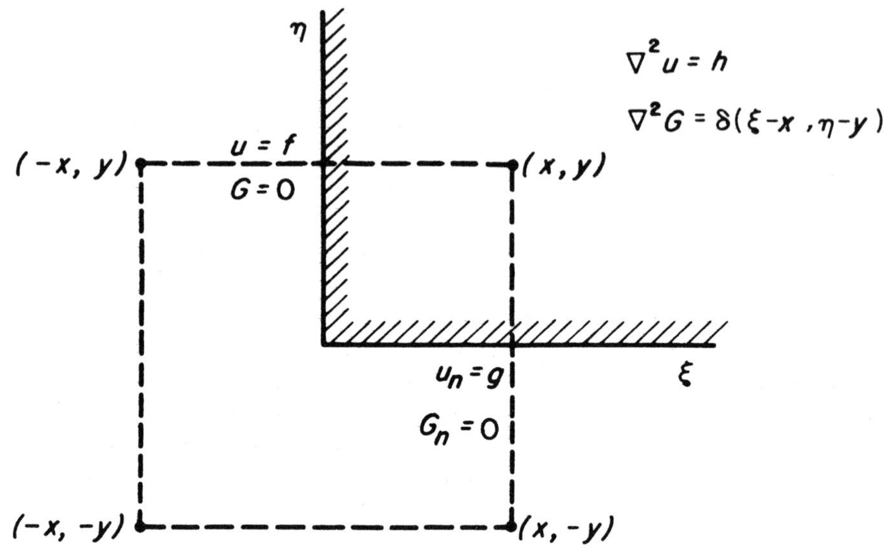

Figure 10.6.2

Let $(-x,y)$, $(-x,-y)$ and $(x,-y)$ be the three image points of the source point (x,y). Then by inspection, we can immediately construct Green's function

$$G = \frac{1}{4\pi}\log\frac{\left[(\xi-x)^2+(\eta-y)^2\right]\left[(\xi-x)^2+(\eta+y)^2\right]}{\left[(\xi+x)^2+(\eta-y)^2\right]\left[(\xi+x)^2+(\eta+y)^2\right]} \tag{10.6.8}$$

This function satisfies $\nabla^2 G = 0$ except at the source point, and $G=0$ on $\xi=0$ and $G_\eta=0$ on $\eta=0$.

The solution from Eq. (10.4.2) is thus

$$\begin{aligned} u(x,y) &= \int\int_D Gh\,d\xi\,d\eta + \int_B (Gu_n - uG_n)\,ds \\ &= \int_0^\infty\int_0^\infty Gh\,d\xi\,d\eta + \int_0^\infty g(\xi)G(\xi,0;x,y)\,d\xi \\ &\quad + \int_0^\infty f(\eta)G_\xi(0,\eta;x,y)\,d\xi \end{aligned}$$

10.7 Method of Eigenfunctions

In this section, we will apply the method of eigenfunctions, described in Chapter 9, to obtain Green's function.

We consider the boundary value problem

$$\begin{aligned} \nabla^2 u &= h \qquad \text{in } D \\ u &= f \qquad \text{on } B \end{aligned} \tag{10.7.1}$$

For this problem, G must satisfy

$$\begin{aligned} \nabla^2 G &= \delta(\xi - x, \eta - y) \qquad \text{in } D \\ G &= 0 \qquad \text{on } B \end{aligned} \tag{10.7.2}$$

and hence the associated eigenvalue problem is

$$\begin{aligned} \nabla^2 \phi + \lambda\phi &= 0 \qquad \text{in } D \\ \phi &= 0 \qquad \text{on } B \end{aligned} \tag{10.7.3}$$

Let ϕ_{mn} be the eigenfunctions and λ_{mn} be the corresponding eigenvalues. We then expand G and δ in terms of the eigenfunctions ϕ_{mn}. Consequently, we write

$$G(\xi, \eta; x, y) = \sum_m \sum_n a_{mn}(x, y)\phi_{mn}(\xi, \eta) \tag{10.7.4}$$

$$\delta(\xi - x, \eta - y) = \sum_m \sum_n b_{mn}(x, y)\phi_{mn}(\xi, \eta) \tag{10.7.5}$$

where

$$\begin{aligned} b_{mn} &= \frac{1}{\|\phi_{mn}\|^2} \int\int_D \delta(\xi - x, \eta - y)\phi_{mn}(\xi, \eta)\, d\xi\, d\eta \\ &= \frac{\phi_{mn}(x, y)}{\|\phi_{mn}\|^2} \end{aligned} \tag{10.7.6}$$

in which

$$\|\phi_{mn}\|^2 = \iint_D \phi_{mn}^2\, d\xi\, d\eta$$

Now substituting Eqs. (10.7.4) and (10.7.5) into Eq. (10.7.2) and using the relation from Eq. (10.7.3) that

$$\nabla^2 \phi_{mn} + \lambda_{mn}\phi_{mn} = 0$$

we obtain

$$-\sum_m \sum_n \lambda_{mn} a_{mn}(x, y)\phi_{mn}(\xi, \eta) = \sum_m \sum_n \frac{\phi_{mn}(x, y)\phi_{mn}(\xi, \eta)}{\|\phi_{mn}\|^2}$$

Hence

$$a_{mn}(x, y) = -\frac{\phi_{mn}(x, y)}{\lambda_{mn}\|\phi_{mn}\|^2} \tag{10.7.7}$$

and the Green's function is therefore given by

$$G(\xi, \eta; x, y) = -\sum_m \sum_n \frac{\phi_{mn}(x, y)\phi_{mn}(\xi, \eta)}{\lambda_{mn}\|\phi_{mn}\|^2} \tag{10.7.8}$$

EXAMPLE 10.7.1. As a particular example, consider the Dirichlet problem in a rectangular domain

$$\nabla^2 u = h \qquad \text{in } D$$
$$u = 0 \qquad \text{on } B$$

The eigenfunctions can be obtained explicitly by the method of separation of variables. We assume a solution in the form

$$u(\xi, \eta) = X(\xi)\, Y(\eta)$$

Substitution of this in

$$\nabla^2 u + \lambda u = 0 \qquad \text{in } D$$
$$u = 0 \qquad \text{on } B$$

yields

$$X'' + \alpha^2 X = 0$$
$$Y'' + (\lambda - \alpha^2)\, Y = 0$$

With the homogeneous boundary conditions $X(0) = X(a) = 0$ and $Y(0) = Y(b) = 0$, X and Y are found to be

$$X_m(\xi) = A_m \sin \frac{m\pi\xi}{a}$$
$$Y_n(\eta) = B_n \sin \frac{n\pi\eta}{b}$$

We have then

$$\lambda_{mn} = \pi^2 \left(\frac{m^2}{a^2} + \frac{n^2}{b^2} \right) \qquad \text{with} \qquad \alpha = \frac{m\pi}{a}$$

Thus, we obtain the eigenfunctions

$$\phi_{mn}(\xi, \eta) = \sin \frac{m\pi\xi}{a} \sin \frac{n\pi\eta}{b}$$

Knowing ϕ_{mn}, we compute $\|\phi_{mn}\|$ and obtain

$$\|\phi_{mn}\|^2 = \int_0^a \int_0^b \sin^2 \frac{m\pi\xi}{a} \sin^2 \frac{n\pi\eta}{b}\, d\xi\, d\eta$$
$$= \frac{ab}{4}$$

We thus obtain from Eq. (10.7.8) the Green's function

$$G(\xi, \eta; x, y) = -\frac{4ab}{\pi^2} \sum_{m=1}^{\infty} \sum_{n=1}^{\infty} \frac{\sin \frac{m\pi x}{a} \sin \frac{n\pi y}{b} \sin \frac{m\pi\xi}{a} \sin \frac{n\pi\eta}{b}}{(m^2 b^2 + n^2 a^2)}$$

10.8 Higher Dimensional Problems

The Green's function method can be easily extended for applications in three and higher dimensions. Since most of the problems encountered in the physical sciences are in three dimensions we will illustrate some examples suitable for practical application.

First of all let us extend our definition of Green's function in three dimensions.

The Green's function for the Dirichlet problem involving the Laplace operator is the function that satisfies

a.

$$\begin{aligned} \nabla^2 G &= \delta(x-\xi, y-\eta, z-\zeta) \qquad \text{in } R \qquad (10.8.1)\\ G &= 0 \qquad \text{on } S \qquad (10.8.2) \end{aligned}$$

b.

$$G(x,y,z;\xi,\eta,\zeta) = G(\xi,\eta,\zeta;x,y,z) \qquad (10.8.3)$$

c.

$$\lim_{\epsilon\to 0} \int\int_{S_\epsilon} \frac{\partial G}{\partial n}\, dS = 1 \qquad (10.8.4)$$

where n is the outward unit normal to the surface

$$S_\epsilon : (x-\xi)^2 + (y-\eta)^2 + (z-\zeta)^2 = \epsilon^2$$

Proceeding as in the two-dimensional case, the solution of the Dirichlet problem

$$\begin{aligned} \nabla^2 u &= h \qquad \text{in } R\\ u &= f \qquad \text{on } S \qquad (10.8.5) \end{aligned}$$

is

$$u(x,y,z) = \int\int_R\int Gh\, dR + \int\int_S fG_n\, dS \qquad (10.8.6)$$

Again we let

$$G(\xi,\eta,\zeta;x,y,z) = F(\xi,\eta,\zeta;x,y,z) + g(\xi,\eta,\zeta;x,y,z)$$

where

$$\nabla^2 F = \delta(x-\xi, y-\eta, z-\zeta) \qquad \text{in } R$$

and

$$\begin{aligned} \nabla^2 g &= 0 \qquad \text{in } R\\ g &= -F \qquad \text{on } S \end{aligned}$$

EXAMPLE 10.8.1. We consider a spherical domain with radius a. We must have

$$\nabla^2 F = 0$$

except at the source point. For

$$r=\left[(\xi-x)^2+(\eta-y)^2+(\zeta-z)^2\right]^{1/2}>0,$$

with (x,y,z) as the origin, we have

$$\nabla^2 F=\frac{1}{r^2}\frac{d}{dr}\left(r^2\frac{dF}{dr}\right)=0$$

Integration then yields

$$F=A+\frac{B}{r} \qquad \text{for} \qquad r>0$$

Applying the condition (10.8.4) we obtain

$$\lim_{\epsilon\to 0}\int\int_{S_\epsilon} G_n\,dS=\lim_{\epsilon\to 0}\int\int_{S_\epsilon} F_r\,dS=1$$

Consequently, $B=-1/4\pi$ and A is arbitrary. If we set $A=0$ for convenience,[16] we have

$$F=-\frac{1}{4\pi r} \tag{10.8.7}$$

We apply the method of images to obtain the Green's function. If we draw a three-dimensional diagram analogous to Fig. 10.6.1, we will have a relation similar to (10.6.1), namely,

$$r'=\frac{a}{\rho}r \tag{10.8.8}$$

where r' and ρ are measured in three-dimensional space. Thus, we seek Green's function

$$G=\frac{-1}{4\pi r}+\frac{a/\rho}{4\pi r'} \tag{10.8.9}$$

which is harmonic everywhere in R except at the source point and is zero on the surface S.

In terms of spherical coordinates

$$\begin{aligned} \xi&=\tau\cos\psi\sin\alpha & x&=\rho\cos\phi\sin\theta\\ \eta&=\tau\sin\psi\sin\alpha & y&=\rho\sin\phi\sin\theta\\ \zeta&=\tau\cos\alpha & z&=\rho\cos\theta \end{aligned}$$

G can be written in the form

$$G=\frac{-1}{4\pi(\tau^2+\rho^2-2\tau\rho\cos\gamma)^{1/2}}+\frac{1}{4\pi\left[\dfrac{\tau^2\rho^2}{a^2}+a^2-2\tau\rho\cos\gamma\right]^{1/2}} \tag{10.8.10}$$

[16]This is the boundedness condition at infinity for exterior problems.

where γ is the angle between r and r'. Now differentiating G, we have

$$\left[\frac{\partial G}{\partial \tau}\right]_{\tau=a}=\frac{a^2-\rho^2}{4\pi a\left(a^2+\rho^2-2a\rho\cos\gamma\right)^{3/2}}$$

Thus, the solution of the Dirichlet problem for $h=0$ is

$$u(\rho,\theta,\phi)=\frac{a\left(a^2-\rho^2\right)}{4\pi}\int_0^{2\pi}\int_0^{\pi}\frac{f(\alpha,\psi)\sin\alpha\,d\alpha\,d\psi}{\left(a^2+\rho^2-2a\rho\cos\gamma\right)^{3/2}} \tag{10.8.11}$$

where $\cos\gamma=\cos\alpha\cos\theta+\sin\alpha\sin\theta\cos(\psi-\phi)$. This integral is called the *three-dimensional Poisson integral formula.*

For the exterior problem where the outward normal is radially inward towards the origin, the solution can be simply obtained by replacing $(a^2-\rho^2)$ by (ρ^2-a^2) in Eq. (10.8.11).

EXAMPLE 10.8.2. Another example involving the Helmholtz operator is the three-dimensional radiation problem

$$\nabla^2 u+\kappa^2 u=0$$

$$\lim_{r\to\infty} r(u_r+i\kappa u)=0$$

where $i=\sqrt{-1}$; the limit condition is called the radiation condition, and r is the field point distance.

In this case, the Green's function must satisfy

$$\nabla^2 G+\kappa^2 G=\delta(\xi-x,\eta-y,\zeta-z)$$

Since the point source solution is dependent only on r, we write the Helmholtz equation

$$G_{rr}+\frac{2}{r}G_r+\kappa^2 G=0 \qquad \text{for} \qquad r>0$$

Note that the source point is taken as the origin. If we write the above equation in the form

$$(Gr)_{rr}+\kappa^2(Gr)=0 \qquad \text{for} \qquad r>0$$

then the solution can easily be seen to be

$$Gr=Ae^{i\kappa r}+Be^{-i\kappa r}$$

or

$$G=A\frac{e^{i\kappa r}}{r}+B\frac{e^{-i\kappa r}}{r}$$

In order for G to satisfy the radiation condition

$$\lim_{r\to\infty} r(G_r+i\kappa G)=0$$

$A=0$ and G thus takes the form

$$G=B\frac{e^{-i\kappa r}}{r}$$

To determine B we have

$$\lim_{\epsilon\to 0}\int\int_{S_\epsilon}\frac{\partial G}{\partial n}\,dS=-\lim_{\epsilon\to 0}\int\int_{S_\epsilon}B\frac{e^{-i\kappa r}}{r}\left(\frac{1}{r}+i\kappa\right)dS=1$$

from which we obtain $B=-1/4\pi$, and consequently, G becomes

$$G=-\frac{e^{-i\kappa r}}{4\pi r}$$

Note that this reduces to $-1/4\pi r$ when $\kappa=0$.

10.9 Neumann Problem

We have noted in the chapter on boundary value problems that the Neumann problem requires more attention than Dirichlet's problem, because additional condition is necessary for the existence of a solution of the Neumann problem.

Let us now consider the Neumann problem

$$\nabla^2 u+\kappa^2 u=h \qquad \text{in } R$$

$$\frac{\partial u}{\partial n}=0 \qquad \text{on } S$$

By the divergence theorem, we have

$$\int\int_R\int \nabla^2 u\,dR=\int\int_S\frac{\partial u}{\partial n}\,dS$$

Thus, if we integrate the Helmholtz equation and use the preceding result, we obtain

$$\kappa^2\int\int_R\int u\,dR=\int\int_R\int h\,dR$$

In the case of Poisson's equation where $\kappa=0$, this relation is satisfied only when

$$\int\int_R\int h\,dR=0$$

If we consider a heat conduction problem, this condition may be interpreted as the requirement that the net generation of heat be zero. This is physically reasonable since the boundary is insulated in such a way that the net flux across it is zero.

If we define Green's function G, in this case, by

$$\nabla^2 G+\kappa^2 G=\delta(\xi-x,\eta-y,\zeta-z) \qquad \text{in } R$$

$$\frac{\partial G}{\partial n}=0 \qquad \text{on } S$$

Then we must have

$$\kappa^2\int\int_R\int G\,dR=1$$

which cannot be satisfied for $\kappa=0$. But, we know by physical reasoning that a solution exists if

$$\int\int_R\int h\,dR=0$$

Hence we will modify the definition of Green's function, so that

$$\frac{\partial G}{\partial n}=C \qquad \text{on } S$$

where C is a constant. When integrating $\nabla^2 G=\delta$ over R, we obtain

$$C\int\int_S dS=1$$

It is not difficult to show that G remains symmetric if

$$\int\int_S G\,dS=0$$

Thus, under this condition, if we take C to be the reciprocal of the surface area, the solution of the Neumann problem for Poisson's equation is

$$u(x,y,z)=C^*+\int\int_R\int G(x,y,z;\xi,\eta,\zeta)h(\xi,\eta,\zeta)\,d\xi\,d\eta\,d\zeta$$

where C^* is a constant.

We should remark here that the method of Green's functions provides the solution in integral form. This is made possible by replacing a problem involving nonhomogeneous boundary conditions with a problem of finding Green's function G with homogeneous boundary conditions.

Regardless of methods employed, the Green's function of a problem with a nonhomogeneous equation and homogeneous boundary conditions is the same as the Green's function of a problem with a homogeneous equation and nonhomogeneous boundary conditions since one problem can be transferred to the other without difficulty. To illustrate, consider the problem

$$\begin{aligned} Lu&=f && \text{in } R\\ u&=0 && \text{on } \partial R \end{aligned}$$

where ∂R denotes the boundary of R.

If we let $v=w-u$, where w satisfies $Lw=f$ in R, then the problem becomes

$$\begin{aligned} Lv&=0 && \text{in } R\\ v&=w && \text{on } \partial R \end{aligned}$$

Conversely, if we consider the problem

$$\begin{aligned} Lu&=0 && \text{in } R\\ u&=g && \text{on } \partial R \end{aligned}$$

we can easily transform this problem into

$$\begin{aligned} Lv &= Lw \equiv w^* && \text{in } R \\ v &= 0 && \text{on } \partial R \end{aligned}$$

by putting $v = w - u$ and finding w that satisfies $w = g$ on ∂R.

In fact, if we have

$$\begin{aligned} Lu &= f && \text{in } R \\ u &= g && \text{on } \partial R \end{aligned}$$

we can transform this problem into either one of the above problems.

Exercises

1. If L denotes the operator

$$Lu = Au_{xx} + Bu_{xy} + Cu_{yy} + Du_x + Eu_y + Fu$$

and if M denotes the adjoint operator

$$Mv = (Av)_{xx} + (Bv)_{xy} + (Cv)_{yy} - (Dv)_x - (Ev)_y + Fv,$$

show that

$$\int\int_R (vLu - uMv)\,dx\,dy = \int_{\partial R} [U\cos(n,x) + V\cos(n,y)]\,ds$$

where

$$\begin{aligned} U &= Avu_x - u(Av)_x - u(Bv)_y + Duv \\ V &= Bvu_x + Cvu_y - u(Cv)_y + Euv \end{aligned}$$

and ∂R is the boundary of a region R.

2. Prove that the Green's function for a region, if it exists, is unique.

3. Determine the Green's function for the exterior Dirichlet problem for a unit circle

$$\begin{aligned} \nabla^2 u &= 0 && \text{in} && r > 1 \\ u &= f && \text{on} && r = 1 \end{aligned}$$

4. Prove that for $x = x(\xi, \eta)$ and $y = y(\xi, \eta)$

$$\delta(x - x_0)\delta(y - y_0) = \frac{1}{|J|}\delta(\xi - \xi_0)\delta(\eta - \eta_0)$$

where J is the Jacobian and (x_0, y_0) corresponds to (ξ_0, η_0). Hence show that for polar coordinates

$$\delta(x - x_0)\delta(y - y_0) = \frac{1}{r}\delta(r - r_0)\delta(\theta - \theta_0)$$

5. Determine, for an infinite wedge, the Green's function that satisfies

$$\nabla^2 G+\kappa^2 G=\frac{1}{r}\,\delta(r-r_0,\theta-\theta_0)$$
$$G=0 \qquad \theta=0 \qquad \text{and} \qquad \theta=\alpha$$

6. Determine, for the Poisson equation, the Green's function which vanishes on the boundary of a semicircular domain of radius R.

7. Find the solution of the Dirichlet problem

$$\nabla^2 u=0 \qquad 0<x<a \qquad 0<y<b$$
$$u(0,y)=u(a,y)=u(x,b)=0$$
$$u(x,0)=f(x)$$

8. Determine the solution of Dirichlet's problem

$$\nabla^2 u=f(r,\theta) \qquad \text{in } D$$
$$u=0 \qquad \text{on } B$$

where B is the boundary of a circle D of radius R.

9. Determine the Green's function for the semiinfinite region $\zeta>0$ for

$$\nabla^2 G+\kappa^2 G=\delta(\xi-x,\eta-y,\zeta-z)$$
$$G=0 \qquad \text{on} \qquad \zeta=0$$

10. Determine the Green's function for the semiinfinite region $\zeta>0$ for

$$\nabla^2 G+\kappa^2 G=\delta(\xi-x,\eta-y,\zeta-z)$$
$$\frac{\partial G}{\partial n}=0 \qquad \text{on} \qquad \zeta=0$$

11. Find the Green's function in the quarter plane $\xi>0,\eta>0$ which satisfies

$$\nabla^2 G=\delta(\xi-x,\eta-y)$$
$$G=0 \qquad \text{on} \qquad \xi=0 \qquad \text{and} \qquad \eta=0$$

12. Find the Green's function in the quarter plane $\xi>0,\eta>0$ which satisfies

$$\nabla^2 G=\delta(\xi-x,\eta-y)$$
$$G_\xi(0,\eta)=0$$
$$G(\xi,0)=0$$

13. Find the Green's function in the plane $0<x<\infty,-\infty<y<\infty$, for the problem

$$\nabla^2 u=f \qquad \text{in } R$$
$$u=0 \qquad \text{on} \qquad x=0$$

14. Determine the Green's function which satisfies

$$\begin{aligned} &\nabla^2 G=\delta(x-\xi,y-\eta) && \text{in } R: 0<x<a, \quad 0<y<\infty \\ &G=0 && \text{on } \partial R: x=0, \quad x=a, \quad y=0 \end{aligned}$$

G is bounded at infinity

15. Find the Green's function that satisfies

$$\nabla^2 G=\frac{1}{r}\delta(r-\rho,\theta-\beta) \qquad 0<\theta<\frac{\pi}{3} \qquad 0<r<1$$

$$G=0 \quad \text{on} \quad \theta=0 \quad \text{and} \quad \theta=\frac{\pi}{3}$$

$$\frac{\partial G}{\partial n}=0 \quad \text{on} \quad r=1$$

16. Solve the boundary value problem

$$\frac{1}{r}\frac{\partial}{\partial r}\left(r\frac{\partial u}{\partial r}\right)+\frac{\partial^2 u}{\partial z^2}+\kappa^2 u=0 \qquad r\geqslant 0, \qquad z>0$$

$$\frac{\partial u}{\partial z}=\begin{cases} 0 & r>a, \quad z=0 \\ C & r<a, \quad z=0, \end{cases} \qquad C=\text{constant}$$

17. Obtain the solution of

$$\nabla^2 u=0 \qquad 0<r<\infty \qquad 0<\theta<2\pi$$

$$u(r,0+)=u(r,2\pi-)=0$$

18. Determine the Green's function for the equation

$$\nabla^2 u-\kappa^2 u=0$$

vanishing on all sides of the rectangle $0\leqslant x\leqslant a, 0\leqslant y\leqslant b$.

19. Determine the Green's function of the Helmholtz equation

$$\nabla^2 u+\kappa^2 u=0 \qquad 0<x<a \qquad -\infty<y<\infty$$

vanishing on $x=0$ and $x=a$.

20. Solve the exterior Dirichlet problem

$$\nabla^2 u=0 \quad \text{in} \quad r>1$$

$$u(1,\theta,\phi)=f(\theta,\phi)$$

21. By the method of images, determine the potential due to a point charge q near a conducting sphere of radius R with potential V.

22. By the method of images, show that the potential due to a conducting sphere of radius R in a uniform electric field E_0 is given by

$$U = -E_0\left(r - \frac{R^2}{r^2}\right)\cos\theta$$

where r, θ are the polar coordinates with origin at the center of the sphere.

23. Determine the potential in a cylinder of radius R and length l. The potential on the ends is zero while the potential on the cylindrical surface is prescribed to be $f(\theta, z)$.

CHAPTER ELEVEN

Integral Transforms

11.1 Fourier Transforms

For problems involving regions of infinite or semiinfinite extent, it is often necessary to resort to the method of integral transforms. In Chapter 5, we have described Fourier series for functions which are periodic with period 2π in the interval $(-\infty,\infty)$. However, functions which are not periodic cannot be represented by a Fourier series. In many problems it is desirable to develop an integral representation for such a function that is analogous to a Fourier series.

We have seen in Sec. 5.10 that the Fourier series for $f(x)$ in the interval $[-l,l]$ is

$$f(x)=\frac{a_0}{2}+\sum_{k=1}^{\infty}\left(a_k\cos\frac{k\pi x}{l}+b_k\sin\frac{k\pi x}{l}\right) \tag{11.1.1}$$

where

$$a_k=\frac{1}{l}\int_{-l}^{l}f(t)\cos\frac{k\pi t}{l}\,dt \qquad k=0,1,2,\ldots \tag{11.1.2}$$

$$b_k=\frac{1}{l}\int_{-l}^{l}f(t)\sin\frac{k\pi t}{l}\,dt \qquad k=1,2,3,\ldots \tag{11.1.3}$$

Substituting (11.1.2) and (11.1.3) into (11.1.1), we have

$$\begin{aligned}f(x)&=\frac{1}{2l}\int_{-l}^{l}f(t)\,dt+\frac{1}{l}\sum_{k=1}^{\infty}\left[\int_{-l}^{l}f(t)\cos\frac{k\pi t}{l}\cdot\cos\frac{k\pi x}{l}\,dt\right]\\&\quad+\left[\int_{-l}^{l}f(t)\sin\frac{k\pi t}{l}\cdot\sin\frac{k\pi x}{l}\,dt\right]\\&=\frac{1}{2l}\int_{-l}^{l}f(t)\,dt+\frac{1}{l}\sum_{k=1}^{\infty}\int_{-l}^{l}f(t)\cos\left[\frac{k\pi}{l}(t-x)\right]dt\end{aligned} \tag{11.1.4}$$

Suppose that $f(x)$ is absolutely integrable, that is,

$$\int_{-\infty}^{\infty}|f(x)|dx$$

converges. Then

$$\frac{|a_0|}{2}=\frac{1}{2l}\left|\int_{-l}^{l}f(t)\,dt\right|$$
$$\leqslant\frac{1}{2l}\int_{-\infty}^{\infty}|f(t)|\,dt$$

which approaches zero as $l\to\infty$. Thus, holding x fixed, as l approaches infinity, Eq. (11.1.4) becomes

$$f(x)=\lim_{l\to\infty}\frac{1}{l}\sum_{k=1}^{\infty}\int_{-l}^{l}f(t)\cos\left[\frac{k\pi}{l}(t-x)\right]dt$$

Now let

$$\alpha_k=\frac{k\pi}{l}\qquad\Delta\alpha=\alpha_{k+1}-\alpha_k=\frac{\pi}{l}$$

Then $f(x)$ can be written as

$$f(x)=\lim_{l\to\infty}\sum_{k=1}^{\infty}F(\alpha_k)\Delta\alpha$$

where

$$F(\alpha)=\frac{1}{\pi}\int_{-l}^{l}f(t)\cos[\alpha(t-x)]\,dt$$

If we plot $F(\alpha)$ against α, we can clearly see that the sum

$$\sum_{k=1}^{\infty}F(\alpha_k)\Delta\alpha$$

is an approximation to the area under the curve $y=F(\alpha)$(see Fig. 11.1.1). As $l\to\infty,\Delta\alpha\to0$, and the sum formally approaches the definite integral. We therefore have

$$f(x)=\int_0^{\infty}\left[\frac{1}{\pi}\int_{-\infty}^{\infty}f(t)\cos\alpha(t-x)dt\right]d\alpha \qquad (11.1.5)$$

which is the *Fourier integral* representation for the function $f(x)$. Its convergence to $f(x)$ is suggested but by no means established by the preceding arguments. We shall now prove that this representation is indeed valid if $f(x)$ satisfies certain conditions.

Lemma 11.1.1 *If f is piecewise smooth in the interval* $[0,b]$, *then for* $b>0$,

$$\lim_{\lambda\to\infty}\int_0^b f(x)\frac{\sin\lambda x}{x}dx=\frac{\pi}{2}f(0+)$$

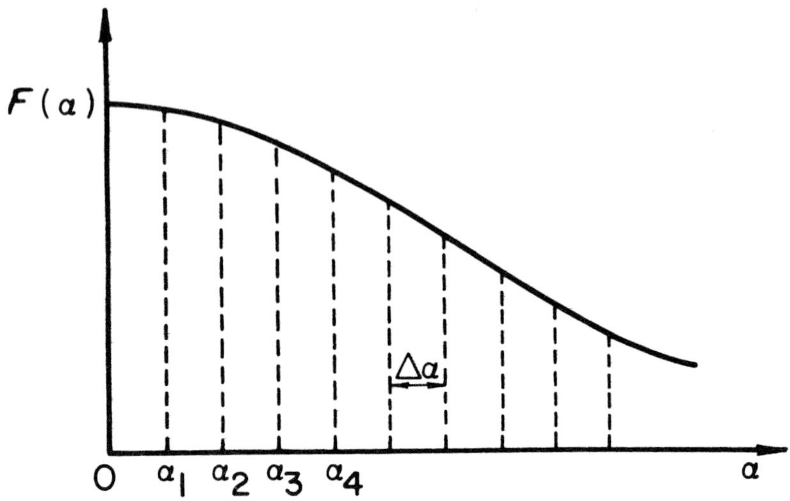

PROOF.

$$\int_0^b f(x)\frac{\sin\lambda x}{x}\,dx=\int_0^b f(0+)\frac{\sin\lambda x}{x}\,dx+\int_0^b\frac{f(x)-f(0+)}{x}\sin\lambda x\,dx$$

$$=f(0+)\int_0^{\lambda b}\frac{\sin t}{t}\,dt+\int_0^b\frac{f(x)-f(0+)}{x}\sin\lambda x\,dx$$

Since f is piecewise smooth, the integrand of the last integral is bounded as $\lambda\to\infty$, and thus by the Riemann-Lebesque lemma, the last integral tends to zero as $\lambda\to\infty$. Hence

$$\lim_{\lambda\to\infty}\int_0^b f(x)\frac{\sin\lambda x}{x}\,dx=\frac{\pi}{2}f(0+) \qquad (11.1.6)$$

since

$$\int_0^\infty\frac{\sin t}{t}\,dt=\frac{\pi}{2}. \qquad \square$$

Theorem 11.1.1 (Fourier Integral Theorem) *If f is piecewise smooth in every finite interval, and absolutely integrable on $(-\infty,\infty)$, then*

$$\frac{1}{2}[f(x+)+f(x-)]=\frac{1}{\pi}\int_0^\infty\left[\int_{-\infty}^\infty f(t)\cos\alpha(t-x)\,dt\right]d\alpha$$

PROOF. Noting that $|\cos\alpha(t-x)|\leqslant 1$ and that by hypothesis

$$\int_{-\infty}^\infty f(t)\,dt<\infty$$

we see that the integral

$$\int_{-\infty}^{\infty} f(t)\cos\alpha(t-x)\,dt$$

converges independently of α and x. It therefore follows that in the double integral

$$I=\int_0^{\lambda}\left[\int_{-\infty}^{\infty} f(t)\cos\alpha(t-x)\,dt\right]d\alpha$$

the order of integration may be interchanged. We then have

$$\begin{aligned} I&=\int_{-\infty}^{\infty}\left[\int_0^{\lambda} f(t)\cos\alpha(t-x)\,d\alpha\right]dt\\ &=\int_{-\infty}^{\infty} f(t)\left[\frac{\sin\lambda(t-x)}{(t-x)}\right]dt\\ &=\left[\int_{-\infty}^{-M}+\int_{-M}^{x}+\int_{x}^{M}+\int_{M}^{\infty}\right]f(t)\frac{\sin\lambda(t-x)}{(t-x)}\,dt \end{aligned}$$

If we substitute $u=t-x$, we have

$$\int_x^M f(t)\frac{\sin\lambda(t-x)}{(t-x)}\,dt=\int_0^{M-x} f(u+x)\frac{\sin\lambda u}{u}\,du$$

which is equal to $\pi f(x+)/2$ in the limit $\lambda\to\infty$, by Lemma 11.1.1. Similarly, the second integral tends to $\pi f(x-)/2$ when $\lambda\to\infty$. If we take M sufficiently large, the absolute values of the first and the last integrals are each less than $\varepsilon/2$. Consequently, as $\lambda\to\infty$

$$\int_0^{\infty}\left[\int_{-\infty}^{\infty} f(t)\cos\alpha(t-x)\,dt\right]d\alpha=\frac{\pi}{2}[f(x+)+f(x-)] \tag{11.1.7}$$

If f is continuous at the point x, then

$$f(x+)=f(x-)=f(x)$$

so that Eq. (11.1.7) reduces to

$$f(x)=\frac{1}{\pi}\int_0^{\infty}\left[\int_{-\infty}^{\infty} f(t)\cos\alpha(t-x)\,dt\right]d\alpha \tag{11.1.8}$$

□

We may express the Fourier integral (11.1.8) in complex form. In this case, we substitute

$$\cos\alpha(t-x)=\frac{1}{2}\left[e^{i\alpha(t-x)}+e^{-i\alpha(t-x)}\right]$$

into Eq. (11.1.8) and write it as two integrals

$$f(x)=\frac{1}{2\pi}\int_0^{\infty}\int_{-\infty}^{\infty} f(t)e^{i\alpha(t-x)}\,dt\,d\alpha+\frac{1}{2\pi}\int_0^{\infty}\int_{-\infty}^{\infty} f(t)e^{-i\alpha(t-x)}\,dt\,d\alpha$$

Changing the integration variable from α to $-\alpha$ in the second integral, we obtain

$$f(x)=\frac{1}{\sqrt{2\pi}}\int_{-\infty}^{\infty}\left[\frac{1}{\sqrt{2\pi}}\int_{-\infty}^{\infty}f(t)e^{i\alpha t}\,dt\right]e^{-i\alpha x}\,d\alpha \qquad (11.1.9)$$

Thus, we state the following theorem:

Theorem 11.1.2 *Let $f(x)$ be continuous, piecewise smooth, and absolutely integrable. If*

$$F(\alpha)=\frac{1}{\sqrt{2\pi}}\int_{-\infty}^{\infty}f(t)e^{i\alpha t}\,dt \qquad (11.1.10)$$

then for all x

$$f(x)=\frac{1}{\sqrt{2\pi}}\int_{-\infty}^{\infty}F(\alpha)e^{-i\alpha x}d\alpha \qquad (11.1.11)$$

The function $F(\alpha)$ is called the *Fourier transform* of $f(x)$, and $f(x)$ is called the *inverse Fourier transform* of $F(\alpha)$. One may note that the factor $1/2\pi$ has been split and placed in front of the integrals in (11.1.10) and (11.1.11). Often we find the factor $1/2\pi$ contained in only one of the relations (11.1.10) and (11.1.11). It is not uncommon to find the term $e^{i\alpha t}$ in (11.1.10) replaced by $e^{-i\alpha t}$, and, as a consequence, $e^{-i\alpha x}$ replaced by $e^{i\alpha x}$ in (11.1.11).

EXAMPLE 11.1.1. Find the Fourier transform of the function $f(x)=e^{-|x|}$

$$\begin{aligned}F(\alpha)&=\frac{1}{\sqrt{2\pi}}\int_{-\infty}^{\infty}e^{-|t|}e^{i\alpha t}\,dt\\&=\frac{1}{\sqrt{2\pi}}\int_{-\infty}^{0}e^{t(1+i\alpha)}\,dt+\frac{1}{\sqrt{2\pi}}\int_{0}^{\infty}e^{-t(1-i\alpha)}\,dt\\&=\frac{1}{\sqrt{2\pi}}\left(\frac{1}{1+i\alpha}+\frac{1}{1-i\alpha}\right)\\&=\sqrt{\frac{2}{\pi}}\;\frac{1}{1+\alpha^2}\end{aligned}$$

Analogous to Fourier sine and cosine series, there are Fourier sine and cosine transforms for odd and even functions respectively.

Corollary 11.1.1 *Let $f(x)$ be defined for $0\leqslant x<\infty$. Let $f(x)$ be extended as an odd function in $(-\infty,\infty)$ satisfying the conditions of Fourier integral theorem. If, at the points of continuity,*

$$F_s(\alpha)=\sqrt{\frac{2}{\pi}}\int_{0}^{\infty}f(t)\sin\alpha t\,dt$$

then

$$f(x)=\sqrt{\frac{2}{\pi}}\int_0^\infty F_s(\alpha)\sin\alpha x\,d\alpha$$

PROOF. From Eq. (11.1.8), we have

$$f(x)=\frac{1}{\pi}\int_0^\infty\left[\int_{-\infty}^\infty f(t)(\cos\alpha t\cos\alpha x+\sin\alpha t\sin\alpha x)\,dt\right]d\alpha$$

Since $f(-t)=-f(t)$ for all real t,

$$\int_{-\infty}^\infty f(t)\cos\alpha t\,dt=0$$

$$\int_{-\infty}^\infty f(t)\sin\alpha t\,dt=2\int_0^\infty f(t)\sin\alpha t\,dt$$

and consequently,

$$f(x)=\sqrt{\frac{2}{\pi}}\int_0^\infty\left[\sqrt{\frac{2}{\pi}}\int_0^\infty f(t)\sin\alpha t\,dt\right]\sin\alpha x\,d\alpha$$

This is the assertion. □

Corollary 11.1.2 *Let $f(x)$ be defined for $0\leqslant x<\infty$. Let $f(x)$ be extended as an even function in $(-\infty,\infty)$ satisfying the conditions of the Fourier integral theorem. Then if, at the points of continuity,*

$$F_c(\alpha)=\sqrt{\frac{2}{\pi}}\int_0^\infty f(t)\cos\alpha t\,dt$$

then

$$f(x)=\sqrt{\frac{2}{\pi}}\int_0^\infty F_c(\alpha)\cos\alpha x\,d\alpha$$

PROOF. From Eq. (11.1.8), we have

$$f(x)=\frac{1}{\pi}\int_0^\infty\left[\int_{-\infty}^\infty f(t)(\cos\alpha t\cos\alpha x+\sin\alpha t\sin\alpha x)\,dt\right]d\alpha$$

Since $f(-t)=f(t)$ for all real t,

$$\int_{-\infty}^\infty f(t)\cos\alpha t\,dt=2\int_0^\infty f(t)\cos\alpha t\,dt$$

$$\int_{-\infty}^\infty f(t)\sin\alpha t\,dt=0$$

and consequently,

$$f(x)=\sqrt{\frac{2}{\pi}}\int_0^\infty\left[\sqrt{\frac{2}{\pi}}\int_0^\infty f(t)\cos\alpha t\,dt\right]\cos\alpha x\,d\alpha$$

This completes the proof. □

EXAMPLE 11.1.2. Find the Fourier sine and cosine transforms of $f(x)=e^{-ax}$ for $a>0$.[17]

$$
\begin{aligned}
F_s(\alpha) &= \sqrt{\frac{2}{\pi}} \int_0^\infty e^{-at} \sin \alpha t \, dt \\
&= \sqrt{\frac{2}{\pi}} \left[-e^{-at} \frac{\cos \alpha t}{\alpha} \right]_0^\infty - \sqrt{\frac{2}{\pi}} \int_0^\infty \frac{a}{\alpha} e^{-at} \cos \alpha t \, dt \\
&= \sqrt{\frac{2}{\pi}} \frac{1}{\alpha} - \sqrt{\frac{2}{\pi}} \frac{a}{\alpha} \left[e^{-at} \frac{\sin \alpha t}{\alpha} \right]_0^\infty - \sqrt{\frac{2}{\pi}} \frac{a^2}{\alpha^2} \int_0^\infty e^{-at} \sin \alpha t \, dt
\end{aligned}
$$

Hence

$$
F_s\left(1+\frac{a^2}{\alpha^2}\right) = \sqrt{\frac{2}{\pi}} \frac{1}{\alpha}
$$

and consequently,

$$
F_s = \sqrt{\frac{2}{\pi}} \frac{\alpha}{a^2+\alpha^2}
$$

In a similar manner, $F_c(\alpha)$ is found to be

$$
F_c = \sqrt{\frac{2}{\pi}} \frac{a}{a^2+\alpha^2}
$$

11.2 Properties of Fourier Transform

Theorem 11.2.1 (Linearity) *The Fourier transform is a linear transformation.*

PROOF. We have

$$
\mathcal{F}[f] = \frac{1}{\sqrt{2\pi}} \int_{-\infty}^\infty f(t) e^{i\alpha t} \, dt
$$

Then for any constants a and b,

$$
\begin{aligned}
\mathcal{F}[af+bg] &= \frac{1}{\sqrt{2\pi}} \int_{-\infty}^\infty [af+bg] e^{i\alpha t} \, dt \\
&= \frac{a}{\sqrt{2\pi}} \int_{-\infty}^\infty f(t) e^{i\alpha t} \, dt + \frac{b}{\sqrt{2\pi}} \int_{-\infty}^\infty g(t) e^{i\alpha t} \, dt \\
&= a\mathcal{F}[f] + b\mathcal{F}[g]
\end{aligned}
$$

□

[17]Hereafter abbreviation $[\ \ ]_a^\infty$ will denote $\lim_{b\to\infty} [\ \ ]_a^b$.

Theorem 11.2.2 (Shifting) *Let $\mathfrak{F}[f]$ be a Fourier transform. Then $\mathfrak{F}[f\ \ (t-c)]=e^{i\alpha c}\mathfrak{F}[f(t)]$ where c is a real constant.*

PROOF. From the definition, we have for $c>0$

$$\begin{aligned}\mathfrak{F}[f(t-c)]&=\frac{1}{\sqrt{2\pi}}\int_{-\infty}^{\infty}f(t-c)e^{i\alpha t}\,dt\\&=\frac{1}{\sqrt{2\pi}}\int_{-\infty}^{\infty}f(\xi)e^{i\alpha(\xi+c)}\,d\xi\quad\text{where}\quad\xi=t-c\\&=e^{i\alpha c}\mathfrak{F}[f(t)]\end{aligned}$$

□

Theorem 11.2.3 (Scaling) *If $\mathfrak{F}\ [f]$ is the Fourier transform of f, then $\mathfrak{F}[f(ct)]=(1/|c|)F(\alpha/c)$ where c is a real constant and $c\neq 0$.*

PROOF. For $c\neq 0$,

$$\mathfrak{F}[f(ct)]=\frac{1}{\sqrt{2\pi}}\int_{-\infty}^{\infty}f(ct)e^{i\alpha t}\,dt$$

If we let $\xi=ct$, then

$$\begin{aligned}\mathfrak{F}[f(ct)]&=\frac{1}{|c|}\frac{1}{\sqrt{2\pi}}\int_{-\infty}^{\infty}f(\xi)e^{i(\alpha/c)\xi}\,d\xi\\&=(1/|c|)F(\alpha/c)\end{aligned}$$

□

Theorem 11.2.4 (Differentiation) *Let f be continuous and piecewise smooth in $(-\infty,\infty)$. Let $f(t)$ approach zero as $|t|\to\infty$. If f and f' are absolutely integrable, then*

$$\mathfrak{F}[f']=-i\alpha\mathfrak{F}[f]$$

PROOF.

$$\begin{aligned}\mathfrak{F}[f'(t)]&=\frac{1}{\sqrt{2\pi}}\int_{-\infty}^{\infty}f'(t)e^{i\alpha t}\,dt\\&=\frac{1}{\sqrt{2\pi}}\left[f(t)e^{i\alpha t}\Big|_{-\infty}^{\infty}-i\alpha\int_{-\infty}^{\infty}f(t)e^{i\alpha t}\,dt\right]\\&=-i\alpha\mathfrak{F}[f(t)]\end{aligned}$$

□

This result can be easily extended. If f and its first $(n-1)$ derivatives are continuous, and if its nth derivative is piecewise continuous, then

$$\mathfrak{F}[f^{(n)}(t)]=(-i\alpha)^n\mathfrak{F}[f(t)]\quad n=0,1,2,\ldots\tag{11.2.1}$$

provided f and its derivatives are absolutely integrable. In addition, we assume that f and its first $(n-1)$ derivatives tend to zero as $|t|$ tends to infinity.

EXAMPLE 11.2.1. Find the solution of the Dirichlet problem in the half-plane $y>0$

$$u_{xx}+u_{yy}=0 \qquad -\infty<x<\infty \qquad y>0$$

$$u(x,0)=f(x) \quad -\infty<x<\infty$$

$$u \text{ is bounded as } y\to\infty$$

$$u \text{ and } u_x \text{ vanish as } |x|\to\infty$$

Let $U(\alpha,y)$ be the Fourier transform of $u(x,y)$ in the variable x. Then

$$U(\alpha,y)=\frac{1}{\sqrt{2\pi}}\int_{-\infty}^{\infty} u(x,y)e^{i\alpha x}\,dx$$

By the relation (11.2.1), we have

$$\begin{aligned}\mathcal{F}[u_{xx}]&=(-i\alpha)^2\mathcal{F}[u]\\ &=-\alpha^2U(\alpha,y)\end{aligned} \tag{11.2.2}$$

Note that

$$\begin{aligned}\mathcal{F}[u_{yy}]&=\frac{1}{\sqrt{2\pi}}\int_{-\infty}^{\infty} u_{yy}e^{i\alpha x}\,dx\\ &=\frac{\partial^2}{\partial y^2}\left[\frac{1}{\sqrt{2\pi}}\int_{-\infty}^{\infty} u(x,y)e^{i\alpha x}\,dx\right]\\ &=\frac{\partial^2 U}{\partial y^2}\end{aligned} \tag{11.2.3}$$

Transforming the Laplace equation, we obtain

$$\mathcal{F}[u_{xx}+u_{yy}]=0$$

By Theorem 11.2.1, we have

$$\mathcal{F}[u_{xx}]+\mathcal{F}[u_{yy}]=0 \tag{11.2.4}$$

Substitution of Eqs. (11.2.2) and (11.2.3) in Eq. (11.2.4) yields

$$U_{yy}-\alpha^2U=0$$

This is an ordinary differential equation in y, the solution of which is

$$U(\alpha,y)=A(\alpha)e^{\alpha y}+B(\alpha)e^{-\alpha y}$$

Since u must be bounded as $y\to\infty$, $U(\alpha,y)$ must also be bounded as $y\to\infty$. Thus, for $\alpha>0, A(\alpha)$ must vanish, and

$$U(\alpha,0)=B(\alpha)$$

For $\alpha<0, B(\alpha)$ must vanish, and

$$U(\alpha,0)=A(\alpha)$$

Thus, for any α, we have

$$U(\alpha,y)=U(\alpha,0)e^{-|\alpha|y}$$

Observing that

$$U(\alpha,0)=\mathcal{F}[u(x,0)]$$
$$=\mathcal{F}[f(x)]$$

from the given condition, we find

$$U(\alpha,y)=\frac{1}{\sqrt{2\pi}}\int_{-\infty}^{\infty}f(x)e^{-|\alpha|y}e^{i\alpha x}\,dx$$

The inverse Fourier transform of $U(\alpha,y)$ is therefore given by

$$u(x,y)=\frac{1}{\sqrt{2\pi}}\int_{-\infty}^{\infty}\left[\frac{1}{\sqrt{2\pi}}\int_{-\infty}^{\infty}f(\xi)e^{-|\alpha|y}e^{i\alpha\xi}\,d\xi\right]e^{-i\alpha x}\,d\alpha$$
$$=\frac{1}{2\pi}\int_{-\infty}^{\infty}f(\xi)\,d\xi\int_{-\infty}^{\infty}e^{\alpha[i(\xi-x)]-|\alpha|y}\,d\alpha$$

It can be easily shown that

$$\int_{-\infty}^{\infty}e^{\alpha[i(\xi-x)]-|\alpha|y}\,d\alpha=\frac{2y}{(\xi-x)^2+y^2}$$

Hence the solution of the Dirichlet problem in the half-plane $y>0$ is

$$u(x,y)=\frac{y}{\pi}\int_{-\infty}^{\infty}\frac{f(\xi)}{(\xi-x)^2+y^2}\,d\xi$$

From this solution, we can readily deduce a solution of the Neumann problem in the half-plane $y>0$.

EXAMPLE 11.2.2. Find a solution of the Neumann problem in the half-plane $y>0$

$$u_{xx}+u_{yy}=0 \qquad -\infty<x<\infty \qquad y>0$$
$$u_y(x,0)=g(x) \qquad -\infty<x<\infty$$

$$u \text{ is bounded as } y\to\infty$$
$$u \text{ and } u_x \text{ vanish as } |x|\to\infty$$

Let $v(x,y)=u_y(x,y)$ then

$$u(x,y)=\int_a^y v(x,\eta)\,d\eta+C, \quad C=\text{constant}$$

where a is an arbitrary constant, and the Neumann problem becomes

$$\frac{\partial^2 v}{\partial x^2}+\frac{\partial^2 v}{\partial y^2}=\frac{\partial^2 u_y}{\partial x^2}+\frac{\partial^2 u_y}{\partial y^2}$$
$$=\frac{\partial}{\partial y}(u_{xx}+u_{yy})$$
$$=0$$
$$v(x,0)=u_y(x,0)=g(x)$$

This is the Dirichlet problem, the solution of which is given by

$$v(x,y)=\frac{y}{\pi}\int_{-\infty}^{\infty}\frac{g(\xi)\,d\xi}{(\xi-x)^2+y^2}$$

Thus we have

$$\begin{aligned}u(x,y)&=\frac{1}{\pi}\int_a^y\eta\int_{-\infty}^{\infty}\frac{g(\xi)\,d\xi}{(\xi-x)^2+\eta^2}\,d\eta+C\\&=\frac{1}{2\pi}\int_{-\infty}^{\infty}g(\xi)\log\left[\frac{(\xi-x)^2+y^2}{(\xi-x)^2+a^2}\right]d\xi+C\end{aligned}$$

11.3 Convolution (Fourier Transform)

The function

$$(f*g)(x)=\frac{1}{\sqrt{2\pi}}\int_{-\infty}^{\infty}f(x-\xi)g(\xi)\,d\xi \tag{11.3.1}$$

is called the *convolution* of the functions f and g over the interval $(-\infty,\infty)$.

Theorem 11.3.1 *If $F(\alpha)$ and $G(\alpha)$ are the Fourier transforms of $f(x)$ and $g(x)$ respectively, then the Fourier transform of the convolution $f*g$ is the product $F(\alpha)G(\alpha)$. Putting this statement in another way,*

$$\begin{aligned}\frac{1}{\sqrt{2\pi}}\int_{-\infty}^{\infty}F(\alpha)G(\alpha)e^{-i\alpha x}\,d\alpha&=(f*g)(x)\\&=\frac{1}{\sqrt{2\pi}}\int_{-\infty}^{\infty}f(x-\xi)g(\xi)\,d\xi\end{aligned}$$

PROOF.

$$\begin{aligned}\mathcal{F}[f*g]&=\frac{1}{2\pi}\int_{-\infty}^{\infty}e^{i\alpha x}\,dx\int_{-\infty}^{\infty}f(x-\xi)g(\xi)\,d\xi\\&=\frac{1}{2\pi}\int_{-\infty}^{\infty}g(\xi)e^{i\alpha\xi}\,d\xi\int_{-\infty}^{\infty}f(x-\xi)e^{i\alpha(x-\xi)}\,dx\end{aligned}$$

With the change of variable $\eta=x-\xi$, we have

$$\begin{aligned}\mathcal{F}[f*g]&=\frac{1}{\sqrt{2\pi}}\int_{-\infty}^{\infty}g(\xi)e^{i\alpha\xi}\,d\xi\frac{1}{\sqrt{2\pi}}\int_{-\infty}^{\infty}f(\eta)e^{i\alpha\eta}\,d\eta\\&=F(\alpha)G(\alpha)\end{aligned}$$ □

The convolution has the following properties:

1. $f*g=g*f$ (communtative)
2. $f*(g*h)=(f*g)*h$ (associative)
3. $f*(g+h)=f*g+f*h$ (distributive)

EXAMPLE 11.3.1. Obtain the solution of the initial value problem of heat conduction in an infinite rod

$$u_t - u_{xx} = 0 \qquad -\infty < x < \infty \qquad t > 0$$
$$u(x,0) = f(x) \qquad -\infty < x < \infty$$
$$u(x,t) \text{ is bounded.}$$

The Fourier transform of $u(x,t)$ in the variable x is

$$U(\alpha,t) = \frac{1}{\sqrt{2\pi}} \int_{-\infty}^{\infty} u(x,t) e^{i\alpha x}\, dx$$

Since

$$\begin{aligned}\mathfrak{F}[u_{xx}] &= (-i\alpha)^2 \mathfrak{F}[u] \\ &= -\alpha^2 U(\alpha,t)\end{aligned}$$

and

$$\begin{aligned}\mathfrak{F}[u_t] &= \frac{1}{\sqrt{2\pi}} \int_{-\infty}^{\infty} u_t e^{i\alpha x}\, dx \\ &= U_t\end{aligned}$$

the heat equation transforms into

$$U_t + \alpha^2 U = 0$$

the solution of which is

$$U(\alpha,t) = F(\alpha) e^{-\alpha^2 t} \tag{11.3.2}$$

Transforming the initial condition, we obtain

$$\begin{aligned}U(\alpha,0) &= \frac{1}{\sqrt{2\pi}} \int_{-\infty}^{\infty} u(x,0) e^{i\alpha x}\, dx \\ &= \frac{1}{\sqrt{2\pi}} \int_{-\infty}^{\infty} f(x) e^{i\alpha x}\, dx\end{aligned}$$

In view of the relation

$$U(\alpha,0) = F(\alpha)$$

which follows from Eq. (11.3.2), we have

$$U(\alpha,t) = U(\alpha,0) e^{-\alpha^2 t}$$

and the inverse transform is given by

$$u(x,t) = \frac{1}{\sqrt{2\pi}} \int_{-\infty}^{\infty} \left[F(\alpha) e^{-\alpha^2 t}\right] e^{-i\alpha x}\, d\alpha \tag{11.3.3}$$

The inverse transform of $G(\alpha)=e^{-\alpha^2 t}$ is

$$g(x)=\frac{1}{\sqrt{2\pi}}\int_{-\infty}^{\infty}e^{-\alpha^2 t-i\alpha x}\,d\alpha$$
$$=\frac{1}{\sqrt{2t}}e^{-x^2/4t}\qquad\text{(see Exercise 10)}$$

Thus, by the Convolution Theorem 11.3.1, the right side of Eq. (11.3.3) has the value

$$\frac{1}{\sqrt{2\pi}}\int_{-\infty}^{\infty}f(\xi)g(x-\xi)\,d\xi$$

Consequently, the solution takes the form

$$u(x,t)=\frac{1}{2\sqrt{\pi t}}\int_{-\infty}^{\infty}f(\xi)e^{-(x-\xi)^2/4t}\,d\xi$$

Consider the case where

$$f(x)=\begin{cases}0 & x<0\\ a & x>0\end{cases}$$

Then $u(x,t)$ becomes

$$u(x,t)=\frac{a}{2\sqrt{\pi t}}\int_{0}^{\infty}e^{-(x-\xi)^2/4t}\,d\xi$$

If we introduce the new variable

$$\eta=\frac{\xi-x}{2\sqrt{t}}$$

then the integral transforms into

$$u(x,t)=\frac{a}{\sqrt{\pi}}\int_{-x/2\sqrt{t}}^{\infty}e^{-\eta^2}\,d\eta$$
$$=\frac{a}{\sqrt{\pi}}\int_{-x/2\sqrt{t}}^{0}e^{-\eta^2}\,d\eta+\frac{a}{\sqrt{\pi}}\int_{0}^{\infty}e^{-\eta^2}\,d\eta$$

Since the integral

$$\int_{0}^{\infty}e^{-\eta^2}\,d\eta=\frac{\sqrt{\pi}}{2}$$

we have

$$u(x,t)=\frac{a}{2}+\frac{a}{\sqrt{\pi}}\int_{0}^{x/2\sqrt{t}}e^{-\eta^2}\,d\eta$$
$$=\frac{a}{2}\left[1+\operatorname{erf}\left(\frac{x}{2\sqrt{t}}\right)\right]$$

where $\operatorname{erf}(\alpha)$ is called the error function and is defined to be

$$\operatorname{erf}(\alpha)=\frac{2}{\sqrt{\pi}}\int_{0}^{\alpha}e^{-\eta^2}\,d\eta$$

These functions are widely used and are tabulated.

11.4 Step Function and Impulse Function (Fourier Transform)

In this section, we shall determine the Fourier transforms of the step function and the impulse function, functions which occur frequently in mathematical physics.

A *unit step function* is defined by

$$u_a(x) = \begin{cases} 0 & x < a \\ 1 & x \geqslant a \end{cases} \qquad a \geqslant 0 \tag{11.4.1}$$

as shown in Fig. 11.4.1.

The Fourier transform of the unit step function can be easily determined. We consider first

$$\begin{aligned} \mathcal{F}[u_a(x)] &= \frac{1}{\sqrt{2\pi}} \int_{-\infty}^{\infty} u_a(x) e^{i\alpha x}\, dx \\ &= \frac{1}{\sqrt{2\pi}} \int_{a}^{\infty} e^{i\alpha x}\, dx \end{aligned}$$

This integral does not exist. We avoid this procedure by defining a new function

$$u_a(x) e^{-\beta x} = \begin{cases} 0 & x < a \\ e^{-\beta x} & x \geqslant a \end{cases}$$

This evidently is the unit step function as $\beta \to \infty$. Thus, we find the Fourier

Figure 11.4.1

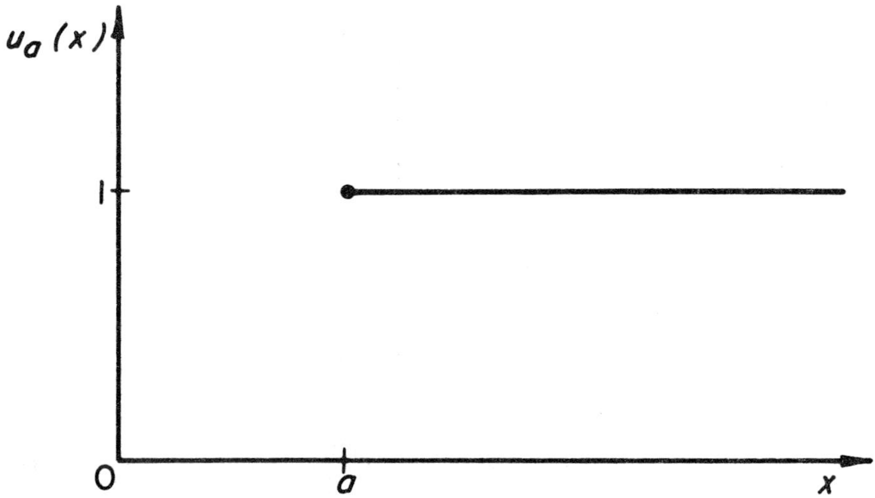

transform of the unit step function as

$$\begin{aligned}\mathcal{F}[u_a(x)] &= \lim_{\beta\to 0}\mathcal{F}[u_a(x)e^{-\beta x}]\\ &= \lim_{\beta\to 0}\frac{1}{\sqrt{2\pi}}\int_{-\infty}^{\infty}u_a(x)e^{-\beta x}e^{i\alpha x}\,dx\\ &= \lim_{\beta\to 0}\frac{1}{\sqrt{2\pi}}\int_{a}^{\infty}e^{-(\beta-i\alpha)x}\,dx\\ &= \frac{ie^{i\alpha a}}{\sqrt{2\pi}\,\alpha}\end{aligned}\tag{11.4.2}$$

For $a=0$,

$$\mathcal{F}[u_0(x)]=(i/\sqrt{2\pi}\,\alpha)\tag{11.4.3}$$

An *impulse function* is defined by

$$p(x)=\begin{cases} h & a-\varepsilon<x<a+\varepsilon\\ 0 & x\leqslant a-\varepsilon,\quad x\geqslant a+\varepsilon\end{cases}$$

where h is large and positive, $a>0$ and ε is a small positive constant, as shown in Fig. 11.4.2. This type of function appears in practical applica-

Figure 11.4.2

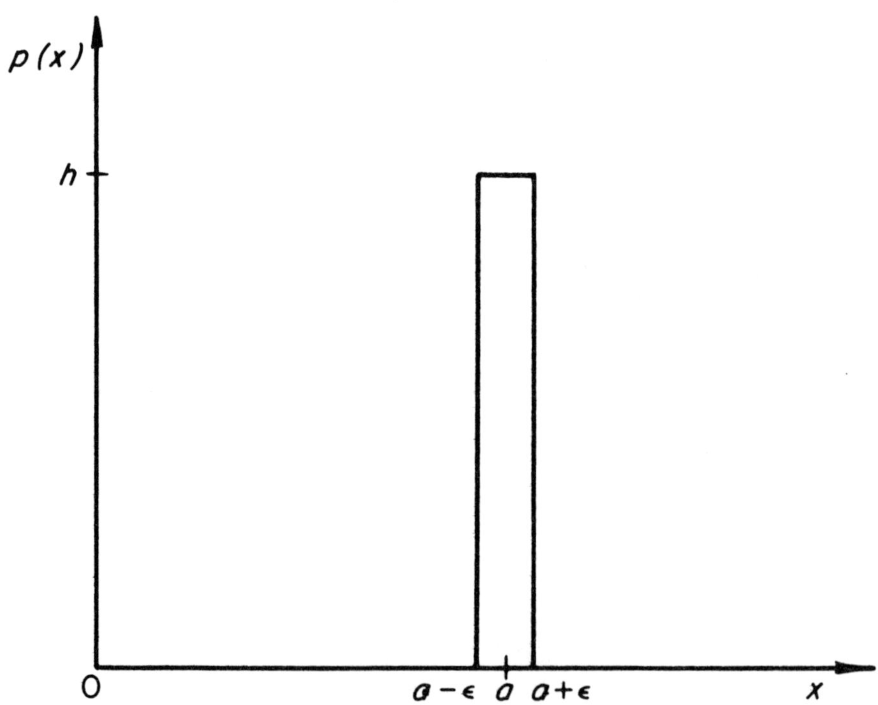

tions, for instance, a force of large magnitude may act over a very short period of time.

The Fourier transform of the impulse function is

$$\begin{aligned}\mathscr{F}[p(x)] &= \frac{1}{\sqrt{2\pi}}\int_{-\infty}^{\infty} p(x)e^{i\alpha x}\,dx\\ &= \frac{1}{\sqrt{2\pi}}\int_{a-\varepsilon}^{a+\varepsilon} he^{i\alpha x}\,dx\\ &= \frac{h}{\sqrt{2\pi}}\,\frac{e^{i\alpha a}}{i\alpha}\left(e^{i\alpha\varepsilon}-e^{-i\alpha\varepsilon}\right)\\ &= \frac{2h\varepsilon}{\sqrt{2\pi}}\,e^{i\alpha a}\,\frac{\sin\alpha\varepsilon}{\alpha\varepsilon}\end{aligned}$$

Now if we choose the value of h to be $1/2\varepsilon$, then the impulse defined by

$$I(\varepsilon) = \int_{-\infty}^{\infty} p(x)\,dx$$

becomes

$$\begin{aligned}I(\varepsilon) &= \int_{a-\varepsilon}^{a+\varepsilon} \frac{1}{2\varepsilon}\,dx\\ &= 1\end{aligned}$$

which is a constant independent of ε. In the limit $\varepsilon\to 0$, this particular function $p_\varepsilon(x)$ with $h = 1/2\varepsilon$ satisfies

$$\lim_{\varepsilon\to 0} p_\varepsilon(x) = 0 \qquad x \neq a$$

$$\lim_{\varepsilon\to 0} I(\varepsilon) = 1$$

Thus, we arrive at the result

$$\begin{gathered}\delta(x-a) = 0 \qquad x\neq a\\ \int_{-\infty}^{\infty} \delta(x-a)\,dx = 1\end{gathered} \tag{11.4.4}$$

This is the *Dirac delta function* which was defined earlier in Chapter 10.

We shall now define the Fourier transform of δ as the limit of the transform of $p_\varepsilon(x)$. We consider then

$$\begin{aligned}\mathscr{F}[\delta(x-a)] &= \lim_{\varepsilon\to 0}\mathscr{F}[p_\varepsilon(x)]\\ &= \lim_{\varepsilon\to 0}\frac{e^{i\alpha a}}{\sqrt{2\pi}}\,\frac{\sin\alpha\varepsilon}{\alpha\varepsilon}\\ &= \frac{e^{i\alpha a}}{\sqrt{2\pi}}\end{aligned}$$

in which we note that by L'Hospital's rule, $\lim_{\varepsilon\to 0}(\sin\alpha\varepsilon/\alpha\varepsilon)=1$. When $a=0$, we obtain

$$\mathcal{F}[\delta(x)]=1/\sqrt{2\pi} \tag{11.4.6}$$

EXAMPLE 11.4.1. (Slowing-down of Neutrons)[18] Consider the problem

$$u_\tau = u_{xx} + \delta(x)\delta(\tau) \tag{11.4.7}$$

$$u(x,0)=\delta(x) \tag{11.4.8}$$

$$\lim_{|x|\to\infty} u(x,\tau)=0 \tag{11.4.9}$$

This is the problem of infinite medium which slows neutrons, in which is located a source of neutrons. Here $u(x,\tau)$ represents the number of neutrons per unit volume per unit time which reach the age τ, and $\delta(x)\delta(\tau)$ represents the source function.

Let $U(\alpha,\tau)$ be the Fourier transform of $u(x,\tau)$. Then the transformation of Eq. (11.4.7) yields

$$\frac{dU}{d\tau}+\alpha^2 U=\frac{1}{\sqrt{2\pi}}\delta(\tau)$$

The solution of which, after applying the condition $U(\alpha,0)=1/\sqrt{2\pi}$, is

$$U(\alpha,\tau)=\frac{1}{\sqrt{2\pi}}e^{-\alpha^2\tau}$$

Hence the inverse transform is given by

$$\begin{aligned} u(x,\tau) &= \frac{1}{\sqrt{2\pi}}\int_{-\infty}^{\infty} e^{-\alpha^2\tau - i\alpha x}\,d\alpha \\ &= \frac{1}{2}\frac{1}{\sqrt{\pi\tau}}e^{-x^2/4\tau} \end{aligned}$$

11.5 Semiinfinite Region

For semiinfinite regions, the Fourier sine and cosine transforms determined in Sec. 11.1 are particularly appropriate in solving boundary value problems. Before we illustrate their applications, we must first prove the differentiation theorem.

---

[18]See I.N. Sneddon, *Fourier Transforms*, McGraw Hill, New York (1951), p. 215.

Theorem 11.5.1 *Let $f(x)$ and its first derivative vanish as $x\to\infty$. If $F_c(\alpha)$ is the Fourier cosine transform, then*

$$\mathcal{F}_c[f''(x)] = -\alpha^2 F_c(\alpha) - \sqrt{\frac{2}{\pi}}\, f'(0)$$

PROOF.

$$\begin{aligned}
\mathcal{F}_c[f''(x)] &= \sqrt{\frac{2}{\pi}} \int_0^\infty f''(x)\cos\alpha x\, dx \\
&= \sqrt{\frac{2}{\pi}}\,[f'(x)\cos\alpha x]_0^\infty + \sqrt{\frac{2}{\pi}}\,\alpha \int_0^\infty f'(x)\sin\alpha x\, dx \\
&= -\sqrt{\frac{2}{\pi}}\, f'(0) + \sqrt{\frac{2}{\pi}}\,\alpha[f(x)\sin\alpha x]_0^\infty \\
&\quad -\sqrt{\frac{2}{\pi}}\,\alpha^2 \int_0^\infty f(x)\cos\alpha x\, dx \\
&= -\sqrt{\frac{2}{\pi}}\, f'(0) - \alpha^2 F_c(\alpha)
\end{aligned}$$

□

In a similar manner, transforms of higher-order derivatives can be obtained.

Theorem 11.5.2 *Let $f(x)$ and its first derivative vanish as $x\to\infty$. If $F_s(\alpha)$ is the Fourier sine transform, then*

$$\mathcal{F}_s[f''(x)] = \sqrt{\frac{2}{\pi}}\,\alpha f(0) - \alpha^2 F_s(\alpha)$$

The proof is left to the reader.

EXAMPLE 11.5.1. Find the temperature distribution in a semiinfinite rod. At the end $x=0$, the heat is supplied at a rate $g(t)$. The initial temperature distribution is zero.

The problem here is

$$\begin{aligned}
u_t &= u_{xx} \quad x>0 \quad t>0 \\
u(x,0) &= 0 \\
u_x(0,t) &= g(t)
\end{aligned}$$

Here we assume that $u(x,t)$ and $u_x(x,t)$ vanish as x approaches infinity.

Let $U(\alpha,t)$ be the Fourier cosine transform of $u(x,t)$. Then the transformation of the heat equation yields

$$U_t + \alpha^2 U = -\sqrt{\frac{2}{\pi}}\, g(t)$$

The solution of this equation is

$$U(\alpha,t)=e^{-\alpha^2 t}\left[-\int_0^t\sqrt{\frac{2}{\pi}}\; g(\tau)e^{\alpha^2\tau}\,d\tau+C\right]$$

Applying the initial condition, we obtain

$$U(\alpha,0)=0=C$$

Thus, $U(\alpha,t)$ becomes

$$U(\alpha,t)=-\sqrt{\frac{2}{\pi}}\;\int_0^t g(\tau)e^{-\alpha^2(t-\tau)}\,d\tau$$

Hence the inverse transform is

$$u(x,t)=-\frac{2}{\pi}\int_0^\infty\left[\int_0^t g(\tau)e^{-\alpha^2(t-\tau)}\,d\tau\right]\cos\alpha x\,d\alpha$$

$$=-\frac{2}{\pi}\int_0^t g(\tau)\,d\tau\int_0^\infty e^{-\alpha^2(t-\tau)}\cos\alpha x\,d\alpha$$

The inner integral is given by (see Exercise 6)

$$\int_0^\infty e^{-\alpha^2(t-\tau)}\cos\alpha x\,d\alpha=\frac{1}{2}\sqrt{\frac{\pi}{t-\tau}}\;e^{-x^2/4(t-\tau)}$$

The solution therefore is

$$u(x,t)=-\frac{1}{\sqrt{\pi}}\int_0^t\frac{g(\tau)}{\sqrt{t-\tau}}\,e^{-x^2/4(t-\tau)d\tau}$$

11.6 Hankel and Mellon Transforms

In addition to Fourier transforms, there are other transforms such as Hankel and Mellon transforms. These transforms can be devised for special types of problems; for instance, the Hankel transform would be very suitable for problems involving Bessel functions.

The *Hankel transform* of order v of the function $f(x)$, denoted by $H(\xi)$ is defined by

$$H(\xi)=\int_0^\infty f(x)J_v(\xi x)x\,dx \qquad v\geqslant-\tfrac{1}{2}$$

and the inverse transform is given by

$$f(x)=\int_0^\infty H(\xi)J_v(\xi x)\xi\,d\xi$$

The *Mellon transform* of the function $f(x)$, denoted by $M(s)$ is defined by

$$M(s)=\int_0^\infty f(x)x^{s-1}\,dx$$

and the inverse transform is given by

$$f(x)=\frac{1}{2\pi i}\int_{c-i\infty}^{c+i\infty} M(s)x^{-s}\,ds$$

The interested reader may refer to Sneddon [37] for further study.

11.7 Laplace Transforms

Due to simplicity, Laplace transforms are frequently used in determining solutions of a wide class of partial differential equations. Like other transforms, Laplace transforms are used to determine particular solutions. In solving partial differential equations, the general solutions are difficult, if not impossible, to obtain. Thus, the transform techniques sometimes offer a useful tool for attaining particular solutions.

The *Laplace transform* of the function $f(t)$, which will be denoted by either of the symbols $F(s)$ or $\mathcal{L}[f(t)]$ is defined by the equation

$$\mathcal{L}[f(t)]=F(s)=\int_0^\infty e^{-st}f(t)\,dt, \quad s>0$$

provided the improper integral converges. s is the transform variable.

The inverse of the Laplace transform shall be denoted by

$$\mathcal{L}^{-1}[F(s)]=f(t)$$

Let us now find the Laplace transforms of some elementary functions.

1. Given $f(t)=c, c$ is a constant.

$$\mathcal{L}[c]=\int_0^\infty e^{-st}c\,dt$$

$$=\left[-\frac{ce^{-st}}{s}\right]_0^\infty$$

$$=c/s$$

2. Given $f(t)=e^{at}$, a is a constant

$$\mathcal{L}[e^{at}]=\int_0^\infty e^{-st}e^{at}\,dt$$

$$=\left[-\frac{e^{-(s-a)t}}{(s-a)}\right]_0^\infty$$

$$=\frac{1}{s-a}\quad s>a$$

3. Given $f(t)=t^2$. Then

$$\mathcal{L}[t^2]=\int_0^\infty e^{-st}t^2\,dt$$

Integration by parts yields

$$=\left[-\frac{t^2e^{-st}}{s}\right]_0^\infty+\int_0^\infty \frac{e^{-st}}{s}2t\,dt$$

Since $t^2e^{-st}\to 0$ as $t\to\infty$, we have

$$\mathcal{L}[t^2]=\frac{2}{s}\left[-\frac{e^{-st}}{s}t\right]_0^\infty+\frac{2}{s}\int_0^\infty \frac{e^{-st}}{s}dt$$

$$=2/s^3$$

4. Given $f(t)=\sin\,\omega t$. Then

$$F(s)=\mathcal{L}[\sin\omega t]=\int_0^\infty e^{-st}\sin\omega t\,dt$$

$$=\left[-\frac{e^{-st}}{s}\sin\omega t\right]_0^\infty+\int_0^\infty \frac{e^{-st}}{s}\omega\cos\omega t\,dt$$

$$=\frac{\omega}{s}\left[-\frac{e^{-st}}{s}\cos\omega t\right]_0^\infty-\frac{\omega}{s}\int_0^\infty \frac{e^{-st}}{s}\omega\sin\omega t\,dt$$

$$F(s)=\frac{\omega}{s^2}-\frac{\omega^2}{s^2}F(s)$$

Thus, solving for $F(s)$, we obtain

$$\mathcal{L}[\sin\omega t]=\omega/(s^2+\omega^2)$$

Theorem 11.7.1 *Let f be piecewise continuous in the interval $[0,T]$ for every positive T, and let f be of exponential order,*[19] *that is, $f(t)=O(e^{at})$ as $t\to\infty$ for some $a>0$. Then the Laplace transform of f exists for $s>a$.*

PROOF. Since f is piecewise continuous and of exponential order, we have

$$|\mathcal{L}[f]|=\left|\int_0^\infty e^{-st}f(t)\,dt\right|$$

$$\leqslant\int_0^\infty e^{-st}|f(t)|\,dt$$

$$\leqslant\int_0^\infty e^{-st}Me^{at}\,dt$$

$$=M\int_0^\infty e^{-(s-a)t}\,dt$$

$$=M/(s-a),\quad s>a$$

[19]A function $f(t)$ is said to be of exponential order as $t\to\infty$ if there exists some $a>0$ such that $e^{-at}|f(t)|$ is bounded for all $t>T$, that is, for $M>a, |f(t)|<Me^{at}$.

Thus

$$\int_0^\infty e^{-st} f(t)\,dt$$

exists for $s > a$. □

11.8 Properties of Laplace Transforms

Theorem 11.8.1 (Linearity) *If $\mathcal{L}[f(t)]$ and $\mathcal{L}[g(t)]$ are the Laplace transforms of f and g respectively, then*

$$\mathcal{L}[af(t)+bg(t)] = a\mathcal{L}[f(t)] + b\mathcal{L}[g(t)]$$

where a and b are constants.

PROOF.

$$\begin{aligned}\mathcal{L}[af(t)+bg(t)] &= \int_0^\infty [af(t)+bg(t)]e^{-st}\,dt \\ &= a\int_0^\infty f(t)e^{-st}\,dt + b\int_0^\infty g(t)e^{-st}\,dt \\ &= a\mathcal{L}[f(t)] + b\mathcal{L}[g(t)]\end{aligned}$$

□

Theorem 11.8.2 (Shifting) *If the Laplace transform of $f(t)$ is $F(s)$, then the Laplace transform of $e^{at}f(t)$ is $F(s-a)$.*

PROOF.

$$\begin{aligned}\mathcal{L}[e^{at}f(t)] &= \int_0^\infty e^{-st}e^{at}f(t)\,dt \\ &= \int_0^\infty e^{-(s-a)t}f(t)\,dt \\ &= F(s-a)\end{aligned}$$

□

EXAMPLE 11.8.1.

a. If $\mathcal{L}[t^2] = 2/s^3$, then $\mathcal{L}[t^2e^t] = 2/(s-1)^3$
b. If $\mathcal{L}[\sin\omega t] = \omega/(s^2+\omega^2)$, then $\mathcal{L}[e^t \sin\omega t] = \omega/[(s-1)^2+\omega^2]$

Theorem 11.8.3 (Scaling) *If the Laplace transform of $f(t)$ is $F(s)$, then the Laplace transform of $f(ct)$ with $c > 0$ is $(1/c)F(s/c)$.*

PROOF.

$$\begin{aligned}\mathcal{L}[f(ct)] &= \int_0^\infty e^{-st}f(ct)\,dt \\ &= \int_0^\infty \frac{1}{c}e^{-(s\xi/c)}f(\xi)\,d\xi \qquad \text{(by substituting } \xi = ct) \\ &= (1/c)F(s/c)\end{aligned}$$

□

EXAMPLE 11.8.2.

a. If $\dfrac{s}{s+1}=\mathcal{L}[\cos t]$, then

$$\frac{1}{\omega}\frac{s/\omega}{(s/\omega)^2+1}=\frac{s}{s^2+\omega^2}=\mathcal{L}[\cos\omega t]$$

b. If $\dfrac{1}{s-1}=\mathcal{L}[e^t]$, then

$$\frac{1}{a}\frac{1}{\frac{s}{a}-1}=\mathcal{L}[e^{at}]$$

$$\frac{2}{s-a}=\mathcal{L}[e^{at}]$$

Theorem 11.8.4 (Differentiation) *Let f be continuous and f' piecewise continuous, in $0\leqslant t\leqslant T$ for all $T>0$. Let f also be of exponential order at $t\to\infty$. Then the Laplace transform of f' exists and is given by*

$$\mathcal{L}[f(t)]=s\mathcal{L}[f(t)]-f(0)$$

PROOF. Consider the integral

$$\int_0^T e^{-st}f'(t)\,dt=\left[e^{-st}f(t)\right]_0^T+\int_0^T se^{-st}f(t)\,dt$$

$$=e^{-st}f(T)-f(0)+s\int_0^T e^{-st}f(t)\,dt$$

Since $|f(t)|\leqslant Me^{at}$ for large t, with $a>0$ and $M>0$,

$$|e^{-st}f(T)|\leqslant Me^{-(s-a)T}$$

In the limit as $T\to\infty, e^{-st}f(T)\to 0$ whenever $s>a$. Hence

$$\mathcal{L}[f'(t)]=s\mathcal{L}[f(t)]-f(0)$$ □

If f' and f'' satisfy the same conditions imposed on f and f' respectively, then the Laplace transform of $f''(t)$ can be obtained immediately by applying the preceding theorem; that is

$$\mathcal{L}[f''(t)]=s\mathcal{L}[f'(t)]-f'(0)$$

$$=s\{s\mathcal{L}[f(t)]-f(0)\}-f'(0)$$

$$=s^2\mathcal{L}[f(t)]-sf(0)-f'(0)$$

Clearly, the Laplace transform of $f^{(n)}(t)$ can be obtained in a similar manner by successive application of the theorem. The result may be written as

$$\mathcal{L}[f^{(n)}(t)]=s^n\mathcal{L}[f(t)]-s^{n-1}f(0)-\ldots-sf^{(n-2)}(0)-f^{(n-1)}(0)$$

Theorem 11.8.5 (Integration) *If $F(s)$ is the Laplace transform of $f(t)$, then*

$$\mathcal{L}\left[\int_0^t f(\tau)\,d\tau\right] = F(s)/s$$

PROOF.

$$\begin{aligned}\mathcal{L}\left[\int_0^t f(\tau)\,d\tau\right] &= \int_0^\infty \left[\int_0^t f(\tau)\,d\tau\right] e^{-st}\,dt \\ &= \left[-\frac{e^{-st}}{s}\int_0^t f(\tau)\,d\tau\right]_0^\infty + \frac{1}{s}\int_0^\infty f(t)e^{-st}\,dt \\ &= F(s)/s \text{ since}\end{aligned}$$

$\int_0^t f(\tau)\,d\tau$ is of exponential order. □

In solving problems by this method, the difficulty arises in finding inverse transforms. Although the inversion formula exists, its use requires a knowledge of complex variables. However, for some problems of mathematical physics, we need not use this inversion formula. We can avoid its use by expanding a given transform by the method of partial fractions in terms of simple fractions in the transform variables. With these simple functions, one refers to the table of Laplace transforms[20] and obtains the inverse transforms. Here, we should note that we use the assumption that there is essentially a one-to-one correspondence between functions and their Laplace transforms. This may be stated as follows:

Theorem 11.8.6 (Lerch) *Let f and g be piecewise continuous functions of exponential order. If there exists a constant s_0, such that $\mathcal{L}[f] = \mathcal{L}[g]$ for all $s > s_0$, then $f(t) = g(t)$ for all $t > 0$ except possibly at the points of discontinuity.*

For a proof, the reader is referred to Kreider et al. [24].

EXAMPLE 11.8.3. Consider the motion of a semiinfinite string with an external force $f(t)$ acting on it. One end is kept fixed while the other end is allowed to move freely in the vertical direction. If the string is initially at rest, the motion of the string is governed by

$$\begin{aligned} u_{tt} &= c^2 u_{xx} + f(t) \qquad 0 < x < \infty \qquad t > 0 \\ u(x,0) &= 0 \\ u_t(x,0) &= 0 \\ u(0,t) &= 0 \\ u_x(x,t) &\to 0 \qquad \text{as } x \to \infty \end{aligned}$$

---

[20]A short table of Laplace transforms is given in the Appendix. For extensive tables, see the references.

Let $U(x,s)$ be the Laplace transform of $u(x,t)$. Transforming the equation of motion and using the initial conditions, we obtain

$$U_{xx}-(s^2/c^2)U=-F(s)/c^2$$

where

$$F(s)=\int_0^{\infty} f(t)e^{-st}\,dt$$

The solution of this ordinary differential equation is

$$U(x,s)=Ae^{sx/c}+Be^{-sx/c}+\left[F(s)/s^2\right]$$

The transformed boundary conditions are

$$U(0,s)=0$$

and

$$\lim_{x\to\infty} U_x(x,s)=0$$

In view of the second condition, we have $A=0$. Now applying the first condition, we obtain

$$U(0,s)=B+\left[F(s)/s^2\right]=0$$

Hence

$$U(x,s)=\left[F(s)/s^2\right]\left[1-e^{-sx/c}\right]$$

a. When $f(t)=f_0$, a constant, then

$$U(x,s)=f_0\left(\frac{1}{s^3}-\frac{1}{s^3}e^{-sx/c}\right)$$

The inverse is given by

$$\begin{aligned} u(x,t)&=\frac{f_0}{2}\left[t^2-\left(t-\frac{x}{c}\right)^2\right] && \text{when} \quad t\geqslant x/c \\ &=(f_0/2)t^2 && \text{when} \quad t\leqslant x/c \end{aligned}$$

b. When $f(t)=\cos\omega t$, where ω is a constant, then

$$\begin{aligned} F(s)&=\int_0^{\infty}\cos\omega t e^{-st}\,dt \\ &=s/(\omega^2+s^2) \end{aligned}$$

Thus, we have

$$U(x,s)=\frac{1}{s(\omega^2+s^2)}\left[1-e^{-sx/c}\right] \tag{11.8.1}$$

By the method of partial fractions, we write

$$\frac{1}{s(s^2+\omega^2)}=\frac{1}{\omega^2}\left[\frac{1}{s}-\frac{s}{s^2+\omega^2}\right]$$

Hence

$$\mathcal{L}^{-1}\frac{1}{s(s^2+\omega^2)}=\frac{1}{\omega^2}(1-\cos\omega t)=\frac{2}{\omega^2}\sin^2\left(\frac{\omega t}{2}\right)$$

If we denote

$$\psi(t)=\sin^2\left(\frac{\omega t}{2}\right)$$

then the inverse of Eq. (11.8.1) may be written in the form

$$\begin{aligned} u(x,t)&=\frac{2}{\omega^2}\left[\psi(t)-\psi\left(t-\frac{x}{c}\right)\right] &&\text{when} \quad t\geqslant x/c\\ &=\frac{2}{\omega^2}\psi(t) &&\text{when} \quad t\leqslant x/c \end{aligned}$$

11.9 Convolution (Laplace Transform)

The function

$$(f*g)(t)=\int_0^t f(t-\xi)g(\xi)\,d\xi \tag{11.9.1}$$

is called the convolution of the functions f and g.

Theorem 11.9.1 (Convolution) *If $F(s)$ and $G(s)$ are the Laplace transforms of $f(t)$ and $g(t)$ respectively, then the Laplace transform of the convolution $f*g$ is the product $F(s)G(s)$.*

PROOF.

$$\begin{aligned} \mathcal{L}[f*g]&=\int_0^\infty e^{-st}\int_0^t f(t-\xi)g(\xi)\,d\xi\,dt\\ &=\int_0^\infty\int_0^t e^{-st}f(t-\xi)g(\xi)\,d\xi\,dt \end{aligned}$$

The region of integration is shown in Fig. 11.9.1. By reversing the order of integration, we have

$$\begin{aligned} \mathcal{L}[f*g]&=\int_0^\infty\int_\xi^\infty e^{-st}f(t-\xi)g(\xi)\,dt\,d\xi\\ &=\int_0^\infty g(\xi)\int_\xi^\infty e^{-st}f(t-\xi)\,dt\,d\xi \end{aligned}$$

If we introduce the new variable $\eta=t-\xi$ in the inner integral, we obtain

$$\begin{aligned} \mathcal{L}[f*g]&=\int_0^\infty g(\xi)\int_0^\infty e^{-s(\xi+\eta)}f(\eta)\,d\eta\,d\xi\\ &=\int_0^\infty g(\xi)e^{-s\xi}\,d\xi\int_0^\infty f(\eta)e^{-s\eta}\,d\eta\\ &=F(s)G(s) \end{aligned}$$

□

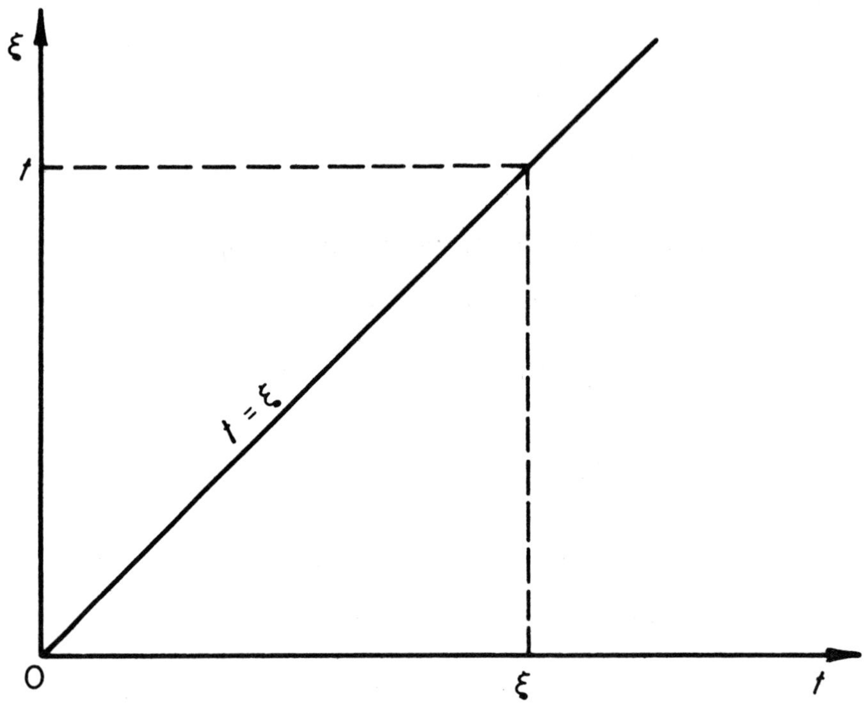

Figure 11.9.1

The convolution has the following properties:

1. $f*g=g*f$ (commutative)
2. $f*(g*h)=(f*g)*h$ (associative)
3. $f*(g+h)=f*g+f*h$ (distributive)

EXAMPLE 11.9.1. Find the temperature distribution in a semiinfinite radiating rod. The temperature is kept constant at $x=0$ while the other end is kept at zero temperature. If the initial temperature distribution is zero, the problem is governed by

$$u_t = ku_{xx} - hu \qquad 0<x<\infty \qquad t>0 \qquad h=\text{constant}$$

$$u(x,0)=0$$

$$u(0,t)=u_0, \qquad t>0, \quad u_0=\text{constant}$$

$$u(x,t)\to 0 \quad \text{as} \quad x\to\infty$$

Let $U(x,s)$ be the Laplace transform of $u(x,t)$. Then transformation with respect to t yields

$$U_{xx}-\left(\frac{s+h}{k}\right)U=0$$
$$U(0,s)=u_0/s$$
$$\lim_{x\to\infty} U(x,s)=0$$

The solution of this equation is

$$U(x,s)=Ae^{\sqrt{\frac{s+h}{k}}x}+Be^{-\sqrt{\frac{s+h}{k}}x}$$

The boundary condition at infinity requires that $A=0$. Applying the remaining boundary condition, we have

$$U(0,s)=B=u_0/s$$

Hence the solution takes the form

$$U(x,s)=(u_0/s)e^{-\sqrt{\frac{s+h}{k}}x}$$

We find that (by using tables)

$$\mathcal{L}^{-1}\left[\frac{u_0}{s}\right]=u_0$$

and

$$\mathcal{L}^{-1}\left[e^{-\sqrt{\frac{s+h}{k}}x}\right]=\frac{xe^{-ht-\frac{x^2}{4kt}}}{2\sqrt{\pi k t^3}}$$

Thus the inverse Laplace transform of $U(x,s)$ is

$$u(x,t)=\mathcal{L}^{-1}\left[\frac{u_0}{s}\cdot e^{-\sqrt{\frac{s+h}{k}}x}\right]$$

By the Integration Theorem 11.8.5, we have

$$u(x,t)=\int_0^t \frac{u_0 x e^{-h\tau-x^2/4k\tau}}{2\sqrt{\pi k}\ \tau^{3/2}}\,d\tau$$

Substitution of the new variable $\eta=x/2\sqrt{k\tau}$ yields

$$u(x,t)=\frac{2u_0}{\sqrt{\pi}}\int_{x/2\sqrt{kt}}^{\infty} e^{-\left(\eta^2+\frac{hx^2}{4k\eta^2}\right)}\,d\eta$$

For the case $h=0$, $u(x,t)$ becomes

$$
\begin{aligned}
u(x,t) &= \frac{2u_0}{\sqrt{\pi}} \int_{x/2\sqrt{kt}}^{\infty} e^{-\eta^2} f\eta \\
&= \frac{2u_0}{\sqrt{\pi}} \int_0^{\infty} e^{-\eta^2} d\eta - \frac{2u_0}{\sqrt{\pi}} \int_0^{x/2\sqrt{kt}} e^{-\eta^2} d\eta \\
&= u_0\left[1 - \operatorname{erf}\left(\frac{x}{2\sqrt{kt}}\right)\right] \\
&= u_0 \operatorname{erfc}\left(\frac{x}{2\sqrt{kt}}\right)
\end{aligned}
$$

where $\operatorname{erfc}(\alpha)$ is called the *complementary error function*. It is related to the error function by

$$\operatorname{erf}(\alpha) + \operatorname{erfc}(\alpha) = 1$$

11.10 Step Function and Impulse Function (Laplace Transform)

We have defined earlier the unit step function. Now, we will find the Laplace transform of it

$$
\begin{aligned}
\mathcal{L}[u_a(t)] &= \int_0^{\infty} e^{-st} u_a(t)\, dt \\
&= \int_a^{\infty} e^{-st}\, dt \qquad (11.10.1) \\
&= e^{-as}/s, \qquad s > 0
\end{aligned}
$$

Theorem 11.10.1 (Second Shifting) *If $F(s)$ is the Laplace transform of $f(t)$, then*

$$\mathcal{L}[u_a(t) f(t-a)] = e^{-as} F(s)$$

PROOF.

$$
\begin{aligned}
\mathcal{L}[u_a(t) f(t-a)] &= \int_0^{\infty} e^{-st} u_a(t) f(t-a)\, dt \\
&= \int_a^{\infty} e^{-st} f(t-a)\, dt
\end{aligned}
$$

Introducing the new variable $\xi = t - a$, we have

$$
\begin{aligned}
\mathcal{L}[u_a(t) f(t-a)] &= \int_0^{\infty} e^{-(\xi+a)s} f(\xi)\, d\xi \\
&= e^{-as} \int_0^{\infty} e^{-\xi s} f(\xi)\, d\xi \\
&= e^{-as} F(s)
\end{aligned}
$$

□

EXAMPLE 11.10.1. a. Given that

$$f(t)=\begin{cases}0 & t<2\\ t-2 & t\geqslant 2\end{cases}$$

find the Laplace transform of $f(t)$.

$$\begin{aligned}\mathcal{L}[f(t)]&=\mathcal{L}[u_2(t)(t-2)]\\&=e^{-2s}\mathcal{L}[t]\\&=e^{-2s}/s^2\end{aligned}$$

b. Find the inverse transform of

$$F(s)=\frac{1+e^{-2s}}{s^2}$$

$$\begin{aligned}\mathcal{L}^{-1}[F(s)]&=\mathcal{L}^{-1}\left[\frac{1}{s^2}+\frac{e^{-2s}}{s^2}\right]\\&=\mathcal{L}^{-1}\left[\frac{1}{s^2}\right]+\mathcal{L}^{-1}\left[\frac{e^{-2s}}{s^2}\right]\\&=t+u_2(t)(t-2)\\&=\begin{cases}t & 0\leqslant t<2\\ 2(t-1) & t\geqslant 2\end{cases}\end{aligned}$$

The Laplace transform of the impulse function $p(t)$ is given by

$$\begin{aligned}\mathcal{L}[p(t)]&=\int_0^\infty e^{-st}p(t)\,dt\\&=\int_{a-\varepsilon}^{a+\varepsilon}he^{-st}\,dt\\&=h\left[-\frac{e^{-st}}{s}\right]_{a-\varepsilon}^{a+\varepsilon}\\&=\frac{he^{-as}}{s}(e^{\varepsilon s}-e^{-\varepsilon s})\\&=2\frac{he^{-as}}{s}\sinh\varepsilon s\end{aligned}\tag{11.10.2}$$

If we choose the value of h to be $1/2\varepsilon$, then the impulse is given by

$$\begin{aligned}I(\varepsilon)&=\int_{-\infty}^{\infty}p(t)\,dt\\&=\int_{a-\varepsilon}^{a+\varepsilon}\frac{1}{2\varepsilon}\,dt\\&=1\end{aligned}$$

Thus, in the limit this particular impulse function satisfies

$$\lim_{\varepsilon\to 0} p_\varepsilon(t)=0 \qquad t\neq a$$

$$\lim_{\varepsilon\to 0} I(\varepsilon)=1$$

From this result, we obtain the Dirac delta function which satisfies

$$\delta(t-a)=0 \qquad t\neq a$$

$$\int_{-\infty}^{\infty}\delta(t-a)\,dt=1 \tag{11.10.3}$$

Thus, we may define the Laplace transform of δ as the limit of the transform of $p_\varepsilon(t)$.

$$\begin{aligned}\mathcal{L}[\delta(t-a)] &= \lim_{\varepsilon\to 0}\mathcal{L}[p_\varepsilon(t)]\\ &= \lim_{\varepsilon\to 0} e^{-as}\frac{\sinh \varepsilon s}{\varepsilon s}\\ &= e^{-as}\end{aligned} \tag{11.10.4}$$

Hence if $a=0$, we have

$$\mathcal{L}[\delta(t)]=1 \tag{11.10.5}$$

One of the very useful results that can be derived is the integral of the product of the delta function and any continuous function $f(t)$.

$$\begin{aligned}\int_{-\infty}^{\infty}\delta(t-a)f(t)\,dt &= \lim_{\varepsilon\to 0}\int_{-\infty}^{\infty}p_\varepsilon(t)f(t)\,dt\\ &= \lim_{\varepsilon\to 0}\int_{a-\varepsilon}^{a+\varepsilon}\frac{f(t)}{2\varepsilon}\,dt\\ &= \lim_{\varepsilon\to 0}\frac{1}{2\varepsilon}\cdot 2\varepsilon f(t^*) \qquad a-\varepsilon<t^*<a+\varepsilon\\ &= f(a)\end{aligned} \tag{11.10.6}$$

Suppose that $f(t)$ is a periodic function with period T. Let f be piecewise continuous on $[0,T]$. Then the Laplace transform of f is

$$\begin{aligned}\mathcal{L}[f(t)] &= \int_0^{\infty}e^{-st}f(t)\,dt\\ &= \sum_{n=0}^{\infty}\int_{nT}^{(n+1)T}e^{-st}f(t)\,dt\end{aligned}$$

If we introduce the new variable $\xi=t-nT$, then

$$\begin{aligned}\mathcal{L}[f(t)] &= \sum_{n=0}^{\infty}e^{-nTs}\int_0^{T}e^{-s\xi}f(\xi)\,d\xi\\ &= \sum_{n=0}^{\infty}e^{-nTs}F_1(s)\end{aligned}$$

where $F_1(s)=\int_0^T e^{-s\xi} f(\xi)\,d\xi$ is the transform of the function f over the first period. Since the series is the geometric series, we obtain for the transform of the periodic function

$$\mathcal{L}[f(t)]=\frac{F_1(s)}{1-e^{-Ts}} \tag{11.10.7}$$

EXAMPLE 11.10.2. Find the Laplace transform of the function

$$f(t)=\begin{cases} h & 0<t<c \\ -h & c<t<2c \end{cases}$$

and

$$f(t+2c)=f(t)$$

shown in Fig. 11.10.1.

$$\begin{aligned} F_1(s) &= \int_0^{2c} e^{-s\xi} f(\xi)\,d\xi \\ &= \int_0^{c} e^{-s\xi} h\,d\xi + \int_c^{2c} e^{-s\xi}(-h)\,d\xi \\ &= \frac{h}{s}(1-e^{-cs})^2 \end{aligned}$$

Thus, the Laplace transform of $f(t)$ is

$$\begin{aligned} \mathcal{L}[f(t)] &= \frac{F_1(s)}{1-e^{-2cs}} \\ &= \frac{h(1-e^{-cs})^2}{s(1-e^{-2cs})} \\ &= \frac{h(1-e^{-cs})}{s(1+e^{-cs})} \\ &= \frac{h}{s}\tanh\frac{cs}{2} \end{aligned}$$

EXAMPLE 11.10.3. A uniform bar of length l is fixed at one end. Let the force

$$f(t)=\begin{cases} f_0 & t>0 \\ 0 & t<0 \end{cases}$$

be suddenly applied at the end $x=l$. If the bar is initially at rest, find the longitudinal displacement for $t>0$.

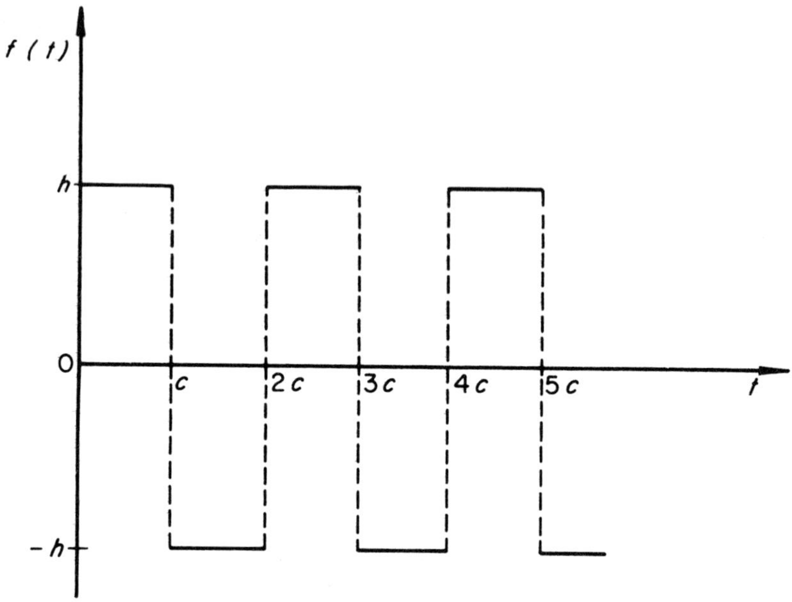

Figure 11.10.1

The motion of the bar is governed by

$$u_{tt} = a^2 u_{xx}, \qquad 0 < x < l, \qquad t > 0, \qquad a = \text{constant}$$
$$u(x,0) = 0$$
$$u_t(x,0) = 0$$
$$u(0,t) = 0$$
$$u_x(l,t) = f_0/E \qquad \text{where } E \text{ is a constant.}$$

Let $U(x,s)$ be the Laplace transform of $u(x,t)$. Then $U(x,s)$ satisfies

$$U_{xx} - \frac{s^2}{a^2} U = 0$$
$$U(0,s) = 0$$
$$U_x(l,s) = f_0/Es$$

The solution of this differential equation is

$$U(x,s) = Ae^{xs/a} + Be^{-xs/a}$$

Applying the boundary conditions, we have

$$A + B = 0$$
$$\left(\frac{s}{a} e^{ls/a}\right) A + \left(-\frac{s}{a} e^{-ls/a}\right) B = f_0/Es$$

Solving for A and B, we obtain

$$A=-B=\frac{af_0}{Es^2(e^{ls/a}+e^{-ls/a})}$$

Hence the transform is given by

$$U(x,s)=\frac{af_0}{Es^2}\frac{e^{xs/a}-e^{-xs/a}}{e^{ls/a}+e^{-ls/a}}$$

Before finding the inverse transform of $U(x,s)$, multiply the numerator and denominator by $(e^{-ls/a}-e^{-3ls/a})$. Thus, we have

$$U(x,s)=\frac{af_0}{Es^2}\Big[e^{-(l-x)s/a}-e^{-(l+x)s/a}-e^{-(3l-x)s/a}+e^{-(3l+x)s/a}\Big]\frac{1}{1-e^{-4ls/a}}$$

Since the denominator has the term $(1-e^{-4ls/a})$, the inverse transform $u(x,t)$ is periodic with period $4l/a$. Hence the final solution may be written in the form

$$u(x,t)=\begin{cases} 0 & 0<t<\dfrac{l-x}{a} \\ \dfrac{af_0}{E}\left(t-\dfrac{l-x}{a}\right) & \dfrac{l-x}{a}<t<\dfrac{l+x}{a} \\ \dfrac{af_0}{E}\left[\left(t-\dfrac{l-x}{a}\right)-\left(t-\dfrac{l+x}{a}\right)\right] & \dfrac{l+x}{a}<t<\dfrac{3l-x}{a} \\ \dfrac{af_0}{E}\left[\left(t-\dfrac{l-x}{a}\right)-\left(t-\dfrac{l+x}{a}\right)-\left(t-\dfrac{3l-x}{a}\right)\right] & \dfrac{3l-x}{a}<t<\dfrac{3l+x}{a} \\ \dfrac{af_0}{E}\left[\left(t-\dfrac{l-x}{a}\right)-\left(t-\dfrac{l+x}{a}\right)-\left(t-\dfrac{3l-x}{a}\right)-\left(t-\dfrac{3l+x}{a}\right)\right] & \dfrac{3l+x}{a}<t<\dfrac{4l}{a} \end{cases}$$

which may be simplified to give

$$u(x,t)=\begin{cases} 0 & 0<t<(l-x)/a \\ \dfrac{af_0}{E}\left(t-\dfrac{l-x}{a}\right) & (l-x)/a<t<(l+x)/a \\ \dfrac{af_0}{E}\left(\dfrac{2x}{a}\right) & (l+x)/a<t<(3l-x)/a \\ \dfrac{af_0}{E}\left(-t+\dfrac{3l+x}{a}\right) & (3l-x)/a<t<(3l+x)/a \\ 0 & (3l+x)/a<t<4l/a \end{cases}$$

This result can clearly be seen in Fig. 11.10.2.

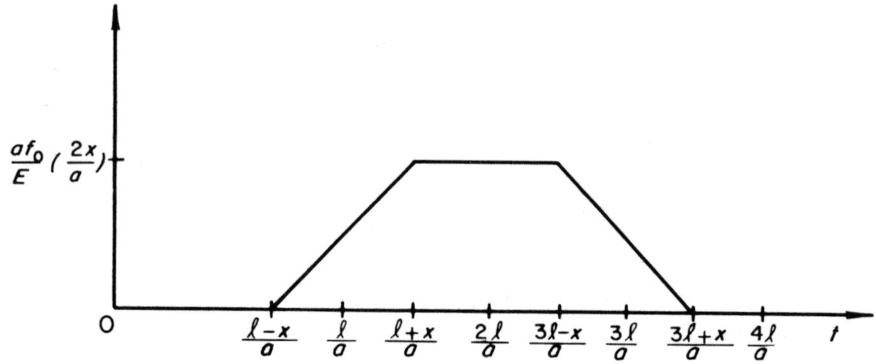

EXAMPLE 11.10.4. Consider a semiinfinite string fixed at the end $x=0$. The string is initially at rest. Let there be an external force

$$f(x,t)=-f_0\delta\left(t-\frac{x}{v}\right)$$

acting on the string. This is a concentrated force f_0 acting at the point $x=vt$.

The motion of the string is governed by

$$u_{tt}=c^2u_{xx}-f_0\delta\left(t-\frac{x}{v}\right)$$

$$u(x,0)=0$$

$$u_t(x,0)=0$$

$$u(0,t)=0$$

$$u(x,t) \text{ is bounded as } x\to\infty$$

Let $U(x,s)$ be the Laplace transform of $u(x,t)$. Transforming the wave equation and using the initial conditions, we obtain

$$U_{xx}-\frac{s^2}{c^2}U=\frac{f_0}{c^2}e^{-xs/v}$$

The solution of this equation is

$$U(x,s)=Ae^{sx/c}+Be^{-sx/c}+\begin{cases}\dfrac{f_0v^2e^{-sx/v}}{(c^2-v^2)s^2} & \text{for } v\neq c\\[2ex] -\dfrac{f_0xe^{-sx/v}}{2cs} & \text{for } v=c\end{cases}$$

The condition that $u(x,t)$ must be bounded at infinity requires that $A=0$.

Application of the condition

$$U(0,s)=0$$

yields

$$B=\begin{cases} \dfrac{-f_0 v^2}{(c^2-v^2)s^2} & \text{for} \quad v\neq c \\ 0 & \text{for} \quad v=0 \end{cases}$$

Hence the transform is given by

$$U(x,s)=\begin{cases} \dfrac{f_0 v^2(e^{-xs/v}-e^{-xs/c})}{(c^2-v^2)s^2} & \text{for} \quad v\neq c \\ -\dfrac{f_0 x e^{-xs/c}}{2cs} & \text{for} \quad v=c \end{cases}$$

The inverse transform is therefore

$$u(x,t)=\begin{cases} \dfrac{f_0 v^2}{(c^2-v^2)}\left[\left(t-\dfrac{x}{v}\right)u\left(t-\dfrac{x}{v}\right)-\left(t-\dfrac{x}{c}\right)u\left(t-\dfrac{x}{c}\right)\right], & v\neq c \\ \dfrac{-f_0 x}{2c}u\left(t-\dfrac{x}{c}\right) & \text{for} \quad v=c \end{cases}$$

In the illustrations of the application of the Laplace transform, only problems that yielded simple transforms were considered. The transforms could be separated into partial fractions, and the inverses were determined by referring to the tables. In general, problems require the use of the Laplace inversion formula. If $F(s)$ is the Laplace transform of a real function $f(t)$, with the transform variable s complex, the inversion integral is given by

$$f(t)=\frac{1}{2\pi i}\int_{\gamma-i\infty}^{\gamma+i\infty}F(s)e^{st}\,ds$$

For readers who are interested in extended application of the Laplace transform, some useful references are listed at the end of this book.

11.11 Application of Green's Function

One of the interesting methods of solving nonhomogeneous equations involves the application of the transform technique in the method of Green's function. We shall illustrate the method by the following example.

EXAMPLE 11.11.1. Consider the diffusion problem in a semiinfinite rod generating heat

$$u_t-u_{xx}=h(x,t) \qquad x>0, \quad t>0 \tag{11.11.1}$$

$$u(x,0)=f(x) \qquad x\geqslant 0 \tag{11.11.2}$$

$$u(0,t)=p(t) \qquad t\geqslant 0 \tag{11.11.3}$$

In this case, we see from Ex. 1, Chap. 10, that

$$Mv = -v_t - v_{xx}$$

so that, again by the superposition technique used in the preceding chapter, F must satisfy

$$MF = -F_\tau - F_{\xi\xi} = \delta(\xi - x, \tau - t)$$

in $\tau > 0, |\xi| < \infty$. (We are here taking ξ to be the field point.)

Note that F is not symmetric about the singular point (x,t), and consequently, F does not depend only on the radial distance as in the case of the Laplace operator.

To solve this problem by the Green's function method, let us consider the equation

$$LF = F_\tau - F_{\xi\xi} = \delta(\xi - x, \tau - t)$$

with the condition

$$F(\xi, \tau; x, t) = 0 \qquad \text{for} \qquad \tau < t$$

We may interpret this physically as the heat distribution due to a unit heat source at (x,t). It is then evident that the effect of the heat source does not exist for $\tau < t$. Hence F must represent heat diffusion for $\tau > t$.

Now our problem

$$MF = -F_\tau - F_{\xi\xi} = \delta(\xi - x, \tau - t) \tag{11.11.4}$$

$$F(\xi, \tau; x, t) = 0 \qquad \text{for} \qquad \tau > t \tag{11.11.5}$$

is the same as the above problem except that the time is reversed.

We will first find the free-space Green's function. Applying the Fourier transformation to Eq. (11.11.4), we obtain

$$-F_\tau^* + \alpha^2 F^* = \delta(\tau - t)\frac{e^{i\alpha x}}{\sqrt{2\pi}}$$

with α as the transform variable. The solution of this equation, in view of the condition (11.11.5), is

$$F^*(\alpha, \tau; x, t) = \frac{H(t-\tau)}{\sqrt{2\pi}} e^{i\alpha x - \alpha^2(t-\tau)}$$

where H is the Heaviside step function defined by

$$H(t-\tau) = \begin{cases} 1 & \text{for} \quad t > \tau \\ 0 & \text{for} \quad t < \tau \end{cases} \tag{11.11.6}$$

The inverse transformation then yields

$$\begin{aligned} F(\xi, \tau; x, t) &= \frac{H(t-\tau)}{2\pi} \int_{-\infty}^{\infty} e^{i\alpha(x-\xi) - \alpha^2(t-\tau)} \, d\alpha \\ &= \frac{H(t-\tau)}{\sqrt{4\pi(t-\tau)}} e^{-(x-\xi)^2/4(t-\tau)} \end{aligned} \tag{11.11.7}$$

To determine the solution $u(x,t)$, we put $v=G$ in problem 1 of the Exercises in Chapter 10.

$$\int\int_R [GLu - uMG]\,dR = \int_{\delta R} [(uG_\xi - u_\xi G)\cos(n,\xi) + (uG)\cos(n,\tau)]\,ds$$

Writing $Lu = u_\tau - u_{\xi\xi} = h$ and $MG = -G_\tau - G_{\xi\xi} = \delta(\xi - x, \tau - t)$ for the semi-infinite region, the above equation becomes

$$u(x,t) = \int_0^t \int_0^\infty Gh\,d\xi\,d\tau + \int_0^\infty f(\xi)G(\xi,0;x,t)\,d\xi$$
$$+ \int_0^t p(\tau)G_\xi(0,\tau;x,t)\,d\tau$$

where we have used the fact that $G=0$ and hence $G_\xi=0$ for $\tau>t$.

We assume G to be of the form

$$G(\xi,\tau;x,t) = F(\xi,\tau;x,t) + g(\xi,\tau;x,t)$$

where g can be obtained simply by placing a negative unit heat source at $(-x,t)$. Thus, we have

$$g = -F(\xi,\tau;-x,t)$$

Since $(-x,t)$ is outside the domain, it is quite obvious from Eq. (11.11.7) that

$$G(\xi,\tau;x,t) = F(\xi,\tau;x,t) - F(\xi,\tau;-x,t)$$

satisfies

$$MG = \delta(\xi - x, \tau - t)$$
$$G = 0 \qquad \text{on} \qquad \xi = 0$$

As a final remark, we should mention that the transform technique used in the method of Green's function needs careful attention. We must be well aware of the conditions imposed in the integral theorems.

At times it is quite convenient to apply two different transforms consecutively on a problem.

For a wealth of problems treated by transform methods, one should consult Sneddon [37]. A clear, concise and systematic treatment of Green's functions can be found in Greenberg [16].

Exercises

1. Find the Fourier transform of $f(x) = e^{-ax^2}$ where a is a constant.

2. Find the Fourier transform of

$$f(x) = \begin{cases} 1 & |x| < a \\ 0 & |x| \geqslant a \end{cases} \qquad a \text{ is a positive constant}$$

3. Find the Fourier transform of

$$f(x)=\frac{1}{|x|}$$

4. Find the Fourier transform of

a. $f(x)=\sin(x^2)$

b. $f(x)=\cos(x^2)$

5. Show that

$$I=\int_0^{\infty} e^{-a^2x^2}\,dx=\sqrt{\pi}\,/2a, \qquad a>0$$

by noting

$$I^2=\int_0^{\infty}\int_0^{\infty} e^{-a^2(x^2+y^2)}\,dx\,dy=\int_0^{\pi/2}\int_0^{\infty} e^{-a^2r^2}r\,dr\,d\theta$$

6. Show that

$$\int_0^{\infty} e^{-a^2x^2}\cos bx\,dx=(\sqrt{\pi}\,/2a)e^{-b^2/4a^2}, \qquad a>0$$

[Hint: Let $I(a,b)=\int_0^{\infty}e^{-a^2x^2}\cos bx\,dx$

$$\therefore \frac{dI}{db}=-\frac{b}{2a^2}I$$

$$\therefore I=Ce^{-b^2/4a^2}$$

Since

$$I(a,0)=C=\int_0^{\infty} e^{-a^2x^2}\,dx=\frac{\sqrt{\pi}}{2a}$$

$$I(a,b)=\frac{\sqrt{\pi}}{2a}e^{-b^2/4a^2}$$

7. Show that

$$\mathfrak{F}[f(at-b)]=\frac{1}{|a|}e^{i\alpha b/a}F(\alpha/a)$$

8. Prove that

$$\mathfrak{F}[f^{(n)}(t)]=(-i\alpha)^n\mathfrak{F}[f(t)]$$

9. Show that

a. $f*0=0*f=0$

b. $f*1\neq f$

10. Show that

$$\frac{1}{\sqrt{2\pi}}\int_{-\infty}^{\infty} e^{-\alpha^2 t - i\alpha x}\,d\alpha = \frac{1}{\sqrt{2t}}\,e^{-x^2/4t}$$

11. Determine the solution of

$$\begin{aligned} u_{tt} &= c^2 u_{xx} \qquad -\infty < x < \infty \qquad t>0 \\ u(x,0) &= f(x) \\ u_t(x,0) &= g(x) \end{aligned}$$

12. Solve

$$\begin{aligned} u_t &= u_{xx} \qquad x>0 \qquad t>0 \\ u(x,0) &= f(x) \\ u(0,t) &= 0 \end{aligned}$$

13. Solve

$$\begin{aligned} u_t &= u_{xx} \qquad x>0 \quad t>0 \\ u(x,0) &= f(x) \quad x>0 \\ u_x(0,t) - hu(0,t) &= 0 \qquad t>0, \quad h>0 \end{aligned}$$

14. Solve

$$\begin{aligned} u_{tt} + c^2 u_{xxxx} &= 0 \qquad -\infty < x < \infty \qquad t>0 \\ u(x,0) &= f(x) \\ u_t(x,0) &= 0 \end{aligned}$$

15. Solve

$$\begin{aligned} u_{tt} + c^2 u_{xxxx} &= 0 \qquad x>0 \qquad t>0 \\ u(x,0) &= 0 \\ u_t(x,0) &= 0 \\ u(0,t) &= g(t) \qquad t>0 \\ u_{xx}(0,t) &= 0 \end{aligned}$$

16. Solve

$$\phi_{xx} + \phi_{yy} = 0 \qquad -a<x<a \qquad 0<y<\infty$$

$$\phi_y(x,0) = \begin{cases} \delta_0 & 0<|x|<a \\ 0 & |x|>a \end{cases}$$

$$\phi(x,y) \to 0 \text{ uniformly in } x \text{ as } y\to\infty$$

17. Solve

$$\begin{aligned} u_t &= u_{xx} + tu \qquad -\infty < x < \infty \qquad t > 0 \\ u(x,0) &= f(x) \\ &u(x,t) \text{ is bounded} \end{aligned}$$

18. Solve

$$\begin{aligned} &u_t - u_{xx} + hu = \delta(x)\delta(t) \qquad -\infty < x < \infty \qquad t > 0 \\ &u(x,t) \to 0 \text{ uniformly in } t \text{ as } |x| \to \infty \\ &u(x,0) = 0 \end{aligned}$$

19. Solve

$$\begin{aligned} &u_t - u_{xx} + h(t)u_x = \delta(x)\delta(t) \qquad 0 < x < \infty \qquad t > 0 \\ &u(x,0) = 0 \\ &u_x(0,t) = 0 \\ &u(x,t) \to 0 \text{ uniformly in } t \text{ as } x \to \infty \end{aligned}$$

20. Solve

$$\begin{aligned} u_{xx} + u_{yy} &= 0 \qquad 0 < x < \infty \quad 0 < y < \infty \\ u(x,0) &= f(x) \quad 0 \leqslant x < \infty \\ u_x(0,y) &= g(y) \quad 0 \leqslant y < \infty \end{aligned}$$

$u(x,y) \to 0$ uniformly in x as $y \to \infty$ and uniformly in y as $x \to \infty$

21. Solve

$$\begin{aligned} u_{xx} + u_{yy} &= 0 \qquad -\infty < x < \infty \quad 0 < y < a \\ u(x,0) &= f(x) \quad -\infty < x < \infty \\ u(x,a) &= 0 \end{aligned}$$

$$u(x,y) \to 0 \quad \text{uniformly in } y \text{ as } |x| \to \infty$$

22. Solve

$$\begin{aligned} u_t &= u_{xx} + u_{yy} \qquad 0 < x < \infty \qquad 0 < y < l \qquad t > 0 \\ u(x,y,0) &= 0 \\ u(0,y,t) &= 0 \\ u(x,0,t) &= 0 \\ u(x,l,t) &= t \sin x, \qquad 0 \leqslant x < \infty, \qquad t \geqslant 0 \end{aligned}$$

23. Solve

$$\begin{aligned} u_t &= u_{xx} \qquad x > 0 \qquad t > 0 \\ u(x,0) &= 0 \\ u(0,t) &= f(t) \\ &u(x,t) \text{ bounded} \end{aligned}$$

24. Solve

$$\begin{aligned} &u_{xx}+u_{yy}=0 \qquad x>0 \qquad 0<y<1 \\ &u(x,0)=f(x) \\ &u(x,1)=0 \\ &u(0,y)=0 \\ &u(x,y)\to 0 \text{ uniformly in } y \text{ as } x\to\infty \end{aligned}$$

25. Solve

$$\begin{aligned} &u_t=u_{xx} \qquad -\infty<x<\infty \qquad t>0 \\ &u(x,0)=e^{-x^2} \\ &u(x,t) \text{ is bounded} \end{aligned}$$

Prove that the problem is well posed.

26. Find the Laplace transform of each of the following functions:

a. t^n b. $\cos\omega t$
c. $\sinh kt$ d. $\cosh kt$
e. te^{at} f. $e^{at}\sin\omega t$
g. $e^{at}\cos\omega t$ h. $t\sinh kt$
i. $t\cosh kt$ j. $\sqrt{\dfrac{1}{t}}$
k. $\sqrt{t}$ l. $\dfrac{\sin at}{t}$

27. Find the inverse transform of each of the following functions:

a. $\dfrac{s}{(s^2+1)(s^2+2)}$ b. $\dfrac{1}{(s^2+1)(s^2+2)}$
c. $\dfrac{1}{(s-1)(s-2)}$ d. $\dfrac{1}{s(s+1)^2}$
e. $\dfrac{1}{s(s+1)}$ f. $\dfrac{s-4}{(s^2+4)^2}$

28. Find the Laplace transform of each of the following functions:

a. $f(t)=\begin{cases} t & 0<t<b \\ 2b-t & b<t<2b \end{cases} \qquad f(t+2b)=f(t)$

b. $f(t)=\begin{cases} h & 0<t<b \\ 0 & b<t<2b \end{cases} \qquad f(t+2b)=f(t)$

c. $f(t)=\begin{cases} \sin t & 0<t<\pi \\ 0 & \pi<t<2\pi \end{cases} \qquad f(t+2\pi)=f(t)$

d. $f(t)=\dfrac{ht}{b} \qquad 0<t<b \qquad f(t+b)=f(t)$

e. $f(t)=hnb \qquad nb<t<(n+1)b \quad n=0,1,2,3,\ldots$

29. Prove the following properties of convolution:

a. $f*g=g*f$
b. $f*(g*h)=(f*g)*h$
c. $f*(g+h)=f*g+f*h$
d. $f*0=0*f$

30. Obtain the solution of the problem

$$\begin{aligned} &u_{tt}=c^2u_{xx} \qquad 0<x<\infty \qquad t>0\\ &u(x,0)=f(x)\\ &u_t(x,0)=0\\ &u(0,t)=0\\ &u(x,t)\to 0 \text{ uniformly in } t \text{ as } x\to\infty \end{aligned}$$

31. Solve

$$\begin{aligned} &u_{tt}=c^2u_{xx} \qquad 0<x<l \qquad t>0\\ &u(x,0)=0\\ &u_t(x,0)=0\\ &u(0,t)=0\\ &u_x(l,t)=\sin^2\omega t, \qquad t\geqslant 0 \end{aligned}$$

32. Solve

$$\begin{aligned} &u_{tt}=c^2u_{xx} \qquad 0<x<\infty \qquad t>0\\ &u(x,0)=0\\ &u_t(x,0)=0\\ &u(0,t)=\sin^2 t, \qquad t\geqslant 0\\ &u(x,t)\to 0 \text{ uniformly in } t \text{ as } x\to\infty \end{aligned}$$

33. Solve

$$\begin{aligned} &u_{tt}=c^2u_{xx} \qquad 0<x<1 \qquad t>0\\ &u(x,0)=0\\ &u_t(x,0)=0\\ &u(0,t)=f(t) \qquad t\geqslant 0\\ &u(1,t)=0 \end{aligned}$$

34. Solve

$$\begin{aligned} &u_t=ku_{xx} \qquad 0<x<\infty \qquad t>0\\ &u(x,0)=f_0 \qquad 0<x<\infty\\ &u(0,t)=f_1 \qquad t>0\\ &u(x,t)\to f_0 \text{ uniformly in } t \text{ as } x\to\infty \end{aligned}$$

35. Solve

$$\begin{aligned} &u_t = ku_{xx} && 0<x<\infty \quad t>0 \\ &u(x,0)=x && x>0 \\ &u(0,t)=0 && \\ &u(x,t)\to x \text{ uniformly in } t \text{ as } x\to\infty \end{aligned}$$

36. Solve

$$\begin{aligned} &u_t = ku_{xx} && 0<x<\infty \quad t>0 \\ &u(x,0)=0 && \\ &u(0,t)=t^2 && t\geqslant 0 \\ &u(x,t)\to 0 \text{ uniformly in } t \text{ as } x\to\infty \end{aligned}$$

37. Solve

$$\begin{aligned} &u_t = ku_{xx} - hu, && 0<x<\infty, \quad t>0, \quad h=\text{constant} \\ &u(x,0)=f_0, && x>0 \\ &u(0,t)=0 && \\ &u_x(0,t)\to 0 \text{ uniformly in } t \text{ as } x\to 0 \end{aligned}$$

38. Solve

$$\begin{aligned} &u_{tt} = c^2u_{xx} + bu_{xxt} && 0<x<\infty \quad t>0 \\ &u(x,0)=\sin x, && 0\leqslant x<\infty \\ &u_t(x,0)=0 && \\ &u(0,t)=0 && \\ &u(x,t) \text{ is bounded as } x\to\infty \end{aligned}$$

39. Solve

$$\begin{aligned} &u_t = ku_{xx} && 0<x<\infty \quad t>0 \\ &u(x,0)=0 && \\ &u(0,t)=f_0 && t>0 \\ &u(x,t)\to 0 \text{ uniformly in } t \text{ as } x\to\infty \end{aligned}$$

40. Solve

$$\begin{aligned} &u_{tt} + au_t + bu = c^2u_{xx} && 0<x<\infty \quad t>0 \\ &u(x,0)=0 && \\ &u_t(x,0)=0 && \\ &u(0,t)=\sin^2\omega t && t\geqslant 0 \\ &u(x,t)\to 0 \text{ uniformly in } t \text{ as } x\to\infty \end{aligned}$$

41. Solve

$$u_{tt}=c^2u_{xx} \qquad 0<x<l \qquad t>0$$
$$u(x,0)=0$$
$$u_t(x,0)=0$$
$$u(0,t)=0$$
$$u_x(l,t)=hf(t) \qquad t>0, \qquad h=\text{constant}$$

42. Solve

$$u_{tt}=c^2u_{xx} \qquad 0<x<\infty \qquad t>0$$
$$u(x,0)=0$$
$$u_t(x,0)=f_0 \qquad t>0$$
$$u(0,t)=0$$
$$u_x(x,t)\to 0 \text{ uniformly in } t \text{ as } x\to\infty$$

43. Solve

$$u_{tt}=c^2u_{xx} \qquad 0<x<\infty \qquad t>0$$
$$u(x,0)=f(x)$$
$$u_t(x,0)=0$$
$$u(0,t)=0$$
$$u_x(x,t)\to 0 \text{ uniformly in } t \text{ as } x\to\infty$$

44. Solve

$$u_t=ku_{xx} \qquad 0<x<\infty \qquad t>0$$
$$u(x,0)=e^{-x} \qquad 0\leqslant x<\infty$$
$$u(0,t)=0 \qquad t\geqslant 0$$
$$u(x,t)\to 0 \text{ uniformly in } t \text{ as } x\to\infty$$

45. Find the free-space Green's function for

a. $MF=c^2F_{\xi\xi}-F_{\tau\tau}=\delta(\xi-x,\tau-t)$
for $\tau>0$, $|\xi|<\infty$.

b. $MF=F_{\xi\xi}+\dfrac{1}{\kappa}F_\tau-\dfrac{1}{c^2}F_{\tau\tau}=\delta(\xi-x,\tau-t)$
for $\tau>0$, $|\xi|<\infty$

Appendix

I Gamma Function

The Gamma function is defined by the improper integral

$$\Gamma(x)=\int_0^{\infty} t^{x-1}e^{-t}\,dt \tag{1}$$

for any $x>0$. This integral is continuous and converges for all $x>0$ (see Crowder and McCuskey [8]).

From the definition, it follows immediately that

$$\begin{aligned}\Gamma(x+1)&=\int_0^{\infty} t^{x}e^{-t}\,dt\\ &=\left[-t^{x}e^{-t}\right]_0^{\infty}+x\int_0^{\infty} t^{x-1}e^{-t}\,dt \qquad (2)\\ &=x\Gamma(x)\end{aligned}$$

since the first term on the right vanishes.

We note that

$$\Gamma(1)=\int_0^{\infty} e^{-t}\,dt=1 \tag{3}$$

and consequently, we obtain, using relation (2),

$$\begin{aligned}\Gamma(2)&=1\cdot\Gamma(1)=1\cdot1=1!\\ \Gamma(3)&=2\cdot\Gamma(2)=2\cdot1=2!\\ \Gamma(4)&=3\cdot\Gamma(3)=3\cdot2\cdot1=3!\end{aligned}$$

In general,

$$\Gamma(n+1)=n! \qquad \text{for } n=0,1,2,\ldots \tag{4}$$

From Eqs. (3) and (4), we obtain the value of 0!

$$\Gamma(1)=0!=1$$

If we write Eq. (2) in the form

$$\Gamma(x)=(1/x)\Gamma(x+1) \tag{6}$$

we obtain, by repeated application of (6),

$$\Gamma(x)=\frac{\Gamma(x+k)}{x(x+1)(x+2)\cdots(x+k-1)} \tag{7}$$

for $k=1,2,3,\ldots$. We see that $\Gamma(x)$ is infinite for all negative integers.

The values of $\Gamma(x)$ are tabulated for $1<x<2$.[21] From these values, one can find, for example,

$$\Gamma(3.5)=(2.5)(1.5)\Gamma(1.5)$$

and

$$\Gamma(-1.4)=\frac{\Gamma(1.6)}{(-1.4)(-.4)(.6)}$$

where $\Gamma(1.5)=0.88623$ and $\Gamma(1.6)=0.89352$.

II Table of Fourier Transforms

$f(x)$	$\mathcal{F}[f(x)]$
$c \quad a\leqslant x\leqslant b$	$\frac{ic}{\sqrt{2\pi}\,\alpha}(e^{i\alpha a}-e^{i\alpha b})$
0 otherwise	
$x^n \quad 0\leqslant x\leqslant a$	$\frac{1}{\sqrt{2\pi}}n!(-i\alpha)^{-(n+1)}-\frac{e^{i\alpha a}}{\sqrt{2\pi}}\sum_{k=0}^{n}\frac{n!}{k!}(-i\alpha)^{k-(n+1)}a^k$
0 otherwise	$n=1,2,3,\ldots$
$\sqrt{\frac{\pi}{2}}\,e^{-\lvert x\rvert}$	$1/(1+\alpha^2)$
e^{-ax^2}	$e^{-\alpha^2/4a}/\sqrt{2a}$
$1/\lvert x\rvert$	$1/\lvert\alpha\rvert$
$e^{-a\lvert x\rvert}/\lvert x\rvert^{1/2}$	$[(a^2+\alpha^2)^{1/2}+a]^{1/2}/(a^2+\alpha^2)^{1/2}$
	$\sqrt{\frac{\pi}{2}} \quad \lvert\alpha\rvert<b$
$(\sin bx)/x$	$0 \quad \lvert\alpha\rvert>b$
$\sinh ax/\sinh \pi x$	$\sqrt{\frac{1}{2\pi}}\,[\sin a/(\cos a+\cosh\alpha)], -\pi<a<\pi$
$\cosh ax/\cosh \pi x$	$\sqrt{\frac{2}{\pi}}\,[\cos\frac{a}{2}\cosh\frac{\alpha}{2}/(\cos a+\cosh\alpha)], -\pi<a<\pi$

[21]For the table of Gamma function, see H. B. Dwight, *Tables of Integrals and Other Mathematical Data*, *Macmillan*, 1955.

III Table of Laplace Transforms

$f(t)$	$\mathfrak{L}[f(t)]$	
c	$c/s,$	c is a constant
t^n	$n!/s^{n+1},$	$n=1,2,3,\ldots$
e^{at}	$1/(s-a)$	
te^{at}	$1/(s-a)^2$	
$\sin \omega t$	$\omega/(s^2+\omega^2)$	
$\cos \omega t$	$s/(s^2+\omega^2)$	
$\sinh \kappa t$	$\kappa/(s^2-\kappa^2)$	
$\cosh \kappa t$	$s/(s^2-\kappa^2)$	
$\sqrt{t}$	$\sqrt{\pi}\ /2\sqrt{s^3}$	
$1/\sqrt{t}$	$\sqrt{\pi}\ /\sqrt{s}$	
$e^{at}\sin \omega t$	$\omega/[(s-a)^2+\omega^2]$	
$e^{at}\cos \omega t$	$(s-a)/[(s-a)^2+\omega^2]$	
$\operatorname{erf}(a\sqrt{t}\ )$	$a/s\sqrt{s^2+a^2}$	
$\operatorname{erfc}(a/2\sqrt{t}\ )$	$e^{-a\sqrt{s}}\ /s$	
$t^{\kappa-1}$	$\Gamma(\kappa)/s^{\kappa}$	$k>0$
$J_0(at)$	$1/\sqrt{s^2+a^2}$	
$I_0(at)$	$1/\sqrt{s^2-a^2}$	
$(\sin\ \omega t)/t$	$\tan^{-1}(\omega/s)$	
$t\sinh \kappa t$	$2\kappa s/(s^2-\kappa^2)^2$	
$t\cosh \kappa t$	$(s^2+\kappa^2)/(s^2-\kappa^2)^2$	
$t\sin \omega t$	$2\omega s/(s^2+\omega^2)^2$	
$t\cos \omega t$	$(s^2-\omega^2)/(s^2+\omega^2)^2$	

Bibliography

Partial Differential Equations

1. H. Bateman, *Partial Differential Equations of Mathematical Physics*, Cambridge U.P., Cambridge (1959).
2. A. Broman, *Introduction to Partial Differential Equations: from Fourier Series to Boundary Value Problems*, Addison-Wesley, Reading, Massachusetts (1970).
3. P. Berg and J. McGregor, *Elementary Partial Differential Equations*, Holden-Day, New York (1966).
4. M. L. Boas, *Mathematical Methods in the Physical Sciences*, Wiley, New York (1966).
5. H. S. Carslaw and J. C. Jaeger, *Conduction of Heat in Solids*, Oxford U. P., Oxford (1959).
6. R. V. Churchill, *Fourier Series and Boundary Value Problems*, McGraw-Hill, New York (1963).
7. R. Courant and D. Hilbert, *Methods of Mathematical Physics, Volume* 2, Interscience, New York (1962).
8. H. K. Crowder and S. W. McCuskey, *Topics in Higher Analysis*, Macmillan, New York (1964).
9. A. F. Danese, *Advanced Calculus, Volume* 2, Allyn and Bacon (1965).
10. R. Dennemeyer, *Partial Differential Equations and Boundary Value Problems*, McGraw-Hill, New York (1968).
11. G. F. D. Duff, *Partial Differential Equations*, University of Toronto, Toronto (1956).
12. G. F. D. Duff and D. Naylor, *Differential Equations of Applied Mathematics*, Wiley, New York (1966).
13. B. Epstein, *Partial Differential Equations*, McGraw-Hill, New York (1962).
14. P. Frank and R. Von Mises, *Die Differential und Integralgleichungen der Mechanik und Physik*, Rosenberg (1943).
15. P. Garabedian, *Partial Differential Equations*, Wiley, New York (1964).

16. M. Greenberg, *Applications of Green's Functions in Science and Engineering*, Prentice-Hall, Englewood Cliffs, New Jersey (1971).
17. G. Grunberg, A new method of solution of certain boundary problems for equations of mathematical physics permitting of a separation of variables, *J. Phys.* **10** 301–320 (1946).
18. J. Hadamard, *Lectures on Cauchy's Problem in Linear Partial Differential Equations*, Dover, New York (1952).
19. G. Hellwig, *Partial Differential Equations*, Blaisdell (1964).
20. J. Irving and N. Mullineux, *Mathematics in Physics and Engineering*, Academic, New York (1959).
21. J. Jeffreys and B. Jeffreys, *Methods of Mathematical Physics*, Cambridge U. P., Cambridge (1956).
22. F. John, *Partial Differential Equations*, Lecture Notes, NYU (1953), unpublished.
23. N. S. Koshlyakov, M. M. Smirnov and E. B. Gliner, *Differential Equations of Mathematical Physics*, Holden-Day, New York (1964).
24. D. Kreider, R. Kuller, D. Ostberg, and F. Perkins, *An Introduction to Linear Analysis*, Addison-Wesley, Reading, Massachusetts (1966).
25. E. Kreyszig, *Advanced Engineering Mathematics*, Wiley, New York (1967).
26. P. D. Lax, *Partial Differential Equations*, Lecture Notes, NYU (1951), unpublished.
27. S. G. Mikhlin, *Linear Equations of Mathematical Physics*, Holt, Rinehart and Winston, New York (1967).
28. K. S. Miller, *Partial Differential Equations in Engineering Problems*, Prentice-Hall, Englewood Cliffs, New Jersey (1953).
29. P. M. Morse and H. Feshbach, *Methods of Theoretical Physics*, Volume 1 and 2, McGraw-Hill, New York (1953).
30. I. Petrovsky, *Lectures on Partial Differential Equations*, Interscience, New York (1954).
31. L. A. Pipes, *Applied Mathematics for Engineers and Physicists*, McGraw-Hill, New York (1958).
32. H. Sagan, *Boundary and Eigenvalue Problems in Mathematical Physics*, Wiley, New York (1961).
33. R. Seeley, *Introduction to Fourier Series and Integrals*, Benjamin, New York (1966).
34. V. I. Smirnov, *Integral Equations and Partial Differential Equations*, Addison-Wesley, Reading, Massachusetts (1964).
35. M. G. Smith, *Introduction to the Theory of Partial Differential Equations*, Van Nostrand, Princeton, New Jersey (1967).
36. I. N. Sneddon, *Elements of Partial Differential Equations*, McGraw-Hill, New York (1957).
37. I. N. Sneddon, *Fourier Transforms*, McGraw-Hill, New York (1951).
38. I. N. Sneddon, *Mixed Boundary Value Problems in Potential Theory*, North Holland, Amsterdam (1966).
39. S. L. Sobolve, *Partial Differential Equations of Mathematical Physics*, Addison-Wesley, Reading, Massachusetts (1964).
40. I. S. Sokolnikoff, *Mathematical Theory of Elasticity*, McGraw-Hill, New York (1956).

41. I. S. Sokolnikoff and R. M. Redheffer, *Mathematics of Physics and Modern Engineering*, McGraw-Hill, New York (1966).
42. A. Sommerfeld, *Partial Differential Equations in Physics*, Academic, New York (1964).
43. I. Stakgold, *Boundary Value Problems of Mathematical Physics*, Macmillan, New York (1967).
44. A. N. Tikhonov and A. A. Samarskii, *Equations of Mathematical Physics*, Macmillan, New York (1963).
45. A. N. Tychonov and A. A. Samarski, *Partial Differential Equations of Mathematical Physics*, Holden-Day, New York (1964).
46. A. G. Webster, *Partial Differential Equations of Mathematical Physics*, Dover, New York (1955).
47. H. Weinberger, *A First Course in Partial Differential Equations*, Blaisdell (1965).
48. E. T. Whittaker and G. Watson, *A Course on Modern Analysis*, Cambridge U. P., Cambridge (1952).

Fourier Series

49. W. E. Byerly, *Fourier Series*, Dover, New York (1959).
50. H. S. Carslaw, *Introduction to the Theory of Fourier Series and Integrals*, Dover, New York (1950).
51. H. F. Davis, *Fourier Series and Orthogonal Functions*, Allyn and Bacon (1963).
52. D. Jackson, *Fourier Series and Orthogonal Polynomials*, Mathematical Association of American Monograph (1941).
53. D. Kreider, R. Kuller, D. Ostberg, and F. Perkins, *An Introduction to Linear Analysis*, Addison-Wesley, Reading, Massachusetts (1966).
54. W. W. Rogosinski, *Fourier Series*, Chelsea (1950).
55. G. Sansone, *Orthogonal Functions*, Interscience, New York (1959).
56. E. C. Titchmarsh, *Introduction to the Theory of Fourier Integrals*, Oxford U. P., Oxford (1962).
57. G. P. Tolstov, *Fourier Series*, Prentice-Hall, Englewood Cliffs, New Jersey (1962).

Eigenvalue Problems and Green's Function

58. G. Birkhoff and R-C. Rota, *Ordinary Differential Equations*, Blaisdell (1962).
59. E. A. Coddington and N. Levinson, *Theory of Ordinary Differential Equations*, McGraw-Hill, New York (1955).
60. R. Cole, *Theory of Ordinary Differential Equations*, Appleton-Century-Crofts (1968).
61. L. Collatz, *Eigenwertproblem und ihre Numerische Behandlung*, Academische Verlags Gesellschaft (1945).
62. R. Courant and D. Hilbert, *Methods of Mathematical Physics, Volume* 1, Interscience, New York (1953).
63. F. Hildebrand, *Methods of Applied Mathematics*, Prentice-Hall, Englewood Cliffs, New Jersey (1965).
64. H. Keller, *Numerical Methods for Two Point Boundary Value Problems*, Blaisdell, Waltham, Massachusetts (1968).
65. J. Indritz, *Methods in Analysis*, Macmillan, New York (1963).

66. K. S. Miller, *Linear Differential Equations in the Real Domain*, Norton (1963).
67. T. Myint-U, *Ordinary Differential Equations*, Elsevier North Holland, Inc., New York (1978).
68. H. Sagan, *Boundary and Eigenvalue Problems in Mathematical Physics*, Wiley, New York (1966).
69. E. C. Titchmarsh, *Eigenfunction Expansions Associated with Second-Order Differential Equations*, Oxford U. P., Oxford (1946).

Integral Transforms

70. G. A. Campbell and R. M. Foster, *Fourier Integrals for Practical Applications*, Van Nostrand, New York (1948).
71. H. S. Carslaw and J. C. Jaeger, *Operational Methods in Applied Mathematics*, Oxford U. P., New York (1948).
72. R. V. Churchill, *Modern Operational Mathematics in Engineering*, McGraw-Hill, New York (1944).
73. A. Erdelyi, W. Magnus, F. Oberhettinger and F. Tricomi, *Table of Integral Transforms*, McGraw-Hill, New York (1954).
74. H. Jeffreys, *Operational Methods in Mathematical Physics*, Cambridge U. P., New York (1931).
75. N. W. McLachlan, *Complex Variable and Operational Calculus with Technical Applications*, Cambridge U. P., London (1942).
76. N. W. McLachlan, *Modern Operational Calculus*, Macmillan, London (1948).
77. R. H. Raven, *Mathematics of Engineering Systems*, McGraw-Hill, New York (1966).
78. E. J. Scott, *Transform Calculus with an Introduction to Complex Variables*, Harper, New York (1955).
79. I. N. Sneddon, *Fourier Transforms*, McGraw-Hill, New York (1951).
80. E. C. Titchmarsh, *Theory of Fourier Integrals*, Oxford U. P., New York (1937).
81. C. J. Tranter, *Integral Transforms in Mathematical Physics*, Wiley, New York (1951).
82. G. N. Watson, *Theory of Bessel Functions*, Cambridge U. P., Cambridge (1966).
83. D. V. Widder, *The Laplace Transform*, Princeton U. P., Princeton, New Jersey (1941).

Tables and Formulas

84. A. Erdelyi, W. Magnus, F. Oberhettinger, and F. G. Tricomi, *Higher Transcendental Functions*, Volumes I, II, and III, McGraw-Hill, New York (1953).
85. E. Jahnke, F. Emde, and F. Losch, *Tables of Higher Functions*, McGraw-Hill, New York (1960).
86. W. Magnus and F. Oberhettinger, *Formulas and Theorems for the Special Functions of Mathematical Physics*, Chelsea (1949).

Problem Books

87. B. M. Budak, A. A. Samarskii and A. N. Tikhonov, *A Collection of Problems on Mathematical Physics*, Macmillan, New York (1964).
88. N. N. Levedev, I. P. Skalskaya, and Y. S. Uflyand, *Problems of Mathematical Physics*, Prentice-Hall, Englewood Cliffs, New Jersey (1965).
89. M. M. Smirnov, *Problems on the Equations of Mathematical Physics*, Noordhoff (1967).

Answers to Exercises

Chapter 1

1. a. Linear, nonhomogeneous, second-order; b. quasilinear, first-order; c. nonlinear, first-order; d. linear, homogeneous, fourth-order; e. linear, nonhomogeneous, second-order; f. quasilinear, third-order; g. nonlinear, second-order.
5. $u(x,y)=f(x)\cos y+g(x)\sin y$
6. $u(x,y)=f(x)e^{-y}+g(y)$
7. $u(x,y)=f(x+y)+g(3x+y)$
8. $u(x,y)=f(y+x)+g(y-x)$

Chapter 3

1. a. $x<0$, hyperbolic;

$$u_{\xi\eta}=\frac{1}{4}\left(\frac{\xi-\eta}{4}\right)^4-\frac{1}{2}\frac{1}{\xi-\eta}(u_\xi-u_\eta)$$

$x=0$, parabolic, the given equation is then in canonical form; $x>0$, elliptic,

$$u_{\alpha\alpha}+u_{\beta\beta}=\frac{\beta^4}{16}+\frac{1}{\beta}u_\beta$$

b. $y=0$, parabolic; $y\neq 0$, elliptic,

$$u_{\alpha\alpha}+u_{\beta\beta}=u_\alpha+e^\alpha$$

d. Parabolic everywhere

$$u_{\eta\eta}=\frac{2\xi}{\eta^2}u_\xi+\frac{1}{\eta^2}e^{\xi/\eta}$$

f. Elliptic everywhere for finite values of x and y

$$u_{\alpha\alpha}+u_{\beta\beta}=u-\frac{1}{\alpha}u_{\alpha}-\frac{1}{\beta}u_{\beta}$$

g. Parabolic everywhere

$$u_{\eta\eta}=\frac{1}{1-e^{2(\eta-\xi)}}\left[\sin^{-1}e^{\eta-\xi}-u_{\xi}\right]$$

h. $y=0$, parabolic; $y\neq 0$, hyperbolic,

$$u_{\xi\eta}=\frac{1+\xi-\ln\eta}{\eta}u_{\xi}+u_{\eta}+\frac{1}{\eta}u$$

2. i. $u(x,y)=f(y/x)+g(y/x)e^{-y}$; ii. $u(r,t)=(1/r)f(r+ct)+(1/r)g(r-ct)$;
3. a. $\xi=(y-x)+i\sqrt{2}\,x, \eta=(y-x)-i\sqrt{2}\,x, \alpha=y-x, \beta=\sqrt{2}\,x, u_{\alpha\alpha}+u_{\beta\beta}=-\frac{1}{2}u_{\alpha}-2\sqrt{2}\,u_{\beta}-\frac{1}{2}u+\frac{1}{2}e^{\beta/\sqrt{2}}$.
   b. $\xi=y+x, \eta=y, u_{\eta\eta}=-\frac{3}{2}u$.
   c. $\xi=y-x, \eta=y-4x, u_{\xi\eta}=\frac{7}{9}(u_{\xi}+u_{\eta})-\frac{1}{9}\sin[(\xi-\eta)/3]$.
   d. $\xi=y+ix, \eta=y-ix, \alpha=y, \beta=x$; the given equation is already in canonical form.
   e. $\xi=x, \eta=x-y/2, u_{\xi\eta}=18u_{\xi}+17u_{\eta}-4$.
   f. $\xi=y+x/6, \eta=y, u_{\xi\eta}=6u-6\eta^2$.
   g. $\xi=x, \eta=y$; the given equation is already in canonical form.
   h. $\xi=x, \eta=y$; the given equation is already in canonical form.
4. i. $u(x,y)=f(x+cy)+g(x-cy)$; ii. $u(x,y)=f(x+iy)+g(x-iy)$; iii. $u(x,y)=(x-iy)f_1(x+iy)+f_2(x+iy)+(x+iy)f_3(x-iy)+f_4(x-iy)$; iv $u(x,y)=f(y+x)+g(y+2x)$; v. $u(x,y)=f(y)+g(y-x)$; vi. $u(x,y)=(-y/128)(y-x)(y-9x)+f(y-9x)+g(y-x)$.
5. i. $v_{\xi\eta}=-(1/16)v$; ii. $v_{\xi\eta}=(84/625)v$.

Chapter 4

1. a. $u(x,t)=t$; b. $u(x,t)=\sin x\cos ct+x^2t+\frac{1}{3}c^2t^3$; c. $u(x,t)=x^3+3c^2xt^2+xt$; d. $u(x,t)=\cos x\cos ct+t/e$; e. $u(x,t)=2t+\frac{1}{2}[\log(1+x^2+2cxt+c^2t^2)+\log(1+x^2-2cxt+c^2t^2)]$; f. $u(x,t)=x+(1/c)\sin x\sin ct$.
2. a. $u(x,t)=3t+\frac{1}{2}xt^2$; c. $u(x,t)=5+x^2t+\frac{1}{3}c^2t^3+(1/2c^2)(e^{x+ct}+e^{x-ct}-2e^x)$; e. $u(x,t)=\sin x\cos ct+(e^t-1)(xt+x)-xte^t$; f. $u(x,t)=x^2+t^2(1+c^2)+(1/c)\sin x\cos ct$.
3.

$$s(r,t)=\begin{cases}0 & 0\leqslant t<r-R\\ \dfrac{s_0(r-t)}{2r} & r-R<t<r+R\\ 0 & r+R<t<\infty\end{cases}$$

4. $u(x,y)=\frac{1}{4}\sin(y+x)+\frac{3}{4}\sin(-y/3+x)+y^2/3+xy$
5. $u(x,t)=\sin x\cos at+xt$, $a=$physical constant
6. b.

$$u(x,t)=\tfrac{1}{2}f(xt)+\tfrac{1}{2}tf\left(\frac{x}{t}\right)+\tfrac{1}{4}\sqrt{xt}\int_{xt}^{x/t}\frac{f(\tau)}{\tau^{3/2}}\,d\tau-\tfrac{1}{2}\sqrt{xt}\int_{xt}^{x/t}\frac{g(\tau)}{\tau^{3/2}}\,d\tau$$

19. $u(x,y)=f\left(\sqrt{\dfrac{y^2+x^2-8}{2}}\right)+g\left(\sqrt{\dfrac{x^2-y^2+16}{2}}\right)-f(2)$

22. $u(x,y)=f\left[\dfrac{y-\dfrac{x^2}{2}+4}{2}\right]+g\left(\dfrac{y+x^2/2}{2}\right)-f(2)$

Chapter 5

1. a. Piecewise continuous, $f(1-)=1, f(1+)=0$; c. piecewise continuous, $f(1-)=2, f(1+)=1$; d. $f(x)$ is not piecewise continuous.
3. a. Neither even nor odd; b. even; c. neither even nor odd; d. odd; e. even; f. odd; g. even; h. odd.
6. a. $2\pi/3$; b. 4π; c. 1; d. not periodic; e. 2π; f. not periodic.
12. a. $$f(x)=-\frac{\pi}{4}+\frac{h}{2}+\sum_{k=1}^{\infty}\frac{1}{\pi k^2}\left[1+(-1)^{k+1}\right]\cos kx+\frac{1}{\pi k}\left[h+(h+\pi)(-1)^{k+1}\right]\sin kx$$

 c. $$f(x)=\sin x+\sum_{k=1}^{\infty}\frac{2(-1)^{k+1}}{k}\sin kx$$

 e. $$f(x)=\frac{\sinh\pi}{\pi}\left[1+\sum_{k=1}^{\infty}\frac{2(-1)^k}{1+k^2}(\cos kx-k\sin kx)\right]$$

13. a. $$f(x)=\sum_{k=1}^{\infty}\frac{2}{k}\sin kx$$

 b. $$f(x)=\sum_{k=1}^{\infty}\frac{2}{\pi k}\left[1-2(-1)^k+\cos\frac{k\pi}{2}\right]\sin kx$$

 c. $$f(x)=\sum_{k=1}^{\infty}\left[2(-1)^{k+1}\frac{\pi}{k}+\frac{4}{\pi k^3}\left((-1)^k-1\right)\right]\sin kx$$

 d. $$f(x)=\sum_{k=2}^{\infty}\frac{2k}{\pi}\left[\frac{1+(-1)^k}{k^2-1}\right]\sin kx$$

14. a. $$f(x)=\frac{3}{2}\pi+\sum_{k=1}^{\infty}\frac{2}{\pi k^2}\left[(-1)^k-1\right]\cos kx$$

b. $f(x)=\frac{\pi}{2}+\sum_{k=1}^{\infty}\frac{2}{\pi k^2}[(-1)^k-1]\cos kx$

c. $f(x)=\frac{\pi^2}{3}+\sum_{k=1}^{\infty}\frac{4(-1)^k}{k^2}\cos kx$

d. $f(x)=\frac{2}{3\pi}+\sum_{k=1,2,4,\dots}^{\infty}\frac{6}{\pi}\left[\frac{1+(-1)^k}{9-k^2}\right]\cos kx, \qquad k\neq 3$

15. b. $f(x)=\sum_{k=1}^{\infty}\frac{2}{k\pi}\sin\frac{k\pi}{2}\cos\frac{k\pi x}{6}$

c. $f(x)=\frac{2}{\pi}+\sum_{k=2}^{\infty}\frac{2}{k\pi}\left[\frac{1+(-1)^k}{1-k^2}\right]\cos\frac{k\pi x}{l}$

f. $f(x)=\sum_{k=1}^{\infty}\frac{k\pi}{1+k^2\pi^2}(-1)^{k+1}(e-e^{-1})\sin k\pi x$

16. a. $f(x)=\sum_{k=-\infty}^{\infty}\frac{1}{\pi}\frac{2+ik}{4+k^2}(-1)^k\sinh 2\pi e^{ikx}$

b. $f(x)=\sum_{k=-\infty}^{\infty}\frac{(-1)^k}{\pi(1+k^2)}\sinh\pi e^{ikx}$

d. $f(x)=\sum_{k=-\infty}^{\infty}(-1)^k\frac{i}{k\pi}e^{ik\pi x}$

17. a. $\sin^2 x=\sum_{k=1,3,4,\dots}^{\infty}\frac{4(1-\cos k\pi)}{k\pi(4-k^2)}\sin kx$

b. $\cos^2 x=\sum_{k=1,3,4,\dots}^{\infty}\frac{2}{k\pi}\frac{1-k^2}{4-k^2}(1-\cos k\pi)\sin kx$

c. $\sin x\cos x=\sum_{k=1,3,4,\dots}^{\infty}\frac{2}{\pi}\frac{1-\cos k\pi}{4-k^2}\cos kx$

18. a. $\frac{x^2}{4}=\frac{\pi^2}{12}-\sum_{k=1}^{\infty}\frac{(-1)^{k+1}}{k^2}\cos kx$

c. $\int_0^\infty \ln\left(2\cos\frac{x}{2}\right)dx = \sum_{k=1}^{\infty}(-1)^{k+1}\frac{\sin kx}{k^2}$

e. $\frac{\pi}{2} - \frac{4}{\pi}\sum_{k=1}^{\infty}\frac{\cos(2k-1)x}{(2k-1)^2} = \begin{cases} -x, & -\pi<x<0 \\ x & 0<x<\pi \end{cases}$

19. a. $f(x) = \frac{\pi}{8} + \sum_{k=1}^{\infty}\frac{1}{2\pi k^2}\left[(-1)^k - 1\right]\cos kx + \frac{(-1)^{k+1}}{2k}\sin kx$

20. a. $f(x) = \frac{l^2}{3} + \sum_{k=1}^{\infty}4(-1)^k\left(\frac{1}{k\pi}\right)^2\cos\frac{k\pi x}{l}$

24. a. $f(x) = -\frac{x}{2} + \sin x + \sum_{n=2}^{\infty}\frac{(-1)^n}{n^5-n}\sin nx, \qquad -\pi<x<\pi$

32. a. $f(x,y) = \frac{16}{\pi^2}\sum_{m=1,3,\ldots}^{\infty}\ \sum_{n=1,3,\ldots}^{\infty}\frac{1}{mn}\sin mx\sin ny$

c. $f(x,y) = \frac{\pi^4}{9} + \frac{1}{2}\sum_{m=1}^{\infty}\frac{8}{3}\pi^2\frac{(-1)^m}{m^2}\cos mx$

$+\frac{1}{2}\sum_{n=1}^{\infty}\frac{8}{3}\pi^2\frac{(-1)^n}{n^2}\cos ny$

$+\sum_{m=1}^{\infty}\sum_{n=1}^{\infty}\frac{16(-1)^{m+n}}{m^2n^2}\cos mx\cos ny$

e. $f(x,y) = \sum_{m=1}^{\infty}\frac{2(-1)^{m+1}}{m}\sin mx\sin y$

Chapter 6

1. a. $u(x,t) = \sum_{n=1}^{\infty}\frac{4}{(n\pi)^3}\left[1-(-1)^n\right]\cos n\pi ct\sin n\pi x$

b. $u(x,t) = 3\cos ct\sin x$

2. a. $u(x,t) = \sum_{u=1,3,4,\ldots}^{\infty}\frac{32\left[(-1)^n - 1\right]}{\pi cn^2(n^2-4)}\sin nct\sin nx$

5. $u(x,t)=\sum_{n=1}^{\infty} a_n T_n(t)\sin\frac{n\pi x}{l}$
where
$a_n=\frac{2}{l}\int_0^l f(x)\sin\frac{n\pi x}{l}\,dx$
and

$$T_n(t)=\begin{cases} e^{-at/2}\left(\cosh\alpha t+\frac{a}{2\alpha}\sinh\alpha t\right) & \text{for } \alpha>0\\ e^{-at/2}\left(1+\frac{at}{2}\right) & \text{for } \alpha=0\\ e^{-at/2}\left(\cos\beta t+\frac{a}{2\beta}\sin\beta t\right) & \text{for } \alpha<0\end{cases}$$

in which

$$\alpha=\frac{1}{2}\left[a^2-4\left(b+\frac{n^2\pi^2c^2}{l^2}\right)\right]^{1/2}$$

$$\beta=\frac{1}{2}\left[4\left(b+\frac{n^2\pi^2c^2}{l^2}\right)-a^2\right]^{1/2}$$

6. $u(x,t)=\sum_{n=1}^{\infty} a_n T_n(t)\sin\frac{n\pi x}{l}$
where
$a_n=\frac{2}{l}\int_0^l g(x)\sin\frac{n\pi x}{l}\,dx$
and

$$T_n(t)=\begin{cases} \frac{2e^{-at/2}}{\sqrt{(a^2-\alpha)}}\sinh\frac{\sqrt{(a^2-\alpha)}}{2}t & \text{for } a^2>\alpha\\ te^{-at/2} & \text{for } a^2=\alpha\\ \frac{2e^{-at/2}}{\sqrt{(\alpha-a^2)}}\sin\frac{\sqrt{(\alpha-a^2)}}{2}t & \text{for } a^2<\alpha\end{cases}$$

7. $\theta(x,t)=\sum_{n=1}^{\infty} a_n\cos a\alpha_n t\sin(\alpha_n x+\phi_n)$ where

$$a_n=\frac{2(\alpha_n^2+h^2)}{2h+(\alpha_n^2+h^2)l}\int_0^l f(x)\sin(\alpha_n x+\phi_n)\,dx$$

and

$\phi_n=\arctan\frac{\alpha_n}{h}$; α_n are the roots of the equation

$$\tan\alpha l=\frac{2h\alpha}{\alpha^2-h^2}$$

11. $u(x,t)=v(x,t)+U(x)$ where

$$v(x,t)=\sum_{n=1}^{\infty}\left[\frac{-2}{l}\int_0^l U(\tau)\sin\frac{n\pi\tau}{l}\,d\tau\right]\cos\frac{n\pi ct}{l}\sin\frac{n\pi x}{l}$$

and

$$U(x)=-\frac{A}{c^2}\sinh x+\left(\frac{A}{c^2}\sinh l+k-h\right)\frac{x}{l}+h$$

12. $u(x,t)=\dfrac{A}{6c^2}x^2(1-x)+\displaystyle\sum_{n=1}^{\infty}\frac{12}{(n\pi)^3}(-1)^n\cos n\pi ct\sin n\pi x$

14. $$u(x,t)=-\frac{hx^2}{2k}+\left(2u_0+\frac{h}{2k}\right)x-\frac{4h}{k\pi}e^{-k\pi^2t}\sin\pi x$$
$$+\sum_{n=2}^{\infty}a_ne^{-kn^2\pi^2t}\sin n\pi x$$

where

$$a_n=\frac{2u_0}{n\pi}[1+(-1)^n]+\frac{2u_0n}{(n^2-1)\pi}[1-(-1)^n]+\frac{2h}{k\pi^3n^3}[(-1)^n-1]$$

15. a. $u(x,t)=\displaystyle\sum_{n=1}^{\infty}\frac{2}{n^3\pi^3}[2(-1)^{n+1}-1]e^{-4n^2\pi^2t}\sin n\pi x$

b. $u(x,t)=\displaystyle\sum_{n=1,3,4,\ldots}^{\infty}[(-1)^n-1]\left[\frac{n}{\pi(4-n^2)}-\frac{1}{n\pi}\right]e^{-n^2kt}\sin nx$

16. $u(x,t)=\displaystyle\sum_{n=1}^{\infty}\frac{2l^2}{n^2\pi^2}[1-(-1)^n]e^{-k\left(\frac{n\pi}{l}\right)^2t}\cos\frac{n\pi x}{l}$

18. $$v(x,t)=Ct\left(1-\frac{x}{l}\right)-\frac{Cl^2}{6k}\left[\left(\frac{x}{l}\right)^3-3\left(\frac{x}{l}\right)^2+2\left(\frac{x}{l}\right)\right]$$
$$+\frac{2Cl^2}{\pi^3k}\sum_{k=1}^{\infty}\frac{e^{-n^2\pi^2kt/l^2}}{n^3}\sin\frac{n\pi x}{l}$$

21. $u(x,t)=v(x,t)+w(x)$ where

$$v(x,t)=e^{-kt}\sin x+\sum_{u=1}^{\infty}a_ue^{-u^2kt}\sin ux$$

where

$$a_u = \frac{-u}{(u^2+a^2)}\left[(-1)_e^{u-a\pi}-1\right] + \frac{2A}{a^2k\pi}\left[\frac{1}{u}\{(-1)^u-1\}\right.$$
$$\left. + \frac{(-1)^u}{u}\left[e^{-a\pi}-1\right]\right]$$

$$w(x) = \frac{A}{a^2k}\left[1-e^{-ax}+\frac{x}{\pi}(e^{-a\pi}-1)\right]$$

26. $u(x,t) = \sum_{n=1}^{\infty} \int_0^t e^{-n^2(t-\tau)} a_n(\tau)\,d\tau \sin nx + \sum_{n=1}^{\infty} b_n(0) e^{-n^2 t} \sin nx$

where

$$a_n(t) = \frac{2}{\pi}\int_0^{\pi} g(x,t)\sin nx\,dx$$

$$b_n(0) = \frac{2}{\pi}\int_0^{\pi} f(x)\sin nx\,dx$$

28. $$u(x,t) = \sum_{n=1}^{\infty} \frac{2}{\pi}\sin\left(n-\frac{1}{2}\right)x \int_0^t \int_0^{\pi} e^{-(2n-1)^2(t-\tau)/4}$$
$$\cdot \sin\left(n-\frac{1}{2}\right)\xi g(\xi,\tau)\,d\xi\,d\tau$$

32. $$u(x,t) = \frac{c}{2}\left\{\frac{\sinh\sqrt{1/c}\;x}{\sinh\sqrt{1/c}\;\pi} - \frac{\sin\sqrt{1/c}\;x}{\sin\sqrt{1/c}\;\pi}\right\}\sin t$$
$$+ \frac{2}{\pi c}\sum_{n=1}^{\infty}(-1)^{n+1}\frac{n}{n^4-(1/c)^2}\sin n^2ct\sin nx$$

in which $\sqrt{1/c}$ is not an integer.

Chapter 7

1. a. $\lambda_n = n^2, \phi_n(x) = \sin nx$ *for* $n = 1,2,3,\ldots$
   b. $\lambda_n = ((2n-1)/2)^2, \phi_n(x) = \sin((2n-1)/2)\pi x$ for $n = 1,2,3,\ldots$
   c. $\lambda_n = n^2, \phi_n(x) = \cos nx$ for $n = 1,2,3,\ldots$
2. a. $\lambda_n = 0, n^2\pi^2, \phi_n(x) = 1, \sin n\pi x, \cos n\pi x$ for $n = 1,2,3,\ldots$
   b. $\lambda_n = 0, n^2, \phi_n(x) = 1, \sin nx, \cos nx$ for $n = 1,2,3,\ldots$
   c. $\lambda_n = 0, 4n^2, \phi_n(x) = 1, \sin 2nx, \cos 2nx$ for $n = 1,2,3,\ldots$
3. a. $\lambda_n = -(3/4 + n^2\pi^2), \phi_n(x) = e^{-x/2}\sin n\pi x, n = 1,2,3,\ldots$
5. a. $\lambda_n = 1 + n^2\pi^2, \phi_n(x) = (1/x)\sin(n\pi \ln x), n = 1,2,3,\ldots$
   b. $\lambda_n = \frac{1}{4} + (n\pi/\ln 3)^2, \phi_n(x) = [1/(x+2)^{1/2}]\sin[(n\pi/\ln 3)\ln(x+2)], n = 1,2,3,\ldots$

c. $\lambda_n = \frac{1}{12}[1+(2n\pi/\ln 2)^2], \phi_n(x) = [1/(1+x)^{1/2}]\sin[(n\pi/\ln 2)\ln(1+x)], n = 1, 2, 3, \ldots$

6. a. $\phi(x) = \sin(\sqrt{\lambda}\ln x), \lambda > 0$

b. $\phi(x) = \sin\sqrt{\lambda}\, x, \lambda > 0$

8. $$f(x) \sim \sum_{n=1}^{\infty} \frac{2}{\pi}\left[\frac{(-1)^n - 1}{n^2}\right]\cos nx$$

13. a. $$G(x,\xi) = \begin{cases} x, & x \leqslant \xi \\ \xi, & x > \xi \end{cases}$$

14. a. $$u(x) = -\cos x + \left(\frac{\cos 1 - 1}{\sin 1}\right)\sin x + 1$$

b. $$u(x) = -\frac{2}{5}\cos 2x - \frac{1}{10}\left(\frac{1+2\sin 2}{\cos 2}\right)\sin 2x + \frac{1}{5}e^x$$

18. $$G(x,\xi) = \begin{cases} x^3\xi/2 + x\xi^3/2 - 9x\xi/5 + x & \text{for } 0 \leqslant x < \xi \\ x^3\xi/2 + x\xi^3/2 - 9x\xi/5 + \xi & \text{for } \xi \leqslant x \leqslant 1 \end{cases}$$

Chapter 8

8. a. $$u(r,\theta) = \frac{4}{3}\left(\frac{1}{r} - \frac{r}{4}\right)\sin\theta$$

c. $$u(r,\theta) = \sum_{n=1}^{\infty} a_n \sinh\left[(n\pi/\ln 3)\left(\theta - \frac{\pi}{2}\right)\right]\sin[(n\pi/\ln 3)\ln r]$$

where

$$a_n = \frac{2}{\ln 3\sinh(n\pi^2/2\ln 3)}\left\{\frac{n\pi\ln 3}{n^2\pi^2 + 4(\ln 3)^2}[9(-1)^n - 1] - \frac{4n\pi\ln 3}{n^2\pi^2 + (\ln 3)^2}[3(-1)^n - 1] + \frac{3\ln 3}{n\pi}[(-1)^n - 1]\right\}$$

9. $$u(r,\theta) = \sum_{n=1}^{\infty} a_n(r^{-n\pi/\alpha} - b^{-2n\pi/\alpha}r^{n\pi/\alpha})\sin\frac{n\pi\theta}{\alpha} + \sum_{n=1}^{\infty} b_n \sinh\left[\frac{n\pi}{\ln(b/a)}(\theta - \alpha)\right]\sin\left[\frac{n\pi}{\ln(b/a)}(\ln r - \ln a)\right]$$

where

$$a_n=\frac{2}{\alpha(a^{-n\pi/\alpha}-b^{-2n\pi/\alpha}a^{n\pi/\alpha})}\int_0^\alpha f(\theta)\sin\frac{n\pi\theta}{\alpha}\,d\theta$$

$$b_n=\frac{-2}{\ln(b/a)\sinh[\alpha n\pi/\ln(b/a)]}$$

$$\cdot\int_a^b f(r)\sin[(n\pi/\ln[b/a])(\ln r+\ln a)]\frac{dr}{r}$$

12. $u(r,\theta)=\sum_{n=1}^{\infty}\frac{2}{\alpha J_v(a)}\left[\int_0^\alpha f(\theta)\sin\left(\frac{n\pi\tau}{\alpha}\right)d\tau\right]J_v(r)\sin\frac{n\pi\theta}{\alpha}, v=n\pi/\alpha$

13. $u(r,\theta)=\frac{1}{2}(a^2-r^2)$

14. a. $u(r,\theta)=-\frac{1}{3}\left(r+\frac{4}{r}\right)\sin\theta+\text{constant}$

16. $u(r,\theta)=\sum_{n=1}^{\infty}\frac{1}{nR^{n-1}}r^n\sin n\theta$

18. $u(r,\theta)=\frac{a_0}{2}+\sum_{n=1}^{\infty}r^n(a_n\cos n\theta+b_n\sin n\theta)$

where

$$a_n=\frac{R^{1-n}}{(n+Rh)\pi}\int_0^{2\pi}f(\theta)\cos n\theta\,d\theta, n=0,1,2,\dots$$

$$b_n=\frac{R^{1-n}}{(n+Rh)\pi}\int_0^{2\pi}f(\theta)\sin n\theta\,d\theta, n=1,2,3,\dots$$

20. $u(r,\theta)=c-\frac{r^4}{12}\sin 2\theta+\frac{1}{6}\left(\frac{r_1^6-r_2^6}{r_1^4-r_2^4}\right)r^2\sin 2\theta$

$$+\frac{1}{6}\left(\frac{r_1^2-r_2^2}{r_1^4-r_2^4}\right)r_1^4r_2^4r^{-2}\sin 2\theta$$

22. a. $u(x,y)=\sum_{n=1}^{\infty}\frac{4[1-(-1)^n]}{(n\pi)^3\sinh n\pi}\sin n\pi x\sinh n\pi(y-1)$

c. $u(x,y)=\sum_{n=1}^{\infty}a_n(\sinh n\pi x-\tanh n\pi\cosh n\pi x)\sin n\pi y$ where

$$a_n=\frac{1}{\tanh n\pi}\left[\frac{2n\pi^3}{n^2\pi^4-4}+\frac{(-1)^n-1}{n\pi}\right]$$

23. a. $u(x,y)=c+\sum_{n=1}^{\infty} a_n(\cosh nx-\tanh n\pi \sinh nx)\cos ny$

where $a_n=2[1-(-1)^n]/n^3\pi \tanh n\pi$

c. $u(x,y)=-\dfrac{1}{\tanh\pi}\left[\cosh y-\tanh\pi \sinh y\right]\cos x+C$

25. $u(x,y)=xy(1-x)+\sum_{n=1}^{\infty}\dfrac{4(-1)^n}{n(\pi)^3\sinh n\pi}\sin n\pi x \sinh n\pi y$

27. $u(x,y)=c+(x^2/2)\left(\dfrac{x^2}{3}-y^2\right)+\sum_{n=1}^{\infty}\dfrac{8a^4(-1)^{n+1}}{(n\pi)^3\sinh n\pi}\cosh\dfrac{n\pi x}{a}\cos\dfrac{n\pi y}{a}$

29. $u(x,y)=x[(x/2)-\pi]+\sum_{n=1}^{\infty} a_n \sin\left(\dfrac{2n-1}{2}\right)x\cosh\left(\dfrac{2n-1}{2}\right)y$

where

$$a_n=\frac{2}{A\pi}\int_0^{\pi}\left[f(x)-h\left(\frac{x^2}{2}-\pi x\right)\right]\sin\left(\frac{2n-1}{2}\right)x\,dx$$

with

$$A=\left(\frac{2n-1}{2}\right)\sinh\left(\frac{2n-1}{2}\right)\pi+h\cosh\left(\frac{2n-1}{2}\right)\pi$$

Chapter 9

1. $u(x,y,z)=\dfrac{\sinh\left[(\pi/b)^2+(\pi/c)^2\right]^{1/2}(a-x)}{\sinh\left[(\pi/b)^2+(\pi/c)^2\right]^{1/2}a}\sin\dfrac{\pi y}{b}\sin\dfrac{\pi z}{c}$

2. $u(x,y,z)=\left[\dfrac{\sinh\sqrt{2}\,\pi z}{\sqrt{2}\,\pi}-\dfrac{\cosh\sqrt{2}\,\pi z}{\sqrt{2}\,\pi\tanh\sqrt{2}\,\pi}\right]\cos\pi x\cos\pi y$

4. a. $u(r,\theta,z)=\sum_{m=0}^{\infty}\sum_{n=1}^{\infty}(a_{mn}\cos m\theta+b_{mn}\sin m\theta)J_m(\alpha_{mn}r/a)$

$$\cdot\frac{\sinh\alpha_{mn}(l-z)/a}{\sinh\alpha_{mn}l/a}$$

where

$$a_{mn}=\frac{2}{a^2\pi\epsilon_n[J_{m+1}(\alpha_{mn})]^2}\int_0^{2\pi}\int_0^a f(r,\theta)J_m(\alpha_{mn}r/a)\cos m\theta\, r\,dr\,d\theta$$

$$b_{mn}=\frac{2}{a^2\pi[J_{m+1}(\alpha_{mn})]^2}\int_0^{2\pi}\int_0^a f(r,\theta)J_m(\alpha_{mn}r/a)\sin m\theta\, r\,dr\,d\theta$$

with

$$\epsilon_n=\begin{cases}1 \text{ for } m\neq 0\\ 2 \text{ for } m=0\end{cases}\qquad \text{and } \alpha_{mn} \text{ is the } n\text{th root of}$$

the equation $J_m(\alpha_{mn})=0$.

5. $u(r,\theta)=\frac{1}{3}+(2/3a^2)r^2P_2(\cos\theta)$

7. $$u(r,z)=\sum_{n=1}^{\infty}\frac{a_n\sinh\alpha_n(l-z)/a}{\cosh\alpha_n l/a}J_0(\alpha_n r/a)$$

where

$$a_n=\frac{2qu}{k\alpha_n^2J_x(\alpha_n)}$$

and α_n is the root of the equation $J_0(\alpha_n)=0$ and k is the coefficient of heat conduction.

8. $$u(r,z)=\frac{4u_0}{\pi}\sum_{n=1}^{\infty}\frac{I_0[(2n+1)\pi r/l]}{I_0[(2n+1)\pi a/l]}\frac{\sin(2n+1)\pi z/l}{(2n+1)}$$

9. $$u(r,\theta)=u_2+\frac{u_1-u_2}{2}\sum_{n=1}^{\infty}\left(\frac{2n+1}{n+1}\right)P_{n-1}(0)\left(\frac{r}{a}\right)^nP_n(\cos\theta)$$

11. $u(r,\theta,\phi)=C+\sum_{n=1}^{\infty}\sum_{m=0}^{\infty}r^nP_n^m(\cos\theta)[a_{nm}\cos m\phi+b_{nm}\sin n\phi]$ where

$$a_{nm}=\frac{(2n+1)(n-m)!}{2n\pi(n+m)!}\int_0^{2\pi}\int_0^{\pi}f(\theta,\phi)P_n^m(\cos\theta)\cos m\phi\sin\theta\,d\theta\,d\phi$$

$$b_{nm}=\frac{(2n+1)(n-m)!}{2n\pi(n+m)!}\int_0^{2\pi}\int_0^{\pi}f(\theta,\phi)P_n^m(\cos\theta)\sin m\phi\sin\theta\,d\theta\,d\phi$$

$$a_{n0}=\frac{2n+1}{4n\pi}\int_0^{2\pi}\int_0^{\pi}f(\theta,\phi)P_n(\cos\theta)\sin\theta\,d\theta\,d\phi$$

12. $$u(x,y,t)=\sum_{n=1,3,4,\ldots}^{\infty}\left(-\frac{4}{\pi}\right)\frac{[1-(-1)^n]}{n(n^2-4)}\cos\sqrt{(n^2+1)}\,\pi ct\sin n\pi x\sin n\pi y$$

13. $$u(r,\theta,t)=\sum_{n=0}^{\infty}\sum_{m=1}^{\infty} J_n(\alpha_{mn}r/a)\cos(\alpha_{mn}ct/a)[a_{mn}\cos n\theta+b_{mn}\sin n\theta]$$
$$+\sum_{n=0}^{\infty}\sum_{m=1}^{\infty} J_n(\alpha_{mn}r/a)\sin(\alpha_{mn}ct/a)[c_{mn}\cos n\theta+d_{mn}\sin n\theta]$$

where

$$a_{mn}=\frac{2}{\pi a^2\epsilon_n[J_n'(\alpha_{mn})]^2}\int_0^{2\pi}\int_0^a f(r,\theta)J_n(\alpha_{mn}r/a)\cos n\theta\, r\,dr\,d\theta$$

$$b_{mn}=\frac{2}{\pi a^2[J_n'(\alpha_{mn})]^2}\int_0^{2\pi}\int_0^a f(r,\theta)J_n(\alpha_{mn}r/a)\sin n\theta\, r\,dr\,d\theta$$

$$c_{mn}=\frac{2}{\pi ac\alpha_{mn}\epsilon_n[J_n'(\alpha_{mn})]^2}\int_0^{2\pi}\int_0^a g(r,\theta)J_n(\alpha_{mn}r/a)\cos n\theta\, r\,dr\,d\theta$$

$$d_{mn}=\frac{2}{\pi ac\alpha_{mn}[J_n'(\alpha_{mn})]^2}\int_0^{2\pi}\int_0^a g(r,\theta)J_n(\alpha_{mn}r/a)\sin n\theta\, r\,dr\,d\theta$$

in which α_{mn} is the root of the equation $J_n(\alpha_{mn})=0$ and

$$\epsilon_n=\begin{cases}2; & n=0\\ 1; & n\neq 0\end{cases}$$

15. $$u(r,\theta,t)=\sum_{n=0}^{\infty}\sum_{m=1}^{\infty} J_n(\alpha_{mn}r)e^{-k\alpha_{mn}t}[a_{nm}\cos n\theta+b_{nm}\sin n\theta]$$

where

$$a_{nm}=\frac{2}{\pi\epsilon_n[J_n'(\alpha_{mn})]^2}\int_0^{2\pi}\int_0^1 f(r,\theta)J_n(\alpha_{mn}r)\cos n\theta\, r\,dr\,d\theta$$

$$b_{nm}=\frac{2}{\pi[J_n'(\alpha_{mn})]^2}\int_0^{2\pi}\int_0^1 f(r,\theta)J_n(\alpha_{mn}r)\sin n\theta\, r\,dr\,d\theta$$

where α_{mn} is the root of $J_n(\alpha_{mn})=0$, and

$$\epsilon_n=\begin{cases}1 & \text{for } n\neq 0\\ 2 & \text{for } n=0\end{cases}$$

16. $u(x,y,z,t)=\sin\pi x\sin\pi y\sin\pi z\cos\sqrt{3}\,\pi ct$

18. $$u(r,\theta,z,t)=\sum_{n=0}^{\infty}\sum_{m=1}^{\infty}\sum_{l=1}^{\infty} J_n(\alpha_{mn}r/a)\sin(m\pi z/l)\cos\omega ct$$
$$\cdot[a_{nml}\cos n\theta+b_{nml}\sin n\theta]$$
$$+\sum_{n=0}^{\infty}\sum_{m=1}^{\infty}\sum_{l=1}^{\infty} J_n(\alpha_{mn}r/a)\sin(m\pi z/l)\sin\omega ct$$
$$\cdot[c_{nml}\cos n\theta+d_{nml}\sin n\theta]$$

where

$$a_{nml}=\frac{4}{\pi a^2 l\epsilon_n[J_n'(\alpha_{mn})]^2}\int_0^{2\pi}\int_0^a\int_0^l f(r,\theta,z)$$
$$\cdot J_n(\alpha_{mn}r/a)\sin(m\pi z/l)\cos n\theta\, r\,dr\,d\theta\,dz$$

$$b_{nml}=\frac{4}{\pi a^2 l[J_n'(\alpha_{mn})]^2}\int_0^{2\pi}\int_0^a\int_0^l f(r,\theta,z)$$
$$\cdot J_n(\alpha_{mn}r/a)\sin(m\pi z/l)\sin n\theta\, r\,dr\,d\theta\,dz$$

$$c_{nml}=\frac{4\omega^{-1}}{\pi a^2 l\epsilon_n[J_n'(\alpha_{mn})]^2}\int_0^{2\pi}\int_0^a\int_0^l g(r,\theta,z)$$
$$\cdot J_n(\alpha_{mn}r/a)\sin(m\pi z/l)\cos n\theta\, r\,dr\,d\theta\,dz$$

$$d_{nml}=\frac{4\omega^{-1}}{\pi a^2 l[J_n'(\alpha_{mn})]^2}\int_0^{2\pi}\int_0^a\int_0^l g(r,\theta,z)$$
$$\cdot J_n(\alpha_{mn}r/a)\sin(m\pi z/l)\sin n\theta\, r\,dr\,d\theta\,dz$$

where α_{mn} is the root of the equation $J_n(\alpha_{mn})=0$ and

$$\omega=\left[(m\pi/l)^2+(\alpha_{mn}/a)^2\right]^{1/2},\quad \epsilon_n=\begin{Bmatrix}1 & \text{for } n\neq 0\\ 2 & \text{for } n=0\end{Bmatrix}$$

20. $$u(r,\theta,z,t)=\sum_{n=0}^{\infty}\sum_{m=1}^{\infty}\sum_{p=1}^{\infty}\left[a_{nmp}\cos n\theta+b_{nmp}\sin n\theta\right]$$
$$\cdot J_n(\alpha_{mn}r/a)\sin(p\pi z/l)e^{-\omega t}$$

where

$$a_{nmp}=\frac{4}{\pi a^2 l\epsilon_n[J_n'(\alpha_{mn})]^2}\int_0^{2\pi}\int_0^a\int_0^l f(r,\theta,z)$$
$$\cdot J_n(\alpha_{mn}r/a)\sin(p\pi z/l)\cos n\theta\, r\,dr\,d\theta\,dz$$

$$b_{nmp}\frac{4}{\pi a^2 l[J_n'(\alpha_{mn})]^2}\int_0^{2\pi}\int_0^a\int_0^l f(r,\theta,z)$$
$$\cdot J_n(\alpha_{mn}r/a)\sin(p\pi z/l)\sin n\theta\, r\,dr\,d\theta\,dz$$

in which

$$\epsilon_n=\begin{cases}1 & \text{for } n\neq 0\\ 2 & \text{for } n=0\end{cases}\qquad \text{and } \omega=\left[(p\pi/l)^2+(\alpha_{mn}/a)^2\right]$$

23. $u(x,y,t)=\sum_{m=1}^{\infty}\sum_{n=1}^{\infty} u_{mn}(t)\sin mx\sin ny$ where

$$u_{mn}(t)=\frac{4(-1)^{m+n+1}}{mn\alpha_{mn}c}\left\{\sin\alpha_{mn}ct\left[\frac{\cos(1-\alpha_{mn}c)t-1}{2(1-\alpha_{mn}c)}+\frac{\cos(1+\alpha_{mn}c)t-1}{2(1+\alpha_{mn}c)}\right]\right.$$
$$\left.+\cos\alpha_{mn}ct\left[\frac{\sin(1-\alpha_{mn}c)t}{2(1-\alpha_{mn}c)}+\frac{\sin(1+\alpha_{mn}c)t}{2(1+\alpha_{mn}c)}\right]\right\}$$

and $\alpha_{mn}=(m^2+n^2)^{1/2}$

25. $$u(x,y,t)=\sum_{n=1}^{\infty}\sum_{m=1}^{\infty}\frac{4A}{mn\pi^2}\frac{[(-1)^n-1][(-1)^n-1]}{k(n^2+m^2)}\left[1-e^{-k(n^2+m^2)t}\right]$$
$$\cdot\sin nx(\sin my-m\cos my).$$

27. $$u(x,y,t)=x(x-\pi)\left(1-\frac{y}{\pi}\right)\sin t+\sum_{n=1}^{\infty}\sum_{m=1}^{\infty} v_{mn}(t)\sin nx\sin my$$

where $\alpha_{mn}^2=m^2+n^2$ and

$$v_{mn}(t)=\frac{8e^{-c^2\alpha_{mn}^2t}\left[1-(-1)^n\right]}{\pi^2mn(1+c^4\alpha_{mn}^4)}\left[\frac{c^2}{n^2}(\alpha_{mn}^2-n^2)(\cos te^{-c^2\alpha_{mn}^2t}-1)\right.$$
$$\left.+\left(\frac{1}{n^2}+c^4\alpha_{mn}^2\right)\sin te^{c^2\alpha_{mn}^2t}\right]$$

30. $$u(x,y,t)=\frac{4qb^4}{\pi^5D}\sum_{n=1,3,\ldots}^{\infty}\frac{1}{n^5}\left[1-\frac{v_n(x)}{1+\cosh(n\pi a/b)}\right]\sin(n\pi y/b)$$

where

$$v_n(x)=2\cosh\frac{n\pi a}{2b}\cosh\frac{n\pi x}{b}+\frac{n\pi a}{2b}\sinh\frac{n\pi a}{2b}\cosh\frac{n\pi x}{b}$$
$$-\frac{n\pi x}{b}\sinh\frac{n\pi x}{b}\cosh\frac{n\pi a}{2b}$$

Chapter 10

3. $$u(\rho,\theta)=\frac{1}{2\pi}\int_0^{2\pi}\frac{(\rho^2-1)f(\beta)\,d\beta}{1-\rho^2-2\rho\cos(\beta-\theta)}$$

7. $$u(x,y) = -\frac{2}{b}\sum_{n=1}^{\infty}\frac{\sin(n\pi y/b)}{\sinh(n\pi a/b)}\left[\sinh\frac{n\pi}{b}(a-x)\int_0^x f(\xi)\sinh\frac{n\pi\xi}{b}\,d\xi + \sinh\frac{n\pi x}{b}\int_x^a f(\xi)\sinh\frac{n\pi}{b}(a-\xi)\,d\xi\right]$$

8. $$u(r,\theta) = -\sum_{n=0}^{\infty}\sum_{k=1}^{\infty}(R/\alpha_{nk})^2 J_n(\alpha_{nk}r/R)(A_{nk}\cos n\theta + B_{nk}\sin n\theta)$$

$$A_{0k} = \frac{1}{\pi R^2 J_1^2(\alpha_{0k})}\int_0^R\int_0^{2\pi} rf(r,\theta)J_0(\alpha_{0k}r/R)\,dr\,d\theta$$

$$A_{nk} = \frac{2}{\pi R^2 J_{n+1}^2(\alpha_{nk})}\int_0^R\int_0^{2\pi} rf(r,\theta)J_n(\alpha_{nk}r/R)\cos n\theta\,dr\,d\theta$$

$$B_{nk} = \frac{2}{\pi R^2 J_{n+1}^2(\alpha_{nk})}\int_0^R\int_0^{2\pi} rf(r,\theta)J_n(\alpha_{nk}r/R)\sin n\theta\,dr\,d\theta$$

$n = 1,2,3,\ldots;$ $k = 1,2,3,\ldots$ and α_{nk} are the roots of $J(\alpha_{nk}) = 0$.

9. $$G(r,r') = \frac{e^{ik|r-r'|}}{|r-r'|} - \frac{e^{ik|\rho-r'|}}{|\rho-r'|}$$

where $r = (\xi,\eta,\zeta)$, $r' = (x,y,z)$, $\rho = (\xi,\eta,-\zeta)$

10. $$G(r,r') = \frac{e^{ik|r-r'|}}{|r-r'|} + \frac{e^{ik|\rho-r'|}}{|\rho-r'|}$$

14. $$G = -\frac{4a}{\pi}\sum_{n=1}^{\infty}\int_0^{\infty}\frac{1}{(\alpha^2a^2+n^2\pi^2)}\sin\frac{n\pi x}{a}\sin\frac{n\pi\xi}{a}\sin\alpha y\sin\alpha\eta\,d\alpha$$

16. $$u(r,z) = \frac{2C}{\pi}\int_0^{\infty}\int_0^{\infty}\frac{1}{(\kappa^2-\lambda^2-\beta^2)}J_0(\beta_r)J_1(\beta a)\cos\lambda z\,d\beta\,d\lambda$$

17. $u(r,\theta) = Ar^{1/2}\sin(\theta/2)$

18. $$G = -\frac{2}{a}\sum_{n=1}^{\infty}\frac{\sinh\sigma y'\sinh\sigma(y-b)}{\sigma\sinh\sigma b}\sin\frac{n\pi x}{a}\sin\frac{m\pi x'}{a}$$

$\sigma = \sqrt{(\kappa^2 + (n^2\pi^2)/\alpha^2)}$ $\quad 0 < x' < x < a, \quad 0 < y' < y < b.$

Chapter 11

1. $F(\alpha) = \sqrt{(1/2a)}\ e^{-\alpha^2/4a}$
2. $F(\alpha) = \sqrt{(2/\pi)}\ (\sin\alpha a)/\alpha$
3. $F(\alpha) = 1/|\alpha|$

4. (a) $F(\alpha)=\sqrt{(1/2)}\,\sin\left(\frac{\alpha^2}{4}+\frac{\pi}{4}\right)$

(b) $F(\alpha)=\sqrt{(1/2)}\,\cos\left(\frac{\alpha^2}{4}-\frac{\pi}{4}\right)$

11. $u(x,t)=\frac{1}{2}[f(x+ct)+f(x-ct)]+\frac{1}{2c}\int_{x\pm ct}^{x+ct} g(\tau)\,d\tau$

12. $u(x,t)=\frac{1}{2\sqrt{t}}\int_0^\infty\left[e^{-(x-\xi)^2/4t}-e^{-(x+\xi)^2/4t}\right]f(\xi)\,d\xi$

14. $u(x,t)=\frac{1}{\sqrt{2\pi}}\int_{-\infty}^{\infty} f(\xi)\cos(a\xi^2 t)e^{-i\xi x}\,d\xi$

15. $u(x,t)=\frac{1}{\sqrt{\pi}}\int_{x/\sqrt{2at}}^{\infty} g\left(\frac{t-x^2}{2a\xi^2}\right)\left(\sin\frac{\xi^2}{2}+\cos\frac{\xi^2}{2}\right)d\xi$

16. $u(x,t)=\frac{\phi_0}{2\pi}\int_{-\infty}^{\infty}\frac{\sin a\xi}{\xi}\frac{e^{i\xi x}}{|\xi|}e^{-|\xi|y}\,d\xi$

17. $u(x,t)=\frac{1}{\sqrt{4\pi t}}\int_{-\infty}^{\infty} e^{-(x-\xi)^2/4t}f(\xi)\,d\xi$

20. $$u(x,y)=\frac{2y}{\pi}\int_0^\infty\frac{(x^2+\tau^2+y^2)f(\tau)\,d\tau}{[y^2+(x-\tau)^2][y^2+(x+\tau)^2]}$$
$$-\frac{1}{2\pi}\int_0^\infty g(\tau)\log\frac{[x^2+(y+\tau)^2]}{[x^2+(y-\tau)^2]}\,d\tau$$

23. $u(x,t)=\frac{2}{\pi}\int_0^\infty \xi\sin x\xi\int_0^t e^{-\xi^2(t-\tau)}f(\tau)\,d\tau\,d\xi$

24. $u(x,y)=\frac{2}{\pi}\int_0^\infty\frac{\sin\xi x\sin h\xi(1-y)}{\sin h\xi}\int_0^\infty f(\tau)\sin\xi\tau\,d\tau\,d\xi$

27. a. $\frac{1}{3}(\cos t-\cos 2t)$; b. $\frac{1}{3}\sin t-\frac{1}{6}\sin 2t$; c. $e^{2t}-e^{t}$; d. $1-e^{-t}-te^{-t}$; e. $1-e^{-t}$; f. $t\cos 2t$.

28. a. $\frac{1}{s^2}\tanh\frac{bs}{2}$

b. $\frac{h}{s(1+e^{-bs})}$

c. $\frac{1}{(s^2+1)(1-e^{-\pi s})}$

d. $\frac{h}{bs^2}-\frac{he^{-bs}}{s(1-e^{-bs})}$

e. $\frac{he^{-bs}}{s(1-e^{-bs})}$

30. $u(x,t)=\begin{cases}\frac{1}{2}[f(ct+x)-f(ct-x)] & \text{for } t>x/c\\ \frac{1}{2}[f(x+ct)+f(x-ct)] & \text{for } t<x/c\end{cases}$

33. $u(x,t)=\begin{cases}0 & \text{for } t<x/c\\ f(t-x/c) & \text{for } x/c<t\leqslant(2-x)/c\end{cases}$

34. $u(x,t)=f_0+(f_1-f_0)\,\text{erfc}\left[\sqrt{(x^2/4\kappa t)}\,\right]$

35. $u(x,t)=x-x\,\text{erfc}(x/\sqrt{4\kappa t}\,)$

36. $u(x,t)=2\int_0^t\int_0^\eta \text{erfc}(x/\sqrt{4\kappa\xi}\,)\,d\eta\,d\xi$

37. $u(x,t)=f_0e^{-ht}[1-\text{erfc}(x/\sqrt{4\pi t}\,)]$

39. $u(x,t)=f_0\,\text{erfc}(x/\sqrt{4\pi t}\,)$

42. $u(x,t)=\begin{cases}f_0t & \text{for } t<x/c\\ f_0x/c & \text{for } t>x/c\end{cases}$

43. $u(x,t)=\begin{cases}\frac{1}{2}[f(x+ct)-f(ct-x)], & t>x/c\\ \frac{1}{2}[f(x+ct)+f(x-ct)], & t<x/c\end{cases}$

Index